Qualitative Representations

Qualitative Representations

How People Reason and Learn about the Continuous World

Kenneth D. Forbus

The MIT Press
Cambridge, Massachusetts
London, England

This book was set in Stone Serif by Westchester Publishing Services. Printed and bound in the United States of America.

Library of Congress Cataloging-in-Publication Data

Names: Forbus, Kenneth D., author.
Title: Qualitative representations: how people reason and learn about the continuous world / Kenneth D. Forbus.
Description: Cambridge, MA : MIT Press, [2018] | Includes bibliographical references and index.
Identifiers: LCCN 2018010193 | ISBN 9780262038942 (hardcover : alk. paper)
Subjects: LCSH: Cognition. | Reasoning. | Space perception.
Classification: LCC BF311 .F6297 2018 | DDC 153.4—dc23
LC record available at https://lccn.loc.gov/2018010193

10 9 8 7 6 5 4 3 2 1

To Dedre Gentner, my best collaborator in all aspects of life

Contents

Preface

How we reason about the continuous world around us is one of the central mysteries of cognitive science. This book is based on multiple decades of research in artificial intelligence (AI) on qualitative reasoning by my group but also by many others. Most of the qualitative reasoning literature assumes a strong artificial intelligence background, which is a shame because I believe it is extremely relevant across cognitive science. This book is my attempt to bridge that gap, to make accessible the insights and ideas that qualitative reasoning research has come up with to a broad audience of cognitive scientists. I hope it will also help AI scientists and engineers better understand the connection between AI and the other branches of cognitive science, because gaining insights across disciplines is why cognitive science exists in the first place.

Acknowledgments

This project has benefited from several kinds of support. It was started in 2012 while my wife Dedre Gentner and I were fellows at the Hanse-Wissenschaftskolleg (HWK) in Delmenhorst, Germany. I thank the HWK staff for making our stays there pleasant and productive. Generous financial support was provided by the Alexander von Humboldt Foundation in the form of a Humboldt Research Prize. Multiple U.S. funding agencies have helped sponsor this research, including the Office of Naval Research, the Air Force Office of Scientific Research, the Defense Advanced Research Projects Agency, and the National Science Foundation, the latter through our Spatial Intelligence and Learning Center. IBM also provided generous support.

Another kind of crucial support is the thoughtful feedback from colleagues across multiple disciplines. I thank Johan de Kleer, Pat Hayes, Ian Horswill, Christian Freksa, and Robert Kahlert for their helpful feedback. Lance Rips, Sue Hespos, and Nora Newcombe helped identify multiple blind spots in making things clearer for psychologists and other cognitive scientists, as well as alerting me to other relevant evidence. Ernie Davis went above and beyond the call of duty, in terms of both amount and depth of feedback and advice. Students in my graduate seminar in spring 2014 helped find many problems and suggested many improvements: David Barbella, Joe Blass, Maria Chang, Subu Kandaswamy, Chen Liang, Clifton McFate, Matthew McLure, Grant Sheldon, and Stephen Zeng. Carrie Ost provided invaluable assistance in manuscript preparation. Despite all this help, errors remain, I am sure, and for those I am responsible.

Our cats helped in their own way: Archie kept me pinned down at the keyboard by refusing to budge from my lap, and Nero would come to my study to fetch me when it was late and I should be turning in.

Finally, I want to thank the many students, collaborators, and colleagues whose research is represented here. Science is a team sport, and I am thrilled to be part of such a stimulating community, both here at Northwestern University and more broadly.

Evanston, December 2017

I Introduction and Preliminaries

Part I provides two things. The first is an overview of what this book is about. The second is optional background material to help put everyone on the same footing, computationally speaking.

- Chapter 1 provides the main argument of the book and motivating examples.
- Chapter 2 provides a quick guide to the relevant aspects of knowledge representation that readers will need.
- Chapter 3 provides an introduction to the computational aspects of reasoning.
- Chapter 4 provides an introduction to analogical matching, retrieval, and generalization, which play central roles in this account.

Chapter 1 is the best place to start. Readers who have a lot of experience with knowledge representation and reasoning may wish to just skim chapters 2 and 3, but you will likely find chapter 4 novel. Some readers prefer digging into the background first; others prefer to dip back into it when needed. This organization will hopefully be useful, whichever style you prefer.

1 Introduction

One of the mysteries of human cognition is our ability to easily reason about the world. We have robust, everyday mental models of the physical world that help us cook, navigate, and build artifacts that enable us to structure the environment so that we can function better. We have models of the social world that help us relate to other people (and other animals). We have models of ourselves that help us decide when to put more effort into a task or when to switch gears and do something else. All of these models share a common concern: they require dealing with continuous properties and phenomena. In everyday physical reasoning, we must think about quantities like pressure, temperature, and mass. We harness processes to transform matter during cooking and industrial production. Thinking about space is crucial for thinking about the physical world, and space is fundamentally continuous. In social reasoning, we think about continuous properties such as degree of responsibility for an event and how much two people like or trust each other. In metacognition, we think about how difficult a problem is and how valuable solving it might be to other things we might be doing. In other words, continuous properties, processes, and systems permeate our mental life. And yet most accounts of representations have avoided, or minimized, the role of thinking about continuous things.

Qualitative representations provide discrete, symbolic descriptions of the continuous world. This makes qualitative representations useful in everyday reasoning. For example, we can know that, if we were to pour water on a table, an object underneath it would not immediately become wet. We do not need a concrete picture or numerical data to come to such conclusions. Moreover, we can easily use such knowledge in planning: if a brief period of rain starts during a picnic, we can put food under the table to keep it dry. Exactly where to put the food depends on how much rain there is, the wind, and a host of other factors. But this simple qualitative distinction provides an initial plan and raises a set of relevant questions for further reasoning.

The goal of research on qualitative reasoning is to formalize our intuitive knowledge and methods of how to reason about continuous phenomena and systems. This knowledge ranges from that of the person on the street, who has never taken formal mathematics or physics courses, to that of experts, such as scientists and engineers. The reasoning techniques developed in qualitative reasoning research are intended to provide both computational models of human commonsense reasoning and how scientists and engineers reason in their professional work.

Most of the research to date has focused on the physical world (and hence why it is often also called *qualitative physics* (Bobrow, 1985; Weld & de Kleer, 1990), but the same ideas have been fruitful in social science and in reasoning about strategies and tactics in games. Moreover, the same ideas may extend into our models of mental life, including self-modeling and metacognition. This book argues that qualitative representations provide a central component of human conceptual structure. For example, qualitative spatial representations serve as a bridge between perception and cognition. If we think of the representations used by our minds as a form of currency for exchanges between different mental processes, then qualitative representations appear to be a major form of cognitive currency.

1.1 Some Examples of Everyday Qualitative Reasoning

Let us consider, informally for now, some examples of qualitative reasoning in everyday life. These will help us see how reasoning about continuous systems permeates our lives. These examples are among those considered in more detail, using qualitative representations, in the rest of the book.

1.1.1 Heating Water

Consider a tea kettle partially filled with water, being heated on a stove. What might happen if you leave it unattended for an hour?

We all know that this is a really bad idea. Some might know it vaguely, whereas others have the visceral feeling you get when seeing a mistake being repeated. And yet, you don't know all of the numerical values or equations that are necessary to derive, from traditional physical principles, what will happen. Nevertheless, you know what kinds of things are likely to happen. You know it will heat up and that, after a while, the water will begin to boil. Eventually, all of the water boils away, and the kettle starts to get extremely hot. If the temperature of the stove is high enough, the kettle might even melt. As chapter 7 illustrates, there are even more outcomes possible, given how little we know about this situation. Knowing the kinds of behaviors that might happen is extremely valuable: it lets us know when something might go wrong and tells us where we need to focus to avoid problems. It helps us understand when we might need more knowledge. This knowledge can take many forms. In everyday life, it may be in the form of more experience and observations. In professional practice, it may be mathematical models, numerical simulations, or experiments conducted on physical models. In either case, the initial qualitative analysis provides the framework for subsequent thinking.

1.1.2 Does Cold Water Freeze Faster Than Warm Water?

Here is an experiment you can try. Find two identical ice cube trays. Fill one with cold water, fill the other with warm water, and put them both in a freezer at the same time. Which will freeze first, the warmer water or the cooler water? And why? Try constructing an explanation before reading on.

If you try it, you will find that the warmer water freezes faster. This is counterintuitive, because the difference between the freezing point and its initial temperature is smaller for the cold water than for the warm water. Many people struggling with this problem for the first time try introducing the concept of "thermal momentum," by analogy with motion. The larger temperature difference, in this model, somehow accelerates the cooling of the warmer water. Unfortunately, thermal momentum does not exist. What else does freezing depend on? A smaller amount of water will freeze faster than a larger amount of water because it has less heat to lose. But both trays were filled with the same amount of water, so if this line of explanation is going to work, something must have happened inside the freezer that affected the amount of the warm water tray more than the cold water tray. What processes can affect the amount of water? Evaporation can, and warmer water evaporates faster than cooler water, so that fits, providing a potential explanation for the phenomenon. Part II explores formal versions of the representations and reasoning we have just been using.

1.1.3 The Seasons

Why are there seasons? A surprising number of children, and even Harvard graduates,[1] incorrectly believe that the Earth is nearer the sun in the summer than in the winter. This cannot be the correct explanation, because when it is winter in Chicago in the United States, it is summer in Brisbane, Australia. Again, you might think about this, reconstructing what you know about it, before reading on.

Recall that the Earth's axis of rotation is tilted. This angle remains roughly constant relative to the plane of the Earth's orbit. Portions of the

Earth that are tilted toward the Sun receive more solar radiation per unit area than portions of the Earth that are tilted away from it; hence, the regions where the tilt is toward the Sun will be experiencing summer when those tilting away are experiencing winter. Chapter 17 describes a simulation of conceptual change by Scott Friedman that starts with a naive misconception and moves to the correct model (Friedman, Forbus, & Sherin, 2011b, 2018), as well as capturing intermediate states of knowledge found in a learning sciences study of middle school students by Bruce Sherin and his colleagues (Sherin, Krakowski, & Lee, 2012). The same system has been used to model how children might learn intuitive notions of force and how self-explanation might help in learning by reading.

1.1.4 Will These Collide?

Consider the balls bouncing around in the simple world of figure 1.1. Can they collide?

As in the case of the kettle on the stove, you don't have enough information to really determine what will happen, but you probably were sure that they might. Adding even a little information can radically change your conclusions. For example, suppose they are both eggs—the first collision with the ground will leave a mess but no further possibility to interact. Suppose we say that the ball on the left goes inside the well and never escapes, whereas the ball on the right never goes inside the well. Again, we can determine that they can never collide, because collisions involve being at the same place at the same time. Suppose we now are given all of the parameters needed to make a traditional mathematical model of this situation: the

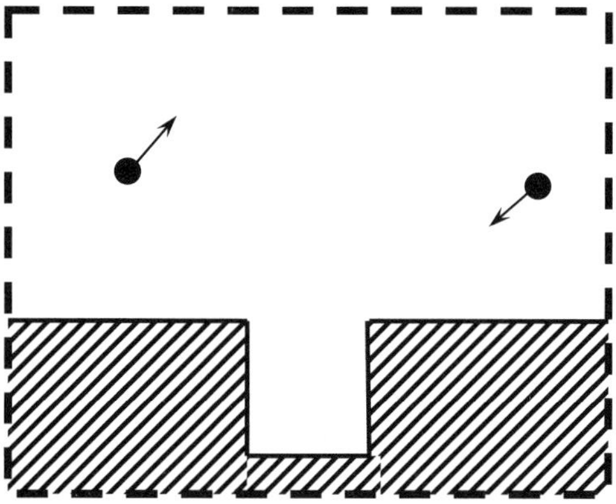

Figure 1.1
Can these balls collide?

exact coordinates of the initial positions of the balls, their velocity, and the coefficients of restitution[2] for the material(s) they are made of, as well as the exact locations of the walls and the assumption that they are perfectly rigid. We could then run the simulation to see if the balls will in fact collide. But this could take a while, and (unless the balls are perfectly inelastic) we might end up seeing that something like the spatial separation we explored as an assumption above actually occurs with the parameters we used. If so, we could stop the simulation, without waiting for the balls to stop, because we already know the answer (Forbus, 1980, 1984). Chapter 14 illustrates how grounding spatial reasoning in metric representations allows qualitative and quantitative reasoning to be combined.

1.1.5 Raven's Progressive Matrices

Figure 1.2 shows an example like those used in the Raven's Progressive Matrices test.[3] The idea is to select which of the eight choices along the bottom is the best fit for the missing element in the 3×3 matrix of figures.

Although this is a visual problem-solving task, it turns out that scores on this test are very highly correlated with *g*, a measure of general intelligence (Raven, Raven, & Court, 2000). Such problems can be solved by combining qualitative spatial representations, automatically computed from digital

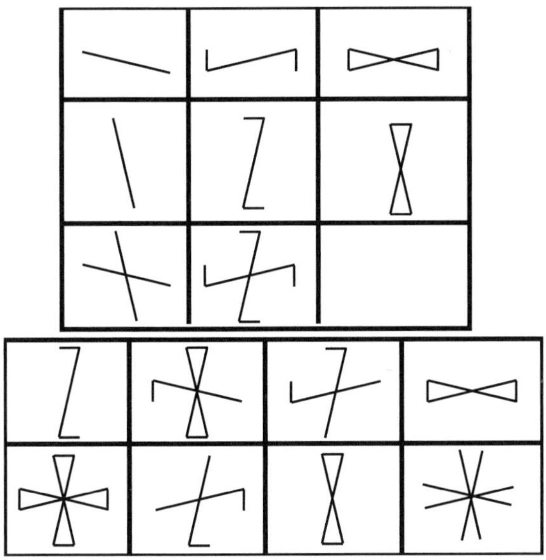

Figure 1.2
The kind of problem found in the Raven's Progressive Matrices test.

ink, with analogical reasoning. As chapter 16 describes, Andrew Lovett's simulation, which uses these ideas, scores better on the real test than most adult Americans (Lovett & Forbus, 2017). Moreover, problems that are harder for people are harder for the simulation, and differences in reaction time between problems are also correctly predicted by the simulation.

1.1.6 Moral Decision Making

One of the important and surprising findings in research on morality in decision making is the existence of sacred or protected values (Ritov & Baron, 1999). In many cases, people make decisions based on utility (e.g., given a choice between two strategies, a company might take the one that appears to be more profitable). The presence of protected values often modifies this behavior. For example, consider this scenario from Ritov and Baron (1999, 83):

> A convoy of food trucks is on its way to a refugee camp during a famine in Africa. (Airplanes cannot be used.) You find that a second camp has even more refugees. If you tell the convoy to go to the second camp instead of the first, you will save 1000 people from death, but 100 people in the first camp will die as a result. Would you send the convoy to the second camp?

When faced with this decision, most participants chose to not redirect the truck, even though more lives would be saved.[4] The intuition is that their action would then be causing the death of the 100, which they viewed as unacceptable. This phenomenon has been explored through a wide range of scenarios, including the infamous and prolific family of trolley problems (Petrinovich, O'Neill, & Jorgensen, 1993; Thomson, 1985). As described in chapter 18, a simulation by Morteza Dehghani (Dehghani, Tomai, Forbus, & Klenk, 2008) uses a qualitative order-of-magnitude representation to capture the effect of protected values and combines first principles and analogical reasoning to detect the presence of protected values.

1.2 The Importance of Qualitative Reasoning in Human Cognition

As these examples illustrate, qualitative reasoning can be both simple and straightforward but also capable of surprising sophistication. It can serve as a bridge between perception and cognition, imposing task-specific distinctions upon the continuous world that carve it into discrete symbols that can be combined into structured, relational representations to support complex reasoning, including causal reasoning. It is central in commonsense reasoning, because the physical world and other continuous phenomena constitute a major portion of what common sense is about. Moreover, I argue that qualitative reasoning is an important component in expert reasoning, grounding expert knowledge in the everyday world, guiding the use of more quantitative knowledge. The importance of qualitative representations in human reasoning must surely be reflected in human language, and in chapter 13, I argue that qualitative reasoning forms an important component of natural language semantics.

In making these arguments, I draw on a number of sources of evidence. In some cases, there is applicable evidence from psychological studies and cognitive simulations. In other cases, I use evidence from artificial intelligence (AI) systems that were designed as performance systems rather than cognitive models per se. This kind of evidence has been less commonly used in cognitive science recently. This is unfortunate because there has been considerable progress in AI over the past twenty years that has mostly been ignored by many other cognitive scientists. One important consideration for cognitive models is ability (Cassimatis, Bello, & Langley, 2008)—that is, can they actually perform the task that they are trying to model, at the levels that people do? Unfortunately, many cognitive models today fail this test, working at best on only a few small hand-coded examples generated by the model builders themselves. (A growing number of cognitive models do better these days, especially those using symbolic modeling techniques.) AI systems, by contrast, are held to ability as a crucial criterion. Even when an AI system was originally designed for performance only, careful scrutiny of the assumptions, representations, and processes involved in that performance can yield valuable insights as to the nature of the reasoning task itself, what Marr (1982) called *computational-level* constraints.[5]

For understanding the nature of cognition, constraints on the information-processing tasks that an organism must solve are just as important as the constraints imposed by biology (what Marr [1982] called the *implementation level*). For understanding higher-order cognition and even high-level visual perception, arguably task constraints are even more important today, given the imperfect and preliminary nature of our current understanding of neural systems. Moreover, careful analysis of how such systems differ from people can provide evidence about the nature of human processing (what Marr [1982] called the *process level*, which concerns algorithms). As described in chapter 12, analyzing how practical qualitative reasoning (QR) systems work has led me to the conclusion that, although the representations developed by the QR community are very reasonable models for human representations, the reasoning techniques that people use are different. I claim that people use analogical reasoning and learning heavily in their qualitative reasoning, and the arguments for that claim, and what this implies, are also explored throughout the book.

1.3 Overview of the Book

The other three chapters in this part lay out some background. Chapter 2 describes the representation conventions assumed here, and chapter 3 provides some background about computation and reasoning. Chapter 4 describes the structure-mapping models of analogical processes that figure heavily in this account. These may be safely skimmed by readers familiar with knowledge representation, reasoning, and analogy, respectively.

Part II discusses *qualitative dynamics*, how to represent and reason about changes in continuous systems that can be described via quantities. This includes qualitative representations for quantities (chapter 5) and how these can be related by forms of qualitative mathematics (chapter 6). How such quantities and relationships can be organized into conceptual structures is discussed, in terms of the *ontologies* that have been explored in qualitative reasoning. We focus on qualitative process theory (Forbus, 1984), describing its representational assumptions and basic inferences supported in chapter 7. This provides the basis for a model of causality in continuous systems, which appears to nicely capture many aspects of human reasoning about them. Chapter 8 illustrates the representational capacity of the theory by a number of examples, including the kettle and ice cube examples from this chapter. Chapter 9 examines the notions of causality introduced in qualitative reasoning in more detail and contrasts them with other recent accounts in cognitive science. The subtleties of reasoning about change, including how time, space, and state interact and the importance of ambiguity, are discussed further in chapter 10. We explore how knowledge about modeling can be organized and used effectively in chapter 11. The problems with the first principles–only account of qualitative reasoning, as well as a more psychologically plausible alternative based on analogical processing, are discussed in chapter 12. How these ideas play out in natural language is explored in chapter 13.

Part III discusses *qualitative spatial reasoning*, that is, how to represent and reason about space qualitatively. We begin in chapter 14 by discussing the basic ideas of qualitative spatial reasoning, including comparing them with independently developed ideas developed at the same time in cognitive psychology (i.e., the *coordinate/categorical* distinction). The synthesis strengthens both. AI research shows how qualitative and quantitative representations are combined to achieve human-level performance in a variety of tasks; cognitive psychologists have explored how such representations may be stored in long-term memory and neural correlates of such processing. Chapter 15 explores the variety of spatial calculi that have been developed by the qualitative spatial reasoning community and how these ideas play out in understanding spatial language. Like dynamics, the specific reasoning processes used by that community may not be broadly psychologically plausible, but the representations seem to capture a variety of useful distinctions. Chapter 16 explores metric, quantitative representations in spatial reasoning, including the roles of diagrams and high-level visual processing. Examples from sketch understanding predominate, given the early state of machine vision.

Part IV focuses on a variety of ways that qualitative reasoning is used in larger-scale tasks and how qualitative representations are learned. Both cognitive development and conceptual change are explored in chapter 17. The combination of qualitative representations and analogical processing suggests a very different model of commonsense reasoning than previously proposed in AI or cognitive science, as chapter 18 explains. This chapter also describes how these ideas have been used to model quantitative estimation (i.e., back of the envelope reasoning), modeling metaphors, and social reasoning (e.g., blame assignment, emotions, and moral reasoning). How qualitative representations have been used to create systems that can perform engineering tasks and scientific modeling is outlined in chapter 19.

Finally, part V wraps it all up. Chapter 20 summarizes the key ideas of the book, and chapter 21 discusses a variety of open questions and exciting new directions for research.

2 Representation: An Overview

Knowledge representation is a deep topic, but there are only a few essentials you need to know to follow the arguments made in this book.

2.1 The Importance of Structured, Relational Representations

Knowledge is a broad term. Our main focus in this book concerns the kinds of mental models that we construct by interacting with the world and via instruction (both formal and informal) about it. Representing such models requires symbolic representational capacities. That is, we have, in some functional sense, tokens that stand for things in the world and compositional means of constructing richer descriptions via combining such tokens. This richer structure is built up via *relations* that connect other tokens. Family relations are familiar examples, as are spatial relations (e.g., above, inside). Relations can also connect events and other relations, as when we invoke causality to explain why something happened. For example, Juliet committing suicide was caused by her belief that Romeo had committed suicide and her love for him. People demonstrably explain things, make plans, and construct hypotheses and models. This book argues that qualitative representations are one of the key materials from which these products of mental life are constructed.

There are some who argue that the notion of representation is fundamentally flawed. Many others (myself included) think of representation as crucial to understanding human cognition.[1] I believe the evidence is heavily in favor of this. Consider the operations you're doing while reading this very sentence, for example. You are decoding these sentences using a complex combination of visual and conceptual processing, including using world knowledge to help with disambiguation even in very early levels of processing (Camblin, Gordon, & Swaab, 2007). If you think you can explain the range of phenomena described in the rest of this book without using the idea of representation, I invite you to try. But the bar is already set very high.

There are others who argue that simple numerical spatial models or purely statistical models or feature vectors can account for much of human cognition. Again, there are many reasons to doubt this. There is evidence that visual comparison uses structured, relational representations (Palmer, 1999; chapters 14 and 16, this volume). Natural language often discusses events, about which information is expressed incrementally. It is hard to see how to accumulate such information without a token to stand for the event in question and relations (i.e., role relations) to link the event to the who, what, when, and where of it (see chapter 13). And of course, there are the "why" questions: explanations are inherently relational, and modeling human causal reasoning about continuous systems is a key question addressed in this book (part II). Hence, we assume structured, relational representations are used in human cognition.

2.2 Logic, Formalism, and Precision

Logic started as an attempt to codify how to think clearly, so it is natural that cognitive scientists, especially AI researchers, would embrace logic in a variety of ways. To understand this book does not require an extensive background in logic. Here we focus on a few essentials, providing perspective as much as background.

Logic is useful in helping ensure that the representations we write have a clear meaning. Our goal is precision, to be clear about the concepts we are discussing—so clear that computational models can automatically reason with the descriptions that we create. Formal representations are one of the tools that can be used to achieve this. Here, logic is used as a tool for helping describe the set of inferences that a representation licenses. Existing formalisms cannot carry that entire burden because computational issues are central in reasoning, and the relationships between logic and computation are still very much a research issue, so we use logic sparingly.[2]

2.2.1 Syntax

Syntax is about culture. Reading is hard, one of those amazing feats of human cognition. Reading formal systems (e.g., representation schemes, programming languages) can be even harder, because we have far less experience with them. Reading requires decoding the marks on the screen (or page), which is a visual skill, and such skills take time to learn. An unfamiliar syntax forces us to do more work to puzzle it out. Visual languages have the same problems, unfortunately, and tend to not scale well: a concept map with more than a dozen nodes, for example, tends to overload viewers.

This book uses a Lisp-like syntax.[3] That is, statements take the following form:

(*<predicate>* . *<arguments>*)

where *<predicate>* is a relation and *<arguments>* are zero or more arguments to that relationship. (The "." is used to indicate an indefinite number of arguments.) For example,

```
(northOf CityOfDelmenhorst CityOfVenice)
```

indicates that Delmenhorst, Germany, is north of Venice, Italy. The advantage of this syntax is that it does not require learning precedence rules, which are conventions for grouping things. In traditional mathematics, for example, the expression $3x+2$ means "the product of 3 and x, plus 2" rather than "the product of 3 times the result of x plus 2." Traditional notations break down when many relationships and functions are required, and making things explicit keeps them clearer. For example,

```
(+ (* 3 x) 2)
```

We follow the common convention of using a fixed-width font for tokens in a representational system to distinguish them from everyday terms more clearly.

We assume the usual logical connectives—`and`, `or`, `implies`, `iff`, `not`—with their usual meanings in logic. Statements can of course be nested. For example,

```
(implies (northOf CityOfDelmenhorst CityOfVenice)
         (< (AverageSummerTemperatureFn
                CityOfDelmenhorst)
            (AverageSummerTemperatureFn CityOfVenice)))
```

indicates that their geographical relationship implies that the average summer temperature of Delmenhorst is lower than that of Venice. Here, the arguments to < are *terms*. Terms provide a means of representing entities in the world (e.g., the cities of Delmenhorst and Venice). Terms can be atomic (e.g., `CityOfDelmenhorst`) or non-atomic (e.g., `(AverageSummerTemperatureFn CityOfDelmenhorst)`). Non-atomic terms are constructed via a function applied to arguments. For example,

(*<function>* . *<arguments>*)

Notice that this is the same syntax used for statements. The difference between statements and non-atomic terms depends on whether the first element of the list is a relation or function. We follow Cyc conventions of capitalizing functions and typically using the suffix "Fn" to make the intended meaning clear to the casual reader.

There are three reasons for choosing this syntax. First, it is widely used in AI and cognitive science research. Part of this is historical, and part of it is that it is really very simple compared to operator scope and other rules needed to make other syntactic conventions work. Second, it scales well to more realistic representations. Most examples in the logic and philosophical literature might use abstract relationships like *P*, but in representing the range of everyday life and professional reasoning, many more concepts and relationships must be invoked. Traditional syntax does not scale well to these demands. Finally, I like it, and I think once you get used to it, you will like it, too.

Constants, like `CityOfDelmenhorst`, are also terms. The fact that this constant is made up of three English words is of interest to human readers but almost never to software.[4] For readability, intuitive names are used whenever possible. Another such convention is to introduce an arbitrary constant by using a concept name followed by an integer (e.g., `Truck18` is intended to be something that is an instance of the concept of a truck). As discussed below, although using intuitive names provides information about the intended meaning of a term, relation, or function, their meaning within the representational system only derives from other statements about them and the computations that they license.

Variables are indicated by prefixing a constant with a "?" (i.e., "`?x`" is a variable, whereas "`x`" is not). Again, this convention is needed because, in realistic representations, it is better to use variable names that are somewhat longer but more meaningful, to support readers, while giving them an unambiguous local cue as to which things are or are not variables. Quantifiers are a special kind of connective that introduces variables into logical statements. The quantifier *forall* indicates universal quantification, and the quantifier *exists* indicates existential quantification. For example,

```
(forall ?day
    (implies (SummerDay ?day)
             (rainingDuringPeriodIn ?day
                                    CityOfDelmenhorst)))
```

That is, it rains every summer day in Delmenhorst (which, although not literally true, certainly seems that way). To contradict this, it would be enough to claim there is a counterexample:

```
(exists ?day
   (and (SummerDay ?day)
        (not (rainingDuringPeriodIn ?day
                                  CityOfDelmenhorst))))
```

which asserts that there is at least one such day, and perhaps more than one, but one does not need to know exactly which day that is. Statements with variables are often called *axioms* or *rules*, and statements without variables are known as *ground statements* or *ground facts*. When writing axioms, it is conventional to leave out universal quantifiers. One could equally say

```
(implies (SummerDay ?day)
         (rainingDuringPeriodIn ?day CityOfDelmenhorst))
```

Notice that there is no explicit quantifier governing the variable `?day` here. A variable that is not governed by a quantifier is called a *free variable*. Most reasoning engines either forbid free variables or interpret them as universally quantified.

Order in logic refers to what sorts of things variables are allowed to quantify over. First-order logic allows the values for variables to be entities. Second-order logical allows variables to quantify over predicates. For example,

```
(forall ?r
  (iff (transitive ?r)
       (forall (?x ?y ?z)
         (implies (and (?r ?x ?y) (?r ?y ?z))
                  (?r ?x ?z)))))
```

defines the notion of a transitive relation.

Why does order matter? One of the properties of a formalism is its *expressiveness*, that is, what sorts of things you can say (or not say) with it. Another property is *completeness*, that is, for those things you can say, can you always determine whether they are true or false? A third property is *tractability*, that is, how hard is it to determine, given a set of statements, whether or not these statements entail another statement (or, equivalently as it happens, if the set of statements and the negation of the other statement are contradictory)? These properties trade off against each other. The more expressive the language, the less tractable it is. The more efficient the reasoning method, the less likely it is to be complete. This is discussed more in chapter 3.

How do we understand the meaning of statements in a logical representation? There are elegant mathematical answers to this, in the form of *model theory*. The idea is to establish a correspondence between the terms and statements of the theory and objects and statements about them in the world that they are intended to represent. In the examples above, `CityOfDelmenhorst` is intended to refer to the city of Delmenhorst, Germany. The accuracy of the predicate calculus statements, when taken as statements about the real-world city, provides a way of evaluating whether or not the statements are correct.

There is an interesting issue lurking here that most readers can ignore. The person, group, or system writing the axioms has some intended meaning in mind. But model theory tells us that any consistent correspondence with some system suffices as a model. For example, the simplified axioms for the blocks world commonly used in AI courses, consisting of around a dozen or so rules, also have as their model ordered pairs of integers, where one integer indicates horizontal position (quantized) on a table, and the second integer indicates the number of blocks below it. So even though people give the predicates names like "Block" and "above" to capture the intended meaning, the existence of radically simpler models tells us that these small sets of axioms are not enough to pin down their meaning to just what was intended. It turns out that adding enough axioms to constrain logical theories to exactly their intended meanings takes substantial effort.[5] It requires far more than a handful of axioms. On the other hand, this makes the knowledge representation enterprise quite different in practice from how many think of it. Knowledge representation is a surprisingly flexible, incremental, and continuous affair: As more axioms are added, the set of possible models is whittled down. This process can be nonlinear—for example, finding out that there are flightless birds and water animals that breathe air radically changes your set of beliefs.

There is an additional complication when representations are used as part of computational systems. All computational systems involve some sort of representation, whether it be a set of numerical parameters or more complex structured data. To some, the ideal world is one in which all the various forms of knowledge that a system has could be written down as logical axioms. Perhaps such a world is attainable, but I personally suspect not. Most AI researchers find it convenient to describe the operation of some aspects of systems in more procedural terms when that language is better suited for expressing the intended meaning. For example, some predicates are grounded in perceptual processes, as in the computation of a visual structure from digital ink (chapter 16). Other predicates are tied to systems that take action, which can be viewed as functional approximations to motor processing. Since such computations help define the causal impacts of representations, a characterization of them is also part of the meaning of the representations. Developing clear descriptions of the computations that use representations is thus an important part of fully specifying their meaning.

If you want to learn more, many fine books have been written about logic and its roles in AI (e.g., Brachman & Levesque, 2004; Genesereth & Nilsson, 1987; Russell & Norvig, 2009) and cognitive science more broadly (e.g., Markman, 1998).

2.3 Schemas, Frames, and Cases

There have been several proposals that knowledge is organized in units that are larger than single statements. An early proposal was Bartlett's (1932) idea of *schemas*, in which a group of interconnected statements describing configurations or entailments is bundled together. By recognizing a situation, a process that involves binding the schema's variables to representations of entities in the current situation, the entailments stated in the schema prototype are then believed to hold in the instance of the schema. Minsky's (1974) idea of *frames* went beyond this idea in several important ways. He proposed that such frames had defaults associated with some of their variables, which could serve as expectations and stand-ins for missing information (e.g., someone's default color of a ball might be red). He argued for hierarchical memory structures, along with links between interrelated systems of frames to represent transformations in visual viewpoints and differential diagnoses (i.e., if you think something is a horse but it has vertical black stripes, consider the zebra).

Hayes (1985a) showed that some aspects of the semantics of these representations could be handily captured by traditional logic. In essence, the schema variables (or frame terminals) can be treated as logical variables, with the statements in the schema or frame being conjunctive conclusions of a (sometimes large) implication. The antecedent can be viewed as a predicate (i.e., the schema name with its variables as arguments represents an instantiation of it). The recognition criteria would then be further implications that imply particular schema statements. This accurately captures some but not all of the original intent of these representations, which also made particular hypotheses about the way memory retrieval works, something about which logic is (deliberately) agnostic. Nevertheless, this transformation is a useful exercise and is commonly used as an implementation technique for building systems using schemas and frames.

On the other hand, the term *cases*, as typically used, describes specific situations, systems, or scenarios. Cases represent a form of experience. They can be thought of as a collection of statements, treated as a unit. Cases are how knowledge is organized in case-based reasoning (Kolodner, 1993; Leake, 2000; Riesbeck & Schank, 1989), where reasoning about a current situation proceeds by retrieving one or more cases from a *case library* and then matching them against the current situation to figure out what to do about it. Case-based reasoning and analogical reasoning share the hypothesis that reasoning from experience is important in human cognition. However, AI research on case-based reasoning has assumed specialized indexing schemes to support retrieval, with both retrieval and matching typically being domain specific. As I argue in chapter 4, one does not need to do this: just as human performance arises from the same cognitive architecture being used across many domains, models of human analogical retrieval and matching provide domain-independent capabilities that can be used for case-based reasoning. Unfortunately, since the turn of the millennium, much work in case-based reasoning has de-evolved to using feature vectors as their representations instead of structured descriptions with relational information. This means, as noted above, that such systems cannot represent explanations, plans, arguments, or proofs—much of what makes human cognition interesting, in other words. I believe that the changing availability of large-scale representation systems, as discussed below, plus improvements in analogical processing capabilities will reverse this.

2.4 Ontologies and Knowledge Bases

Isolated concepts are not of much use to represent the world. In philosophy, ontology is the study of what kinds of things there are, so it was natural that as AI scientists started developing formal ways to organize knowledge, they adopted the term and started giving it more formal bite. For our purposes, we can think of an ontology as a set of concepts and relationships, specified via a small set of *structural relationships*. There is generally some form of conceptual hierarchy supported to make efficient use of knowledge (e.g., things generally true of `Animal` can be stated about it rather than redundantly under `Cat`, `Dog`, `Mouse`, `Elephant`, etc.). The ontology by itself does not contain enough axioms to specify the meaning of the concepts to the degree necessary to reason with them effectively (e.g., that animals need to eat and die after they are born). The additional axioms that provide interrelationships between the concepts (including ground facts, e.g., `(isa Skippy-TVCharacter Kangaroo)`) are, together with the ontology, called a *knowledge base*.

There are already a number of ontologies and knowledge bases around. Some of these are small and carefully crafted to support particular experiments and applications. Some knowledge bases consist of carefully curated axioms (e.g., Cyc), whereas others contain mixtures of structured and unstructured information (e.g., YAGO, which also includes Wikipedia articles as "concepts"). With the widening adoption of the Semantic Web, there has been a veritable explosion of ontologies for particular areas, such as biology, and for specific industries (e.g., the *New York Times* and BBC ontologies). For this book, it is useful to know that large-scale ontologies can and have been built, and they are being populated at scale (e.g., Google's Knowledge Graph, Microsoft's Satori, and Prismatic, the knowledge base learned by reading for IBM's Watson; Fan, Ferrucci, Gondek, & Kalyanpur, 2010). By "at scale," I mean hundreds of millions in the case of Watson and billions in the case of Google's Knowledge Graph. These efforts, driven by intensely practical goals, are excellent demonstrations that symbolic representations can scale in useful ways.

Fortunately, only a few conventions about ontologies are needed to understand this book. We draw them from the Cyc ontology,[6] although they are very common. There are three reasons for using this particular set. First, building good representations takes time and effort—better to reuse prior excellent work than to reinvent the wheel yourself (often badly). Second, using someone else's representations and knowledge bases in a computational model reduces tailorability.[7] Third, the upper ontology in Cyc has been used as a starting point for other large ontologies, such as YAGO and DBPedia, making an acquaintance with it a worthwhile investment.

Concepts in Cyc are represented by *collections*. For example, the collection `DomesticCat` has as members everything that we would recognize as a house cat.[8] To indicate membership, we use the `isa` relation:

```
(isa Nero DomesticCat)
```

Several different kinds of structural relationships hold between collections. The most important is `genls`, which indicates that one collection is a subset of another. For example,

```
(genls DomesticCat Mammal)
```

says that anything that is an instance of `DomesticCat` is also an instance of `Mammal`. Thus, any axiom that we wish to be true of cats, dogs, and giraffes, for example, might be written in terms of `Mammal`.

Information about predicates is itself represented in the knowledge base by structural relations.[9] For our purposes, the structural relationship to know about predicates is `genlPreds`, which is to predicates what `genls` is to collections. For example,

```
(genlPreds connectedViaHinge connectedTo)
```

indicates that when something is connected via a hinge to something else, it is also connected to something else.

One of the important advances in ontologies and knowledge base construction are mechanisms for partitioning the knowledge base into useful subsets and relating them appropriately. The most tested way of doing this is Cyc's concept of *microtheories*. A microtheory is a set of facts that can be viewed as forming a locally consistent set of information about some topic. Microtheories, like collections and predicates, have inheritance relationships among them: if one microtheory M1 inherits from M2, then anything believed in M2 is also believed in M1. All reasoning is done with respect to a microtheory and those it inherits from, which is the *logical environment* for that reasoning. The idea of a logical environment is crucial for any system that has to reason about fictional worlds, alternative hypotheses and theories, and the beliefs of other agents. For example, in reasoning about the possible outcomes of a situation, different microtheories are used to record distinct alternatives.

Cyc is quite large, in terms of its ontology. We use a subset of OpenCyc, plus our own extensions for qualitative reasoning, analogy, and language processing, in our research currently. As of this writing, it contains 1.3 million facts, divided into 1,135 microtheories, with more than 87,000 collections, 26,000 relations, and 5,000 functions. Even so, when using it, one finds that there are many gaps. Cycorp's goal was to build a large enough knowledge base by hand that could then be extended via learning, and there have been several successful experiments in doing this (Curtis et al., 2009; Forbus et al., 2007; Witbrock et al., 2015). OpenCyc is an especially valuable resource because of its size, its contents are freely available, and it has been linked to large resources with ground facts, such as DBPedia (Bizer et al., 2009). Even larger than OpenCyc is ResearchCyc, which has far more axioms than OpenCyc.[10] Cyc itself is larger still.

2.5 Richness and Structure of Predicate Vocabularies

A question of profound psychological importance is how large is the vocabulary of predicates that people use in their mental representations. As language users, it seems likely that lexical-level representations exist and are useful in language processing as well as skim reading. IBM's Watson, for example, demonstrated that a knowledge base consisting predominantly of lexical-level representations sufficed for human-level factoid question-answering in the kind of questions posed in the *Jeopardy!* TV game (Fan, Kalyanpur, Gondek, & Ferruci, 2012). On the other hand, for deeper reasoning about the contents of text, richer conceptual representations of the type found in Cyc have proven useful (Chang & Forbus, 2015; Lockwood & Forbus, 2009). Early models, such as conceptual dependency (Schank, 1972) and the LNR scheme (Norman, Rumelhart, & the LNR research Group, 1975), postulated a small set of primitives that could be combined to provide the semantics for a wide variety of verbs. Subsequent experience suggests that these might indeed provide an abstract summary that is useful for some purposes, but there is also a need for deeper, more specific representations. I believe that the kinds of qualitative representations described here serve as a crucial part of the vocabulary for those deeper representations. Our new-found abilities to do large-scale cognitive systems experiments in building knowledge bases via learning by reading and other modalities may provide fresh insights into these questions.

2.6 Summary: Evaluating Representations

Understanding representations is a central issue in understanding cognition. Representations are the currency of cognition, the inputs and outputs of the processes of cognition. Some of them can be banked (i.e., stored in long-term memory) to guide future operation and learning. This is why so much research in AI and cognitive science has focused on understanding representation.

The breadth of possible representations and the sometimes subtle trade-offs among them make it important to keep in mind some criteria for evaluating representations. Here are three criteria that are useful to keep in mind:

1. What do its elements mean, in terms of intended models? Carefully described theories are obviously better than representational vocabularies where new primitives are pulled out of the air or whose meaning is at best suggestive due to the use of natural-language sounding terms for predicates (McDermott, 1976). As per the discussion of model theory above, this means that good representations are constrained by axioms/rules that use them, and by the computations specified over them.
2. How can descriptions in a representational vocabulary be computed from inputs? In some cases, it might be presumed that they are directly computed by some subsystem in an organism (e.g., a specification of the representations produced by a vision system or natural-language parser).

But in most cases, an account of how these representations might be computed from more basic information needs to be provided.

3. How can the representation be used for reasoning? What kinds of reasoning does it support? What is the computational complexity of that reasoning?

The third criterion is sufficiently important that it merits its own chapter, so we turn to reasoning next.

3 Reasoning: An Overview

To understand human cognition requires, I believe, a broad view of reasoning. Simplistic models like "logic does it all" or "statistics does it all" can be useful approximations for particular investigations, but both leave out vital phenomena. For example, there is a common misconception that using symbolic, relational representations necessarily requires using logic, serial processing and no numerical or statistical information. In reality, the space of options is much broader and is continually being expanded. This chapter provides a brief guide to some of these issues and options. Seasoned cognitive scientists will be familiar with most of the ideas described here, whether or not they agree with any particular idea.

3.1 Computational Complexity and Tractability

Reasoning is an act of computation, carried out by organisms and/or systems that are realized in the physical world. Thus, they consume resources in doing reasoning, and characterizing resource usage is thus an important part of understanding reasoning. This is what is called *computational complexity* in computer science, and some understanding of it needs to be part of every cognitive scientist's toolkit.

To talk about the complexity of a computation, computer scientists use what is called *big-O* notation. This notation describes how the cost of a computation scales relative to the size of the problem it is applied to. Searching for an item in an unordered list of n elements, for example, is $O(n)$ in time (i.e., linear time). As the size of the list (n) doubles, in the worst case, you will have to search twice as long. That cost assumes a serial processor—if you have parallel processors, then you can do it in constant time, but $O(n)$ processors are required. These ideas really do apply to any resource, not just time—processors, storage space, and even energy consumed! But to keep things simple, we use time in serial processing for most of our examples.

One can do better than linear time if the list is sorted. Think about a dictionary in hardcover book form. To search a dictionary, you might start in the middle. If the item you are looking for isn't where you first looked, you know from the ordering if it is in the earlier half or the later half. That means you can recursively apply the same process, searching just one of the two halves, until you find it. This search process takes only $O(\log(n))$ time—if you double the number of items, you only have to do one more search step, on average, so searching an ordered list is logarithmic in its time complexity on a serial machine.

For any task, there are better or worse algorithms. Sorting a set of data, as one might do to get an ordered list and thereby speed subsequent retrievals, provides a nice example. A really simple way to sort is to walk through a list, flipping pairs of adjacent items that are out of order. This algorithm (called *bubble sort*) is $O(n^2)$. But there are much better sorting algorithms whose complexity drops to $O(n*\log(n))$. The difference between $O(n*\log(N))$ and $O(n^2)$ may not seem like much, but it can make an immense difference in practical terms. Can one do better than $O(n*\log(n))$? In fact, it has been proven that, on a serial machine, $O(n*\log(n))$ is the best you can do. Such bounds are useful because they apply to any system, mechanical or biological, that is doing the same task under the same assumptions.

Computational complexity often tends to be about worst-case analyses, to establish bounds on resource usage. If the distribution of types of problems is known in advance, sometimes average-case performance results can be obtained. And if one is measuring an algorithm, concrete estimations of its complexity in practice can also be found. Can this be used to determine whether or not a particular algorithm is being used by people? In some cases, yes, but because assumptions about algorithms, representations, and architecture (e.g., how many processors and of what kinds) are all involved, it has to be done with great care (Barton, Berwick, & Ristad, 1987).

A key issue in complexity is what is called *tractability*. A computation is *polynomial* if there is some polynomial of n such that the property of interest can be described as in the order of that polynomial. A computation is *nonpolynomial* if the expression in big-O notation is something that grows faster than a polynomial, such as an exponential. Computations that are polynomial time are traditionally called *tractable*, and those that aren't are traditionally called *intractable*. This term is often misunderstood in cognitive science. Intractable does not mean impossible. It means that the computational cost grows rapidly as the size of the inputs grows. If the size of the inputs is kept small, such computations are often cheap, especially if one can allocate parallel processing resources for each input. But as the size of the inputs grow, the cost of exponential computations grows quite rapidly. Some important implications of this are discussed below.

3.2 Deduction, Abduction, and Induction

The term *reasoning* is what Marvin Minsky (2007) called a *suitcase term*: a convenient way of speaking about what is actually many different phenomena, stuffed into a single container even though their nature and operation (as it is with clothes and toiletries and shoes, for example) are quite different. We start here with a basic set of distinctions. The first model of reasoning, proposed by philosophers a few millennia ago, is deductive reasoning (Shapiro, 2013). A classic rule of deductive reasoning is called *modus ponens*, or *conditional elimination*. That is, if we believe P implies Q, and we believe P, then we must believe that Q is true. This piece of inference does seem to capture something important about reasoning. There are two ways to build a system of deductive reasoning. The first is to develop a set of rules that expresses valid consequences, given a set of assumptions. (Modus ponens is one such rule.) This is the *natural deduction* approach. Many systems of natural deduction have been developed, exploring different trade-offs in notations. The second approach is to use a single rule, *resolution*, which relies on transforming statements into a more primitive but equivalent form, called clausal form. Both of these approaches can be made equally powerful, in terms of what conclusions they can draw, in theory.

Logicians and mathematicians have come up with some useful theoretical ways to characterize sets of rules. A set of inference rules is *sound* if the conclusions it draws are always correct, given correct starting assumptions. A set of inference rules is *complete* if anything that actually follows from a set of assumptions can be proven by the rules. For propositional reasoning and first-order predicate calculus, one can make systems of rules that are sound and complete. These are important properties of logical systems, to be sure. However, we must also ask about the computational properties such logics have. One property is *decidability*: can an algorithm be found that will always produce correct results? For simple tasks like sorting, the answer is clearly yes. However, first-order logic is *undecidable*, that is, no algorithm will guarantee producing correct proofs of everything that follows from a set of assumptions in the general case. The "in the general case" caveat is important: deduction algorithms are widely used in applications such as scheduling and design verification, showing that they can be useful for many real-world reasoning tasks.

A second property we must be concerned with is, of course, computational complexity. This brings us to one of the reasons why most cognitive scientists believe that logical deduction cannot be the main explanation for human reasoning. Deduction in first-order logic is nonpolynomial in time. This means that it can become impractically slow to perform as the amount of knowledge involved grows. Reasoning in higher-order logics is undecidable (i.e., one cannot guarantee that an algorithm that will always converge to an answer, due to its increased expressive power). Worse, higher-order logics are known to be incomplete, meaning that true statements can be made in them that are true but cannot be proven to be so (this was one of Godel's famous results; Nagel, Newman, & Hofstadter, 2001). Yet all the evidence points to human beings having massive amounts of knowledge, so the size of the inputs (i.e., knowledge to be applied to the current situation) is far from small. Nevertheless, people reason very rapidly and successfully, with massively more knowledge than today's automated reasoning systems that operate on logic can do. And what is especially tantalizing is that, as people know more, they are often *faster* in reasoning than when they know less (Forbus & Gentner, 1997). Understanding how and why is one of the key challenges of cognitive science, and our proposal for explaining this is outlined at the end of chapter 4 and examined further in chapter 12.

It is easy to read too much into these properties, particularly into completeness. Any physically realized computational system will, by its nature, be incomplete: we don't have infinite time and infinite resources. For example, much nonsense has been written about the supposed implications of Godel's work for AI and cognitive science. The only constraint it really imposes is that if people (or machines) operated as purely deductive systems using a second-order (or higher) logic, then they would not be able to prove every statement that actually follows from their assumptions. I personally find the idea that people, or machines, are operating purely deductively to be nonsense, and there are plenty of counterexamples, as discussed below. In fact, none of the smartest systems in AI are purely deductive. All use heuristics that are unsound and exploit statistics as part of their operations where appropriate.

What makes the story complicated is that aspects of deduction are essential in explaining reasoning. Any representation system that cannot express disjunction and negation simply cannot handle many of the kinds of thinking that we do. For instance, knowing that a cat is inside the house means that it is in one of the rooms, and if one has searched all but one room without finding it (carefully closing doors behind you to prevent movement), then it must be in the only room that you haven't searched. Similarly, being able to have general knowledge, containing variables that enable it to be applied to a broad range of circumstances, is also crucial.

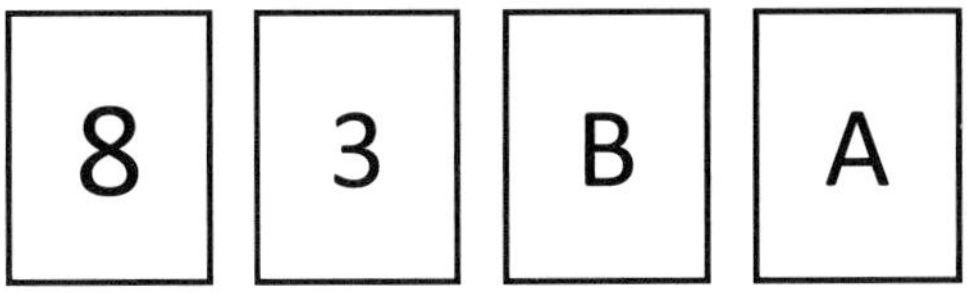

Figure 3.1
An example of a Wason task.

Any reasoning system that does not define a notion of contradiction will not be able to detect errors in its reasoning, so aspects of deduction capture important properties of reasoning. On the other hand, there are several reasons to view deduction as insufficient for a model of human reasoning, in addition to issues of computational complexity. One such reason is that there is ample evidence that people often do not always reason deductively. The classic Wason task (Wason, 1968) from cognitive psychology provides an example. Suppose you have four cards (see figure 3.1). Each card has a letter on one side and a number on the other. Your job is to pick up as many cards as you need to confirm that the cards conform to the following rule: if the number is odd, then the other side of the card is a vowel. Which cards do you need to pick up?

As it happens, most people get this wrong. One must pick up the card marked three to ensure that the other side is a vowel. One must also pick up the card marked B to ensure that the number on the other side is not odd. No other cards need to be picked up. Many variations of this task have been tried. Abstract versions of the task tend to lead to the same performance. On the other hand, concrete framings that rely on common sense do not (e.g., if someone is younger than twenty-one, his or her drink must not contain alcohol), even though they seem logically equivalent to the original task. There have been several explanations for this, including the use of constructing and inspecting concrete models (Johnson-Laird, 1983) and domain-specific schema (Cheng & Holyoak, 1985). Note, however, that although neither of these methods are deductive, they all assume some form of structured representation. That is, they have ways of representing the cards, there are symbols of particular types on one side or another, and they should be governed by a rule.

Another limitation of deductive logic as a model for human reasoning lies in the strength of the axioms. Deductions are taken to be true all the time, without exception. Rules, in human discourse, are slippery things, not the hard-and-fast, universally true statements of deduction. There have been many attempts to provide formalisms that model this character, ranging from fuzzy logic (Zadeh, 1996) to nonmonotonic logics (Antonelli, 2012). We return to a connection between qualitative representations and fuzzy representations in chapter 5. The idea of a nonmonotonic logic is to capture the phenomena of default reasoning. For example, if we know Tweety is a bird, then we know that Tweety can fly: Unless, of course, it is a penguin, baked, dead, or a stuffed toy. Consider the size of the set of consequences that follow from a set of assumptions. Suppose a new assumption is added. If it is logically entailed by the others, then the size of the set of consequences does not change. Otherwise, it grows, based on what can be inferred from the new assumption plus the others. Thus traditional logics are *monotonic*. This is problematic when new information can invalidate default assumptions. Returning to Tweety, when we learn that it is a bird, we assume it can fly, which entails many consequences (e.g., you need to keep windows closed to confine it in a room). Subsequently learning that Tweety is a penguin invalidates those default assumptions—the set of consequences can actually shrink. That is what is meant by *nonmonotonic logic*. Defaults and nonmonotonic reasoning are essential in human reasoning, so understanding how to do nonmonotonic reasoning is important. The formalisms of nonmonotonic logic are still in fairly early stages of development, so they have not been used much outside of those exploring their properties at this writing.

One nonmonotonic predicate is used in this book, `uninferredSentence`. Its meaning is this: (`uninferredSentence` *<p>*) is true exactly when *<p>* cannot be currently proven by the system doing the reasoning. This is commonly known as *negation by failure* in the logic programming community, and it is a key feature of Prolog. It also provides a form of closed-world assumption, in that the failure to derive the statement relies on the available knowledge when the query was made.

Building models of statements is one nondeductive way of using structured representations. Another is *constraint propagation*. One version of this, *spreading activation*, was one of the earliest computational models proposed in cognitive science (Collins & Quillian, 1969). The idea is that, given a question like "Is Tweety an animal?" that activation would spread up links indicating superordinates, and if the node for animal became active, then the answer would be yes (figure 3.2).

Such semantic networks received substantial attention early in cognitive science and have evolved into what are known as *description logics* (Baader, Calvanese, McGuinness, Nardi, & Patel-Schneider, 2010). Description logics are the theoretical foundation for a key Semantic Web technology, the OWL family of representation languages. In this evolution, the idea of propagating numerical values was lost, and a very formal semantics that is crisp,

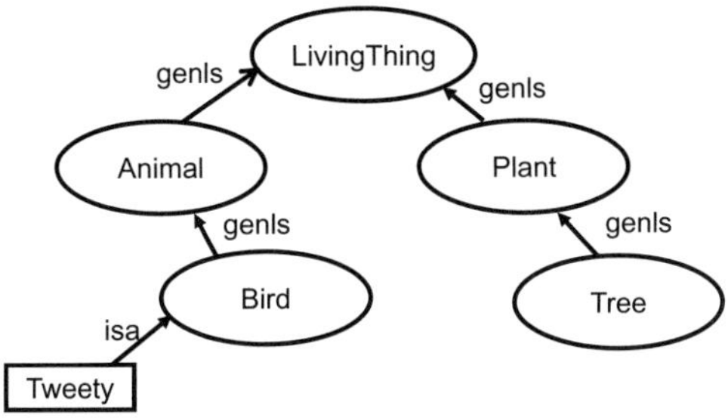

Figure 3.2
A simple semantic network.

but unfortunately somewhat limited, was gained. On the other hand, such spreading activation techniques are still widely used in long-term memory models in cognitive architectures, as discussed below.

So far, we have seen that deduction does capture some aspects of human reasoning. Another kind of reasoning that captures part of human reasoning is *abduction*. Abduction is the process of finding explanations. It is often viewed formally as

```
(implies A B)
B
----
A
```

Here A is an *abductive assumption* (i.e., something assumed to be true in order to support the belief in B). Although unsound, abduction is crucial in plan recognition (e.g., "Why is that person loitering outside the restaurant? Maybe he is waiting for someone.") and for natural-language understanding (e.g., "The brick is hot." typically does not mean that the brick is spicy when used in cooking or sexy, although it is logically possible that either might be valid interpretations in some unusual context). In general, multiple assumptions must be made because the inference might require multiple steps. Moreover, in any rich axiom system, there can be multiple sets of relevant implications. Usually, some notion of simplicity or cost is used to guide a search process to find the simplest (or least costly) set of assumptions to explain the phenomenon (Hobbs, 2004). Abductive methods do not guarantee that their explanations are the correct one, only that, given available evidence, they could be an explanation.

Induction involves learning new concepts and rules from examples or from more concrete rules. Learning to recognize different kinds of animals is an example of an inductive task. Inputs can include positive evidence ("That's a chipmunk.") and negative evidence ("No, that's not a doggie. That's a skunk."). Classification is the focus of many types of traditional machine learning, but it is insufficient to explain human learning. People also learn rules and causal relationships (e.g., if you need to unlock something, find a key). A number of machine-learning models combine logic and statistics to learn such relational information, such as inductive logic programming (Muggleton, 1992) and Markov logic networks (Richardson & Domingos, 2006). The extent to which any of these systems can serve as a model for human cognition is not known at this time. Current formulations of Markov logic networks, for example, use a technique known as *propositionalization* for reasoning, which is exponential in complexity in the size of the axioms and hence extremely slow computationally. The analogical learning techniques discussed in chapter 4, I believe, provide solid models for handling such relational learning, but they have not yet been tested on the range of problems that the inductive logic programming and logic/probability hybrid communities have explored.

3.3 Pattern Matching and Unification

Pattern matching is a way of binding variables, to apply general knowledge to specific situations. Suppose we have an axiom like

```
(implies (Human ?x) (Mortal ?x))
```

And we know

```
(Human Robbie)
```

By matching the antecedent of the implication with this ground fact, we bind `?x` to `Robbie`, and we get the implication that (alas)

```
(Mortal Robbie)
```

Unification is a specific form of pattern matching. It computes the set of variable bindings needed to make two statements identical. A number of other forms of pattern matching have been proposed in AI, but most have dropped away in favor of unification. In discussing analogical matching, we will see quite a different form of pattern matching. As it happens, neither form of pattern matching subsumes the other. Analogical matching allows bindings between constants, which is not allowed in unification. Unification allows multiple variables to be bound to the same value, which is not allowed in analogical matching. An interesting open question is whether analogical matching can completely replace unification in a cognitive architecture.

3.3.1 Storing and Retrieving Knowledge

We assume that people, like other cognitive systems, can be viewed as having a knowledge base, a functional way of storing what they know. How should a cognitive system retrieve knowledge from its knowledge base? The typical AI answer is to find all matching knowledge, relative to a particular logical environment. This can result in a large number of answers, depending on what is known. Some cognitive architectures, taking a clue from psychological theories, add additional constraints in the form of numerical filtering schemes. For example, estimates of the utility of a fact and how recently it has been used are often stored in the knowledge base and updated automatically as processing proceeds. These schemes typically rely on spreading activation, with automatic decay in activation to model effects of recency. Spreading activation models are what are called *data-parallel* algorithms (i.e., each piece of data is assumed to be an active processing unit, which means their scaling properties are quite promising). Implementing them efficiently on serial hardware is a challenge, so little research has been done on scaling to knowledge bases that even remotely approach human size (but see Derbinsky, Laird, & Smith, 2010).

As noted in chapter 2, a number of cognitive theories postulate that knowledge is retrieved in larger units than single statements. In cognitive architectures based on production rules, the declarative data retrieved are chunks, which are equivalent to multiple propositions. Similarly, schemas, frames, and cases can be viewed as sets of propositional statements, grouped based on their utility in being used together. However, case-based reasoning systems rarely use spreading activation schemes. Instead, retrieval in most case-based reasoning systems relies on indexing (i.e., maintaining persistent data structures that link potential matches with easy to compute properties of the current situation).

3.4 Closed-World Assumptions

Rarely does one have all of the knowledge that would be useful to make a decision or to think an issue through. People are constantly using heuristics to overcome gaps in knowledge (Tversky & Kahneman, 1974; Gigerenzer, Todd, & the ABC Research Group, 2000). One important strategy is to make *closed-world assumptions* (Collins, Warnock, Aiello, & Miller, 1975). A closed-world assumption is assuming that the knowledge one has on hand is all that is relevant to the reasoning being performed. Many systems of reasoning use implicit closed-world assumptions. An example is Prolog's negation by failure rule, where if something is not provable by a system of rules, it is deemed to be false. A more powerful way of using closed-world assumptions is to make them explicit. That is, an explicit statement is created during reasoning, and subsequent inferences that are based on that assumption are justified in terms of it (Forbus & de Kleer, 1994). If a problem is found later with a conclusion, the dependence on that closed-world assumption can be used to diagnose what additional information must be gathered (or models learned) to come to more accurate conclusions. Such closed-world assumptions are used heavily in qualitative reasoning.

3.5 Probability

Uncertainty is a fact of life. Probability is a mathematically clean way to model uncertainty. There is ample evidence that people compute implicit statistics on a variety of parameters (Anderson, 2009). However, there is also ample evidence that human reasoning often does not work in ways that follow normative prescriptions from a probabilistic reasoning perspective (Gigerenzer et al., 2000; Kahneman, Slovic, & Tversky, 1982). This did not prevent Bayesian models from becoming extremely popular in cognitive science at the turn of the twenty-first century (e.g., Griffiths, Chater, Kemp, Perfors, & Tenenbaum, 2010), although it has been argued that they are ultimately unsatisfying because they do not provide the kind of mechanistic explanation that cognitive science seeks (Jones & Love, 2011; Marcus & Davis, 2013). Bayesian accounts of causality can be viewed as complementary to the accounts of causality developed in the qualitative reasoning community, as discussed in chapter 9. As the next chapter shows, priors for facts can be derived from experience via analogical generalization. This means that analogical learning provides a powerful substrate for probabilistic reasoning and learning.

4 Analogy

Reasoning, as described so far, operates from first principles. That is, given a set of assumptions, a collection of rules is used deductively (or abductively) to infer conclusions from statements representing the current situation. In the pure first principles account, there is no room for experience. This seems unrealistic. Human experts constantly rely on examples and experience in their professional lives, and all of us rely on examples and experience in our daily lives. A robust model of human reasoning cannot be based on first-principles reasoning alone.

Dedre Gentner and I have come to believe that analogical processing is a core operation of human cognition. That is, processes governed by the laws of structure-mapping theory (Gentner, 1983) seem to be operating from low-level visual processing (Lovett, Gentner, Forbus, & Sagi, 2009; Lovett, Tomai, Forbus, & Usher, 2009) to problem solving (Catrambone, Craig, & Nersessian, 2006) to conceptual change (Gentner et al., 1997). I believe heavy reliance on analogical processing is the solution to the mystery of the effectiveness and scalability of human reasoning outlined above. This chapter outlines the basics of structure-mapping theory; describes our models of analogical matching, retrieval, and generalization; and outlines why we think they are so important in human cognition. These models are used throughout the book, especially in chapters 12, 16, 17, and 18.

4.1 Some Psychologically Motivated Representation Conventions

There are some additional conventions beyond those of chapter 2, drawn from structure-mapping theory, which we will be using in this book.

The first convention is the distinction between *attributes* and *relations*. An attribute is a unary predicate used in making statements about properties of an object and what kind of object it is. For example, the statements

```
(RedObject Truck18)
(FireTruck Truck18)
```

are two attribute statements that indicate that `Truck18` is red and is a fire truck, respectively. In Cyc terms, these would be written via `isa` statements:

```
(isa Truck18 RedObject)
(isa Truck18 FireTruck)
```

These are viewed as a syntactic variation, with identical meaning. Note that the following (assuming sensible axioms to pin down the meaning of the predicates) are all logically equivalent:

```
(RedObject Truck18)
(primaryObjectColor Truck18 RedColor)
(= (ColorFn Truck18) RedColor)
```

However, they are not psychologically equivalent, according to structure-mapping theory. In structure mapping, functions (here, `ColorFn`) are treated differently from predicates because the former denote dimensions of entities, whereas the latter are relationships that can participate directly in other relations. Atomic terms, like `Truck18`, are considered *entities* under structure mapping. Nonatomic terms, such as `(ColorFn Truck18)`, are also considered entities, but more substitutions are allowed in them, as explained below.

Structure mapping also introduces a different notion of order, which describes the level of nesting with statements. Entities have order 0, and the order of a statement is one plus the maximum of the order of its arguments. Thus, our earlier implication about summer being rainy in Delmenhorst is a third-order statement, by this definition, despite it not having any variables. In other words, this structural notion of order, which is useful for thinking about analogy, is completely unrelated to the logician's notion of order. When order is mentioned in this book, it is the structure-mapping notion of order that is meant by default, with the exception of chapters 2 and 3.

4.2 Structure-Mapping Theory

Structure-mapping theory proposes that analogy involves comparing two structured, relational representations. These descriptions can indeed have attribute information as well, of course. These descriptions are traditionally called the *base* and *target*, respectively. Commonly, the base is the description about which more is known, but this is not always so.

Comparison proceeds via *structural alignment*, which computes one or more *mappings*. Each mapping consists of three parts:

1. A set of *correspondences* that identifies what goes with what (i.e., how entities and statements in the base align with entities and statements in the target)
2. A *structural evaluation score*, providing an estimate of match quality
3. A set of *candidate inferences* that indicates how information connected to the mapping in the base can be projected into the target. The set of candidate inferences can be empty, and reverse candidate inferences (i.e., from the target to the base) are also allowed.

The correspondences provide information about how the two descriptions are alike, with the structural evaluation score providing information about how much they are alike. Candidate inferences can be viewed as conjectures about one description based on information projected from the other. They are conjectures because there is no guarantee that they are valid; they must be scrutinized by processes outside the purview of analogical processing. Candidate inferences can also be viewed as information about differences. Candidate inferences can have *skolems* (i.e., new entities postulated by the analogy). An important historical example was the introduction of *caloric* to make the heat/water analogy work. Just as a difference in pressure causes water to flow, a difference in temperature was conjectured to be causing a flow in a fluidlike substance, caloric. Working through the implications of that analogy led to a search for a deeper understanding of heat (i.e., if caloric really were a form of fluid, then one should be able to drain all of it out of a body—something that Count Rumford showed could not be done).

There is psychological evidence that this same process is the heart of human similarity judgments (Markman & Gentner, 1993), difference judgments (Sagi, Gentner, & Lovett, 2012), and many metaphors (Wolff & Gentner, 2011), as well as analogy.

Figure 4.1 provides an example comparing a spring-block oscillator to a pendulum.

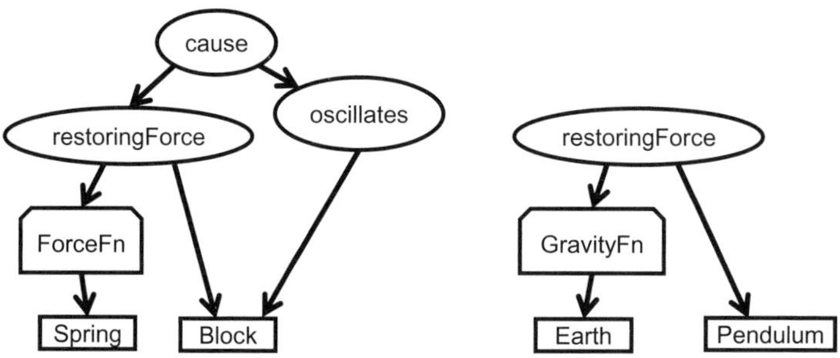

Figure 4.1
Simplified descriptions of a spring-block oscillator and a pendulum.

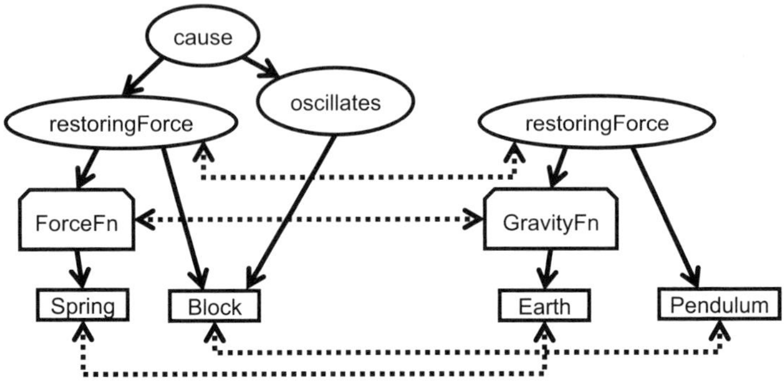

Figure 4.2
Correspondences for a simple analogy.

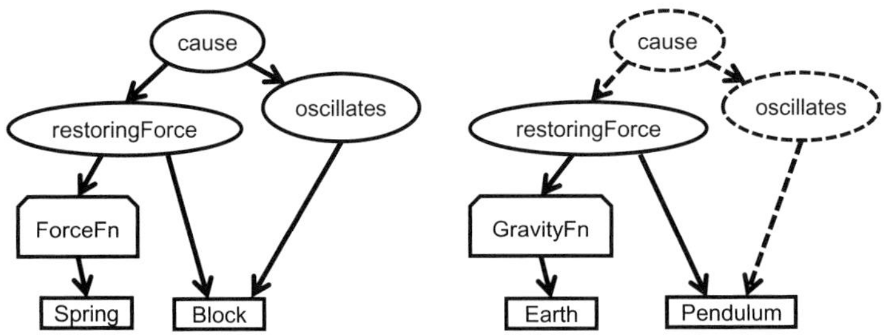

Figure 4.3
Example of a candidate inference.

(For exposition these representations are kept simple; they are a small subset of what would usually be included. Descriptions automatically produced by qualitative reasoning systems are generally several times larger.) The structured representation on the left can be read as, "The restoring force that the spring provides on the block causes the block to oscillate." On the right is an (again very partial) description of the pendulum, which can be read as, "The gravity of the Earth provides a restoring force on the pendulum." Entities are depicted by rectangles, functions by rectangles with cut corners, and relations by ovals. Figure 4.2 shows the mapping that our analogical matching system SME, described below, produces when comparing these two descriptions, using dotted arrows to show what corresponds with what.

Part of the power of analogy comes from its suggestion of how knowledge from one description can be imported, via projection, into the other. Figure 4.3 illustrates the *candidate inference* suggested by these correspondences—namely, that the restorative force on the pendulum will cause it, too, to oscillate.

The process of structural alignment is governed by four constraints:

1. *Identicality*: By default, only identical relations, attributes, and functions are matched. Nonidentical functions can be matched when suggested by higher-order matches (i.e., matching statements that have them as arguments). For example, in figure 4.2, `ForceFn` and `GravityFn` are aligned because of their shared roles as arguments to a `restoring-Force` statement. Similarly, nonidentical relations, which are themselves similar in some way, can be aligned to support a higher-order match.
2. *One-to-one mappings*: Each item can be matched with at most one other in a mapping.
3. *Parallel connectivity*: If two statements match, their corresponding arguments must also match. The combination of one-to-one mappings and parallel connectivity is often referred to as *structural consistency*.
4. *Systematicity*: Mappings involving systems of relations, especially higher-order constraining relations, are preferred.

Why do these constraints matter, cognitively? Identicality is a strong semantic constraint: analogy is not just subgraph isomorphism. Consider the following four statements:

1. `(preferredPreyType Cats Mice)`
2. `(preferredPreyType ConArtists NaiveInvestors)`
3. `(preferredFoodType VeganPeople Vegetable)`
4. `(typicalMaterialType AutomobileBody Metal)`

All four are general statements about concepts. The first statement says that things that are mice are a preferred type of prey for things that are cats. Allowing #1 and #2 to align in an analogy makes sense: con artists can bilk sophisticated people, but naïve investors are easier prey. One can make a case for allowing #1 and #3 to match, because both involve a preference concerning consumption; it is only how the meal is pursued and caught that differs. On the other hand, #1 and #4 do not make sense to align at all, because #4 is a constitutive relationship. Hence, a matcher should always be willing to align #1 and #2, as well as align #1 and #3 if it will yield some larger inference because these statements are themselves part of a larger relational structure. The reason for restricting such nonidentical correspondences to cases where they will fit in a larger structure is that blindly doing all such substitutions could lead to a combinatorial explosion.

There have been several ways of relaxing strict identicality in the literature; hence, it is often referred to as *tiered identicality*. One is *minimal ascension* (Falkenhainer, 1990), where two predicates are allowed to match if they share a close common superordinate. For example, `Dog` and `Cat` might be allowed to match if they are both subconcepts of `HouseholdPet`. Another example, in IBM's Watson, was a structure-mapping matcher that used WordNet similarity measures on a predicate vocabulary that is essentially lexical (Murdock, 2011).

Why are nonidentical functions allowed to match? Functions are ways of specifying non-atomic terms, as described in chapter 2. Psychologically, these provide information about dimensions or parts of some kind. For example, in understanding an animal's immune system, one might find a correspondence like

`(MilitaryFn Country18)` ↔ `(ImmuneSystemFn Animal6)`

that is, that the military of a country is like the immune system of an animal. Other statements in the base might provide the information that the military protects the country from foreign invaders, but it can fail by attacking the country itself. Being able to place a country's military in correspondence with an organism's immune system would then provide a basis for reasoning about the normal function of immune systems and one way they can fail (i.e., autoimmune diseases). Cross-domain analogies often have such nonidentical functions corresponding, as in the famous example of the heat/water analogy, where temperature and pressure are aligned in determining what causes flow.

Structural consistency is important for ensuring coherent candidate inferences. Consider an analogy used in explaining the Enron scandal, one of the largest frauds in American history. A comparison was made to the *Titanic*, which sunk after hitting an iceberg. Lookouts on the *Titanic* warned the captain but were ignored. When the ship sunk, many passengers lost their lives. If we work through the scandal, we end up with the set of correspondences between the entities shown in table 4.1.

Table 4.1
An analogy between the sinking of the *Titanic* and the Enron scandal.

Titanic ↔ Enron
Captain of the *Titanic* ↔ Enron CEO Ken Lay
Iceberg ↔ Accounting fraud
Passengers ↔ Investors
Lookouts ↔ Whistleblowers

As discussed below, these entity correspondences are actually created as a consequence of matches between statements, such as the following (remaining in English for simplicity):

> "Lookouts warn captain but are ignored" ↔ "Whistleblowers warn Key Lay but are ignored"

At one point in the trial, Ken Lay referred to himself as a whistleblower. Suppose we take this claim seriously. Then we also would have

> Lookouts ↔ Ken Lay

which violates the 1:1 constraint. This might not seem to be a problem until one tries to construct a candidate inference:

> Lookouts warn captain but are ignored → Ken Lay warns Ken Lay but is ignored.

This is not the sort of inference that should be sanctioned—the 1:1 constraint helps ensure that candidate inferences are at least coherent.

The importance of parallel connectivity is that it ensures constituents of explanations can be projected. Suppose we ignore parallel connectivity and map the following two statements:

```
(implies (and (LeakingFluidDevice BrakeCylinder2)
              (partOf BrakeCylinder2 Car54))
         (DangerousDevice Car54))
  ↔
(implies (EjectsFlames BattleBot12)
         (DangerousDevice BattleBot12))
```

It seems reasonable that the two `DangerousDevice` statements should align, but what should `BrakeCylinder2` go with? The two explanations aren't very similar, so matching them (and hence the implication) doesn't make sense.

Systematicity is important because descriptions with higher-order constraining relations (in the structure-mapping sense of order) structurally correspond to statements that are well-justified arguments, semantically. Consequently, mappings with higher systematicity are more likely to lead to stronger candidate inferences.

The distinction between attributes and relations suggests that comparisons can be grouped into four different kinds, depending on whether the amount of overlapping attributes or relations is high or low (see table 4.2).

These have different psychological properties. Analogies and overall similarity matches are more likely to produce inferences, and these inferences are more likely to be believed. Surface and overall similarity matches

Table 4.2
Types of matches that fall out of analogical processing.

Type	Attributes	Relations
Anomaly	Low	Low
Analogy	Low	High
Surface (also known as mere appearance)	High	Low
Overall similarity (also known as literal similarity)	High	High

are more likely to be retrieved, even though appearance matches are not viewed as valuable inferentially. We return to this dissociation when discussing retrieval below.

4.3 Psychological Support for Structure-Mapping Theory

A variety of studies support the claims of structure-mapping theory. For example,

- Systematicity and structural consistency influence the interpretation of analogies (Clement & Gentner, 1991).
- Structural consistency influences inference in analogical reasoning (Markman, 1997; Spellman & Holyoak, 1992).
- Structural consistency influences inference in category-based induction (Lassaline, 1996; Wu & Gentner, 1998).
- Systematicity influences inference in analogical reasoning and category-based induction (Bowdle & Gentner, 1997; Clement & Gentner, 1991; Wu & Gentner, 1998).
- Ordinary similarity comparisons use structural alignment and mapping (Gentner, 1989; Gentner & Markman, 1997; Markman & Gentner, 1993; Medin, Goldstone, & Gentner, 1993).
- Relational shift hypothesis: Early in development, object matches win over relational matches, in part because of inadequate relational knowledge (Gentner, 1988; Gentner & Rattermann, 1991; Gentner & Toupin, 1986; Richland, Morrison, & Holyoak, 2006).
- Learning higher-order domain relations enables children to perform relational mappings (Gentner & Ratterman, 1991; Goswami & Brown, 1989; Kotovsky & Gentner, 1996; Ratterman & Gentner, 1998).

The evidence continues to accumulate, suggesting that it provides a strong base upon which to build computational models.

4.4 Computational Models of Analogical Processing

The ideas of structure-mapping theory have been extended to cover analogical retrieval and analogical generalization as well as analogical matching. The computational models for these three processes have been used to model a variety of human phenomena and have also been used as modules in performance-oriented systems (Forbus, 2001). From an AI perspective, together they form the basis for a technology of analogical reasoning and learning, grounded in human cognition. From a broader cognitive science perspective, the fact that these systems can and have been used to perform in a wide range of domains and tasks provides evidence that these ideas may indeed be at the heart of human cognition. We describe our models for analogical matching, retrieval, and generalization in turn. These will play a major role in our account of qualitative reasoning. While they permeate our thinking about how people reason with and learn qualitative representations, they will be especially important in chapters 12 and 16 and in part IV.

4.4.1 Matching

The structure-mapping engine (SME) provides a model of analogical matching (Falkenhainer, Forbus, & Gentner, 1989; Forbus, Ferguson, Lovett, & Gentner, 2017). It takes as input two structured, relational representations, the base and target, as outlined above. These descriptions can be incrementally updated with new information, leading to incremental updating of the match (Forbus, Ferguson, & Gentner, 1994), but we won't discuss incrementality further here. SME produces one or more mappings, as defined by structure-mapping theory. Its default operation is to produce up to three mappings, if the next mappings are within 10 percent of the best mapping in terms of their structural evaluation scores.

Matching graphs can be expensive. For example, detecting that two graphs are identical (subgraph isomorphism) is NP-Complete, meaning that it is in the class of algorithms that are exponential or worse. However, SME uses a strategy that is common in computing: give up guaranteeing an optimal answer in favor of more rapidly computing approximate answers that are good enough in practice. SME operates in polynomial time by exploiting a *greedy merge* algorithm. Greedy algorithms do not guarantee optimality, but they are very fast. The worst-case complexity of the current

SME algorithm is $O(N^2\log(N))$ on a serial processor, making it quite efficient in practice.

How SME works is interesting, both as a novel approach to matching and also because its algorithm makes additional psychological predictions that have held up under laboratory testing. It starts by computing, in parallel, local matches between statements based on identical predicates. For instance, in the description of figure 4.1, the two `restoringForce` statements will be conjectured to match. This set of local matches is expanded, continuing in parallel, to attempt to match the arguments of statements that have been proposed as matching. This expansion is where matches are hypothesized between pairs of entities, pairs of non-atomic terms that have different functions, and pairs of statements with nonidentical predicates. Returning to figure 4.1, if the `restoringForce` statements are to match, by parallel connectivity, then `Block` and `Pendulum` should be matched, as should `(ForceFn Spring)` and `(GravityFn Earth)`. This initial parallel process typically results in an inchoate set of local matches. The next phase, still operating in parallel, does three things: (1) it marks match hypotheses containing unaligned arguments as structurally inconsistent, (2) pairs of match hypotheses that would violate the 1:1 constraint are marked as mutually incompatible (i.e., *nogood*), and (3) evidence is propagated downward from match hypotheses between statements to match hypotheses between their arguments. The downward propagation of evidence provides a means of implementing systematicity: entity correspondences that support large, deeply nested relational structures get more evidence than those that do not. Next, large structurally consistent sets of match hypotheses are identified by starting with the match hypotheses that are structurally consistent and not included in the support for some other match hypothesis. These *kernels* form the basis for constructing global mappings. The score of each kernel is simply the sum of the evidence for each match hypothesis in it. Kernels are sorted by their score and greedily merged to form mappings. The greedy process uses the nogoods to avoid adding kernels that would violate the 1:1 constraint. Candidate inferences are then computed by taking the statements in the base (or target, if reverse inferences are being computed also) that are connected to the mapping and computing their projection via making the substitutions implied by the correspondences. If entities in the projected statement do not have correspondences, placeholders called *analogy skolems* are introduced to represent them.

The SME algorithm can be viewed as a *middle-out* algorithm. Bottom-up analogical matchers, which begin by matching entities (e.g., Winston, 1980), have poor computational complexity, with $O(n!)$ in the number of entities in the base and target. Top-down analogical matchers (e.g., Burstein, 1983; Keane, 1995) attempt to identify a priori what structure in the base ought to be projected and then search for ways to align them. The data-parallel phase of SME is $O(n^2)$ on a serial machine (or $O(n^2)$ processors, if parallel). Entity matches and nonidentical statement and term matches are only proposed when suggested by the data, which is a considerable savings in practice. This algorithm also makes an important prediction about the time course of analogical matching: there should be an initial alignment phase that is symmetric, without any distinction between base and target (i.e., the data-parallel phase), followed by an asymmetric phase (i.e., the construction of candidate inferences, which occurs from base to target by default).

This predication has been tested in two kinds of studies. The first concerns metaphor interpretation (Wolff & Gentner, 2011). Some metaphors only make sense in one direction: people are much more likely to say, "My job is a jail" than "My jail is a job." Suppose you give people a series of sentences and ask if they are good metaphors. If you demand an answer from them before 600 milliseconds, they have trouble distinguishing between forward and reversed metaphors, which is what one would predict if they are interrupted during that first parallel processing phase. If you interrupt them later (say, at 1200 milliseconds), they show the expected preference for forward over reversed direction.

The second test of this prediction concerns difference detection. A paradox of difference detection is that people are faster to detect that two things are different when they are very different, but people are faster to state a difference when two things are very similar (Sagi et al., 2012). Sagi and colleagues' explanation for this paradox relies on how difference detection is rooted in the computation of similarity. When two items are very different, the initial match hypothesis forest constructed during the parallel phase of processing is very small. This alone suffices to say that the two are different, because a small forest means there cannot be a large match. However, when the task is to *state* a specific difference, it helps to have a difference that is psychologically salient. Much evidence supports the idea that psychologically salient differences are *alignable differences*, meaning differences related to the commonalities (e.g., Markman & Gentner, 1993). These, the theory goes, are computed using candidate inferences and thus require the entire SME algorithm to have run to completion.

There are two aspects of analogy and similarity that SME does not currently model. The first is working memory limitations. We have not built in such limitations because the data on what they should be are quite murky.

Evidence from task constraints suggests that between 10 and 100 relational statements are typically needed in cases for a variety of tasks (Forbus et al., 2017). The second is quantitative similarity, which SME currently ignores. We return to this in chapter 18.

4.4.2 Retrieval

Human memory is vast. How do we manage to retrieve relevant experiences and schemas? The many are called/few are chosen (MAC/FAC) model of similarity-based retrieval (Forbus, Gentner, & Law, 1995) uses a two-stage process to provide both scalability and relevance. The inputs to MAC/FAC are a *probe*, which is just a case consisting of structured representations, and a *case library*, which is a (potentially very large) set of cases. The output of MAC/FAC is one or more *remindings*, which are cases from the library that are good matches for the probe, along with the mapping(s) between the probe and the remindings. The first stage, MAC, uses a special nonstructured representation, *content vectors*, which are automatically computed from structured representations. A content vector is a vector whose dimensions are predicates and functions, with the strength in each dimension being a function of the number of statements and terms using those predicates or functions. Figure 4.4 illustrates.

This method of construction means that the dot product of content vectors provides an estimate of the size of the match hypothesis forest that will be computed by SME for the original structured representations. This provides a good heuristic for estimating the likelihood of a strong similarity match for those structured representations. Content vectors are normalized to be unit vectors, to avoid size bias (i.e., overretrieval of cases with more statements). In MAC, a content vector for the probe is compared

```
(cause (restoringForce
          (ForceFn Spring1)
          Block2)
        (oscillates Block))
(cause (and (Spring Spring1)
             (connectedTo Spring1
                          Block2))
        (restoringForce
           (ForceFn Spring1)
            Block2))
(cause (performedBy Hammering3
                     Coyote4)
        (oscillates Coyote4))
```

cause	0.3
connectedTo	0.1
ForceFn	0.1
Oscillates	0.2
performedBy	0.1
restoringForce	0.1
Spring	0.1

Figure 4.4
A simple description and its content vector.

with content vectors for every case in the case library, in parallel, by taking their dot products. The best and up to two additional matches, if they are sufficiently close to the best (usually within 10 percent), are returned as the output of the MAC stage. The FAC stage then uses SME, again in parallel, to compare the cases produced by MAC to the probe, using the structured representations for the cases with the probe as the target. The output of the FAC stage is the case that is most similar to the probe (as measured by the structural evaluation score of the best mapping when they were compared). Up to two additional cases and their mappings can be returned as well if they are sufficiently close to the best (as before, within 10 percent).

Scalability arises from the data-parallel computation of the content vectors in the MAC stage, which is the only place where everything in the case library is tested. Because at most three descriptions will be input to FAC, there is a strict upper bound on the number of comparisons that will be performed, independent of the size of the case library. Relevance arises in two ways. First, the dot product of two content vectors provides an estimate of the size of the match hypothesis forest that SME would compute for the two structured representations. It cannot exactly calculate the size for two reasons. First, it cannot detect when a match between pairs of statements will be ruled out because their arguments can't align, thereby violating parallel connectivity. This also means that it cannot accurately calculate the effects of systematicity, so it will end up overweighing surface commonalities. Second, it cannot detect when statements or terms with nonidentical predicates or functions will be aligned. Nevertheless, it turns out to be a fairly reasonable estimate. Once the candidate matches from MAC are passed to the FAC stage, SME operates over full structural representations to produce accurate numerical similarity scores, plus candidate inferences, which are then useful as conjectures about the current situation.

The MAC/FAC model is agnostic about whether or not cases in human memory are organized as one large case library or are broken up somehow. Our model of analogical generalization, described below, makes some more specific conjectures about this.

MAC/FAC captures two important findings about similarity-based retrieval (Forbus et al., 1995). First, most often, retrievals are overall similarities and surface matches to the probe. This result requires the additional assumption that, in human representations, there is much more surface information than deep structure (the *specificity conjecture*; Forbus & Gentner, 1989). We believe that this is a plausible assumption because we can perceive more than we deeply understand or choose to encode in any particular situation. Second, cross-domain analogies do occur, but they are relatively rare.

Ecologically, this makes sense: the odds of a random pair of descriptions yielding an enlightening match are quite low. On the other hand, overall similarity—in other words, within-domain analogies—are extremely useful. If I want to start my car, for example, I do it the same way I did it last time, without thinking about it.

Analogical matching and retrieval together form a surprisingly powerful combination for reasoning. If there are prior experiences with relevant explanations, suggestions, or caveats in the case library, these will be applied to new situations as candidate inferences. Examples will be found in chapters 16 to 19. For instance, accumulating worked solutions of physics problems enables a system to solve many kinds of novel but similar problems (see chapter 19).

4.4.3 Generalization

Learning by accumulating experience in the form of cases is surprisingly powerful. However, there is much evidence that people also generalize from experience (e.g., Crowley & Siegler, 1999; Elio & Anderson, 1981; Newell, 1994; Shepard, 1987; Tenenbaum & Griffiths, 2001). Human learning is somewhat conservative, often requiring multiple examples to master a new concept. But human learning is often much faster than today's statistical learning algorithms, which can require orders of magnitude more stimuli than people do to master a task (Kuehne, Gentner, & Forbus, 2000). We believe that an important component of human learning is *analogical generalization*. The key idea is that, when a comparison is made, the overlap can be viewed as a form of abstraction, where the constituents of base and target that don't map are deemphasized and eventually eliminated. Via multiple matches, most of the surface information is worn away, leading to more generalized descriptions that can match in a wider variety of circumstances.

SAGE (McLure, Friedman, & Forbus, 2010) is a computational model of analogical generalization that provides a specific framework for this intuition. SAGE stands for Sequential Analogical Generalization Engine. SAGE assumes localized storage of incoming examples into one or more *generalization pools*. A generalization pool has a *trigger*, which is a set of conditions under which something is stored there. For example, if we are learning about animals, we may have a generalization pool for all instances of `Animal` that we see but also for `Dog` and `Cat` and `People`. (We assume an incoming example can be added into more than one generalization pool.) Each generalization pool maintains a set of *examples* and *generalizations*. A generalization is simply a structured representation. When a new example arrives, MAC/FAC is used to retrieve the most similar generalizations and examples by treating these sets as a case library. An *assimilation threshold* indicates how similar a new example has to be in order to be assimilated into that generalization (or form a new generalization if the retrieved item is another example). If nothing retrieved is sufficiently similar, the new example is stored as one of the set of examples in the generalization pool. If the most similar retrieved item is a generalization, then the example is assimilated into that generalization. The assimilation process involves updating frequency information stored with each fact in the generalization, thereby implementing the "wearing away" of infrequent information. Statements whose probabilities fall below a certain threshold are culled to keep the size of generalizations bounded. If the most similar retrieved item is an example, then the new and old examples are combined into a new generalization, with the information aligned in both having their probability be 1.0 and the rest being probability 0.5. Nonidentical entities are also replaced with *generalized entities*, a kind of skolem individual guaranteed to be unique.

The assimilation process has several interesting properties. First, it produces probabilistic representations via keeping track of the frequency with which statements occur in examples. Second, although generalized entities are more abstract than the original entities, they still are not logical variables. To apply a generalization requires using analogical matching. Third, generalizations can still have concrete information associated with them. They are no longer specific examples, but they often are not as abstract as the kind of purely first-principles deductive rule that someone might write by hand.

As a concept-learning system, SAGE has some unique features. At the level of concepts used as triggers, it is performing supervised learning. However, within a generalization pool, it can maintain multiple generalizations, enabling it to more easily learn disjunctive concepts. Because outlier examples are stored, it maintains some of the advantages of nearest-neighbor algorithms. Learning within a generalization pool can be viewed as unsupervised because no a priori information is provided about how many clusters there should be (unlike, say, *k-means* algorithms).

Classification using SAGE works as follows. Suppose the example is to be classified with respect to which of *K* concepts it best fits and that there is a generalization pool for each concept. Classification occurs via MAC/FAC, using the union of the generalization pools as the case library. The generalization pool from which the most similar generalization or example is drawn is considered the concept most appropriate for the new example, with the similarity score indicating the degree of fit to that concept.

SAGE can be used in two ways. The first involves generalization pools stored in long-term memory, as done in modeling word learning, for example. The second involves generalization pools stored in working memory (Kandaswamy, Forbus, & Gentner, 2014). This use of SAGE fits evidence of online abstraction occurring within sequences of stories presented in succession (Day & Gentner, 2007), as well as analogical generalization from two simultaneous visual examples (Christie & Gentner, 2010).

SAGE and its predecessor, SEQL, have been used to model several psychological phenomena. For example, Marcus et al. (1999) showed that infants could rapidly learn spoken patterns, like "Pa Ti Pa," and learn to distinguish between an ABA and AAB pattern after only sixteen trials. Connectionist simulations, such as Seidenberg and Ellman's (1999), required many times more exposure to the same stimuli, plus many epochs of pretraining to teach the network each syllable it was going to be using. By contrast, a SEQL-based simulation, using the same phonetic representation that Seidenberg and Ellman used, was able to learn these patterns in the same number of stimuli as the infants (Kuehne et al., 2000). More examples of how SAGE has been used to model spatial language learning (chapter 16) and aspects of conceptual change (chapter 17) are described later in this book.

4.5 The Centrality of Analogy in Human Cognition

Analogical processing, as defined above, provides capabilities that we think make it a good explanation for much of human reasoning and learning:

- *Analogy allows inference from little information.* Even a single example, if retrieved when appropriate, can provide predictions and/or explanations for a new situation.
- *Analogical reasoning provides an efficient form of abduction.* The inferences made in analogy can be predictions if they concern what might happen next in a situation. But they can also be hypothesized explanations for an observed behavior, projecting information about a prior understood situation into a new one (Blass & Forbus, 2017; Falkenhainer, 1990). The explanations in the prior behavior do not need to have come from a completely articulated first-principles theory: they can be concrete explanations about what happened in a particular prior situation.
- *Analogical processing predicts faster performance as knowledge increases.* As cases and generalizations accumulate, for any new situation, there is more likely to be a close analog and hence a rapid, one-shot answer generated via retrieval plus matching.

- *Analogical learning, although initially conservative, appears to occur at human-like rates.* In most cases, SAGE only requires a dozen or so examples to perform well, whereas machine-learning methods, when they work at all, often require several orders of magnitude more examples. The generalization process automatically produces probability information for each statement, providing contextualized priors that can be used in statistical reasoning. Moreover, analogical generalizations provide a natural, graded form of learning, with more abstract generalizations being applicable to a wider range of circumstances.

One way these ideas are being explored is via the Companion cognitive architecture (Forbus, Klenk, & Hinrichs, 2009; Forbus & Hinrichs, 2017), which incorporates the above models, along with other reasoning capabilities. A number of the models discussed in this book, particularly in part IV, are built on top of the Companion architecture.

II Dynamics

Dynamics concerns representing and reasoning about how continuous properties of things change over time. As we saw in two of the motivating examples from chapter 1 (i.e., what happens when heating water on a stove and whether warm or cold water freezes faster), subtle conclusions can be drawn even without numerical parameters or differential equations. This section describes how qualitative representations can be used to reason about change in continuous systems. It summarizes a rich set of representations and reasoning techniques developed by researchers in qualitative reasoning (and other areas as well). It also explores their implications for understanding commonsense reasoning, aspects of expert reasoning, causality, and natural-language semantics. Specifically,

- Chapter 5 outlines representations for quantities (i.e., continuous properties). These draw upon the rich tradition of mathematics, although I avoid most formal details and theorems because there are ample resources elsewhere for readers who want to dig in deeper. Instead, I outline the trade-offs between these representations, to understand which might play what roles in understanding human cognition.
- Chapter 6 lays out what is called *qualitative mathematics* (i.e., representations of relationships between quantities). These relationships form the links out of which larger causal arguments about continuous systems are made, making them very important for understanding cognition.
- Chapter 7 describes the core ideas of *qualitative process theory*, which provides a framework for formalizing human mental models in a wide range of continuous domains. I argue that its notion of mechanism provides grounding for causal arguments as well as an organizing structure for composable knowledge about continuous systems.
- Chapter 8 illustrates the power of qualitative process theory by demonstrating how it can be used to capture a range of commonsense models.

A domain theory for liquids and gases is presented, including an explanation for the ice cube freezing problem. Multiple models of one-dimensional motion are examined, including Aristotelian, Newtonian, and impetus models, motivated by misconception studies. This chapter also illustrates how interesting dynamic phenomena, such as oscillation and equilibria, can be derived from qualitative representations.

- Chapter 9 returns to causality, providing a framework for thinking about causality. It then compares and contrasts the three accounts of causality developed in qualitative reasoning research (qualitative process [QP] theory, confluences, and causal ordering). Unlike many causal models in psychology, these richer models provide ways of dealing with loops and feedback.
- Chapter 10 examines the notion of qualitative simulation more deeply. The abstractness of qualitative representations that makes them so useful for commonsense reasoning comes at a cost—namely, ambiguity in simulation. This has some surprising consequences, examined here.
- Chapter 11 explores how qualitative representations can be used to organize knowledge about the modeling process itself. Scientists and engineers have massive amounts of both professional and everyday knowledge, which must be marshalled appropriately when solving problems. An extension of QP theory, *compositional modeling*, provides a framework for expressing modeling assumptions and their interrelationships. Automatic model formulation algorithms are also summarized and aspects of their psychological plausibility examined.
- Chapter 12 argues that human qualitative reasoning is mostly performed via analogy. This is a major departure from most work in qualitative reasoning, which focuses on first-principles reasoning. Specifically, I propose that mental simulation is best accounted for by analogical reasoning over qualitative representations derived from experience and cultural knowledge.
- Chapter 13 examines how qualitative representations can be used in natural-language semantics. Constructions used to communicate most aspects of QP theory in English are summarized, showing they are compatible with linguistic models of semantics via linking them to FrameNet. QP theory can be viewed as supplying an inferential component for semantics concerning the continuous world.

These chapters build upon each other and are designed to be read in the order provided. However, not all readers may be interested in every representation in chapter 5 or every example in chapter 8. Also, readers uninterested in modeling professional reasoning can safely skip chapter 11.

5 Quantity

We reason about a wide variety of continuous properties. Weight, level, pressure, heat, and temperature are examples of physical quantities. One-dimensional versions of position, force, acceleration, and velocity are also quantities. More abstractly, prices, quality, stability, difficulty, and simplicity are properties that we often think of as more or less continuous. For some quantities, like weight, instruments can measure reasonably accurate numerical values, calibrated upon some well-agreed-on cultural standard. For others, like depth of friendship, it is far from clear that any specific number can be easily measured, although we can often ascertain that we are closer to one person than another.

To reason qualitatively about such continuous properties requires quantizing their possible values into meaningful units. There are a number of ways to do this, some of which are better than others for particular purposes. Following Minsky (2007), we argue that people use multiple representations of quantities. However, there are many ways to carve up values that are psychologically implausible. Let us examine one of them, to hone our intuitions. Suppose we decided to carve up temperature into ranges, such that every 10 degrees Centigrade is a distinct value. That is, zero to 10 degrees is one value, 11 to 20 is another, and so on. This successfully captures the idea that the difference between 12 and 13 degrees doesn't really matter, and these values can be treated as more or less the same. However, this idea has several problems. If we are thinking about water, then there is a major difference between zero and 1 degree: zero is its freezing point, in degrees Celsius, so there is something special about zero that this representation does not capture. Another problem is that the difference between 10 and 11 degrees really doesn't matter for most purposes, but it does in this representation. Our choice of 10-degree boundaries was completely arbitrary: it does not reflect any deep property about the world or about a task we are performing in it, so this kind of regular discretization is an example of a poor qualitative representation.

What properties should good qualitative representations have? Five properties trade off against each other, which is why multiple representations are needed:

1. *Obtainable.* The representation can be computed by our perceptual system or easily derived from perceptual information.
2. *Relevance.* The distinctions made by a qualitative representation should be meaningful with respect to some task.
3. *Resolution.* Qualitative representations can be fine-grained or coarse, and the number of distinctions can be fixed or varied, depending on different needs.
4. *Operations supported.* Qualitative representations vary in terms of the kinds of operations they can support. Three important categories of operations are as follows:

    a. *Compare* relative magnitudes
    b. *Propagate* information about given values to derive new values
    c. *Combine* terms involving quantities via relationships

5. *Graceful extension.* Sometimes higher-resolution information is needed in the course of solving a problem. Can it be added without invalidating old conclusions?

An important goal in qualitative reasoning research is to understand the trade-offs between these properties. Much of the work has focused on more coarse-grained representations because traditional mathematics provides a solid set of fine-grained representations. As the examples in chapters 7 and 8 illustrate, surprising amounts of reasoning can be done with coarse-grained qualitative representations.

Figure 5.1 illustrates the representations of quantity discussed in this chapter. All have been proposed as psychologically plausible by various researchers. There is reasonable evidence for some of them, whereas others seem less plausible. The rest of this chapter discusses the cases for and against each of them. We start with representations from traditional mathematics and computer science because they are likely to be the most familiar.

5.1 The Reals

The classical efforts to formalize the nature of continuous properties were carried out by mathematicians. We owe the integers, rational numbers (i.e., ratios of integers), and the real numbers (a technical term) to their efforts. In the nineteenth century, the theory of real analysis developed the notion of

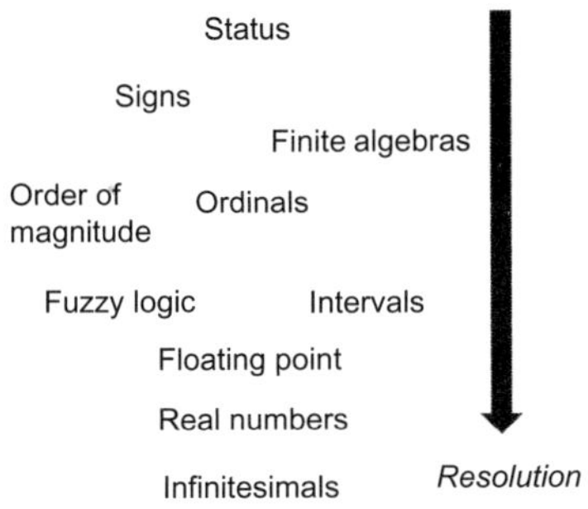

Figure 5.1
Representations for quantities.

real numbers to the point where rigorous proofs of the theorems of calculus could be provided. Because the real numbers representation is a necessity for some branches of mathematics, it is clearly a representation used by professionals.

Could the reals be used as a model for our everyday, intuitive notion of quantity as well? As Hayes (1979) pointed out, the real numbers have some counterintuitive properties. For example, point-set topology tells us that the number line between 0 and 1 has the same number of points as the unit square between (0,0) and (1,1). If we think of lines and areas as being made of real "stuff," this is obviously nonsense. Similarly, the Banach-Tarski theorem (Banach & Tarski, 1924) tells us that we can take a unit sphere and, by some clever manipulations, create multiple spheres of the same size. Again, this is something that cannot happen in the material world. Key to these counterintuitive results is a property called *density*: between any two real numbers, one can always find another real number. Because the new number is itself a real number, this definition applies recursively. This is incompatible with substances made of atoms. If we are thinking of continuous media as ultimately atomic, we must give up on density.

Another important property of the reals is *continuity*. Suppose we have three real numbers, A, B, and C, such that A < B and B < C. Suppose we now think about a quantity whose initial value is A and that is increasing over time. Continuity tells us that its value cannot reach C without having first gone through B. It turns out that continuity is a very important constraint for reasoning about change because it restricts possible behaviors. Intuitively, to go smoothly from A to C, one must go through B. Interestingly, as we shall see, most qualitative representations of quantity satisfy continuity, despite their discreteness.

5.2 Finite Approximations to the Reals

Another problem with the reals as a model for continuous properties is that physical systems cannot compute with them. Each real number contains an infinite amount of information, and the finite size and noise inherent in physical systems means they can only store and manipulate finite amounts of information, so neither people nor computers can be directly computing with real numbers, although some people and some computers are capable of reasoning about them abstractly. Consequently, finite approximations such as *floating-point numbers* are used in computer hardware (32 bits or 64 bits per number are today's popular choices). Even though they are finite, they still contain an enormous amount of information, which is why numerical simulations are so widely used to approximate reality.

The power of numerical simulation has seduced some into proposing that people's ability to mentally simulate physical changes rests on numerical simulation, of the same type found in modern computers (e.g., Battaglia, Hamrick, & Tenenbaum, 2013). When we consider more closely what such simulations entail, this looks extremely unlikely. To derive a predicted behavior via numerical simulation requires three things. First, an accurate mathematical model must be available for the situation. This is much harder than it looks. Good numerical models for many everyday phenomena remain elusive. Second, initializing simulations requires a plethora of accurate data, far beyond what is plausible for people to garner about situations given our sensory capabilities. This includes not just data about external conditions, which can sometimes be easily observed, but also exact data about internal parameters of the system, which often cannot be obtained without laborious experimental work. Finally, carrying out a simulation requires massive amounts of computation, organized in a fairly serial manner that seems quite foreign to neural hardware. And of course, even if it could be done, such a simulation would only capture a single behavior. Running the simulation multiple times (e.g., Monte Carlo techniques) is even more resource intensive, exponentially so in terms of accuracy sought, and still would not guarantee that the results included all of the behaviors that a person would call qualitatively distinct.[1] Consequently, numerical simulation does not appear to be a good model for mental simulation. (For an excellent discussion of the technical issues, please see Davis & Marcus, 2016.)

5.3 Finite Algebras and Fuzzy Logic

One of the first qualitative schemes proposed is the use of a small vocabulary of symbols, interconnected by an ordering relationship. For example, a person's height might be coded as one of these symbols:

`VeryShort, Short, Medium, Tall, VeryTall`

We can extend the definition of > so that every symbol on the right is greater than the symbol on its left (e.g., `Medium` > `Short`). Transitivity is satisfied, of course, as is continuity: if someone is growing, he or she has to be of medium height before he or she grows to be tall. An advantage of such representations is their naturalness. They are common in natural language and can be effective means of reference (e.g., "the tall woman with the long hair and tiny dog"). They have also been used in gathering data in ecosystem research because the sparse and uncertain nature of the available data there makes them a good fit for expressing what has actually been measured (Guerrin, 1995).

There are well-known subtleties with this form of representation. Mapping from perceptual information to one of these categories requires identifying an appropriate reference set. For example, a tall child and a tall basketball player may be very different in height from a tall person on the street. Fuzzy logic (Zadeh, 1996) addresses these problems by using different distributions for different variables. Fuzzy logic also allows graded membership, using overlapping ranges to enable something to be considered a partial member of two categories (e.g., a "medium-short stick"). Advocates of fuzzy logic argue that it provides natural ways of combining values and performing inference. Critics (e.g., Elkan, 1994) argue that it violates basic logical intuitions and that the successes of fuzzy logic stem from the use of continuous parameters in rules rather than overlapping intervals. Nevertheless, fuzzy logic has proven to be useful in building control systems and has been used in some qualitative reasoning systems (Shen & Leitch, 1993).

Although finite symbol vocabularies are clearly obtainable and relevant for linguistic reference, they do have several limitations. First, it is not clear how to combine them in operations. For example, suppose we are trying to predict how long a stick made from two others would be. If we knew the lengths quantitatively, we would add them. But what exactly is `Short + Medium`? Is it still `Medium`? Or will it be `Long`? We know it cannot be `Short` because it must be at least `Medium`. Presumably, it would not be `VeryLong`, but there is no basis for choosing between `Medium` and `Long`. If we define addition over this representation, its results are ambiguous. This is part of the nature of qualitative representations: we can encode "long" more easily (and more accurately) than "3.2 meters," but we can't expect the same precision of results. Nevertheless, it is still a useful representation because it supports rapid comparison. It also supports graceful extension: if we now are given two particular sticks that are `Short` and `Medium`, we can lay them end to end and determine visually if they satisfy our (implicit) criteria for `Medium` versus `Long`.

5.4 Signs

Suppose we abstract away from particular numerical values and keep only their signs. That is, if `A` is a quantity, then we can denote the sign value of `A` via `[A]`. The values it can take on are the following:

`[A]` =− means that `A` is negative

`[A]` = `0` means that `A` is zero

`[A]` =+ means that `A` is positive

Moreover, if we define `[∂A]` as the sign of the derivative of `A`, we get the following:

`[∂A]` =− means that `A` is decreasing

`[∂A]` = `0` means that `A` is not changing

`[∂A]` =+ means that `A` is increasing

Signs can be viewed as a special case of finite algebras, but one where the categories imposed align naturally with conceptual distinctions that people tend to make. Notice that this representation is easily obtainable from perceptual information. People are sensitive to changes, so much so that we have a number of adjectives devoted to them (e.g., rising, falling, steady). Sign values are relevant for many purposes. For example, many natural quantities are always nonnegative (e.g., mass, heat), often with special conditions attaining when they are zero (e.g., when the population of a species is zero, it has become extinct). The values of quantities can often be aligned so that differences in signs correspond to differences in qualitative state (e.g., whether we are broke, in debt, or doing okay financially depends on the sign of the balance in our bank account).

Although there is some ambiguity, sign values can still be composed via operations like addition, as table 5.1 illustrates.

The biggest limitation of sign values is their fixed resolution. That is, they always divide the number line into exactly three parts. For some domains,

Table 5.1
Addition table for sign values.

[A] + [B]	–	0	+
–	–	–	?
0	–	0	+
+	?	+	+

Note: Ambiguous results are indicated by "?" The entries with "?" indicate ambiguity: because we don't know anything about the magnitudes of A and B, there is nothing we can say for certain about the sum of a negative and positive number. Importantly, this ambiguity can be used to provide a signal during reasoning about what information would be helpful: if we knew even just the relative magnitudes of A and B, we would be able to determine the sign of the outcome.

such as analog electronics, this is not a problem. But for others, it can be confining. Consider, for example, trying to use signs to represent the temperature of the coffee in a cup. What should we use as the zero point? Perhaps we could use the idea of there being a perfect temperature for drinking, so that + would correspond to too hot and – would correspond to too cold. But this representation will not tell us what happens if we put the cup into a microwave oven because the idea that it could get hot enough to boil would not be expressible in this representation. So let us choose instead the boiling point of water (which is close enough, even though the other chemicals in coffee might well change this a bit) as the zero point. This gives us a representation that is sensitive to boiling. But now, because it is summer, we decide we want cold coffee and put the cup in the freezer instead. Our new representation won't handle freezing at all. The crux of the problem is that multiple comparisons are relevant in dealing with many types of quantities. One can try to approximate this by adding multiple quantities for every continuous parameter, but then we lose the simplicity of the sign representation. Ordinal relations handle this problem of multiple references, so let us turn to them next.

5.5 Ordinal Relations

Many qualitative distinctions rest on comparisons between values. Whether a piece of water is liquid, ice, or steam depends on its temperature relative to both its freezing point and boiling point. Which way fluids flow in a piping system is determined by their relative pressures. Thus, representing quantities in terms of a set of ordinal relationships with other quantities, called the *quantity space* representation (Forbus, 1984), provides a relevant representation. The quantity space for a quantity Q consists of two things: (1) a set of *limit points* that it is compared against and (2) a set of ordinal relations involving the quantity and its limit points. For example, when considering the temperature of water (`T(W)`) in a kettle sitting on the stove, its quantity space will have three limit points: the freezing point of water (`Tfreeze`), the boiling point of water (`Tboil`),[2] and the temperature of the stove (`Tstove`). Why the temperature of the stove? Because the relative temperatures of the water and the stove tell us whether or not heat will flow between them and, if so, in what direction it will flow. Notice that the set of ordinal relationships does not have to be complete: in our example, we don't actually know the relationship between `Tboil` and `Tstove`.[3]

Quantity spaces satisfy many of our desirable properties for qualitative representations. Ordinal information is derivable via comparison for perceptual quantities and is easy to state in natural language (e.g., hotter, heavier), so it is easily obtainable. Because relative values help determine important properties of state and whether or not processes can occur, they are clearly relevant. By adding new limit points, resolution can be varied. For example, if we wanted to model heat lost to the atmosphere, we could do so by adding another limit point comparing the water's temperature to that of the atmosphere around it. If we have quantitative information about the properties, we can compute ordinal relationships, so this representation exhibits graceful extension. Quantity spaces can be defined for derivatives and, by including zero as a limit point in them, provide the same expressive power for capturing change that signs do. The major limitation is in the operations supported: because there is not a specific token in the representation representing a value, algebras for operations like we did for signs cannot be specified. Instead, qualitative relationships are used to derive new ordinals (especially signs of derivatives, which are crucial summaries of change), thereby using what is known about one part of a system and the nature of the system itself to constrain the rest.

5.6 Numerical Intervals

A commonly used representation in science and engineering is numeric intervals. For example, the desired temperature of a room might be specified to be (68°F, 78°F), and the resistance of a resistor might be specified as 330Ω, ±1%. There is a considerable body of research and practice on the use of intervals for numerical analyses, and the range of operations they support is quite broad. Moreover, they support variable resolution by simply using wider or narrower intervals. However, there are some subtleties. Consider the following constraints:

$A = Z/(X - Y)$

$Z = [1,2]$

$X = [2,4]$

$Y = [3,5]$

What is the value of A? The interval value for the expression $(X - Y)$ is, given the values above, [–3,1]. This includes zero, and dividing by zero is of course undefined. That means our answer must actually consist of multiple intervals, with additional elements introduced to handle what in essence are infinite limits. This means operations that combine them must be done cautiously.

5.7 Order of Magnitude

Sometimes effects are so different in size that the smaller effects can be completely ignored. For example, the effect of evaporation on the level of coffee in a cup can be safely ignored while coffee is being poured into it, but it is worth taking into account if the partially filled cup is abandoned for several days. (Reasoning about such abandonment may also require considering the potential growth of mold, another phenomenon that can safely be ignored when drinking coffee fresh.) For example, in Dehghani et al.'s (2008) version of Dauge's (1993) formalism, three relationships are introduced, based on a sensitivity parameter, K.

- Almost equal: $A =_K B \Leftrightarrow |A - B| \leq K * \mathrm{Max}(|A|,|B|)$
- Greater than: $A \neq_K B \Leftrightarrow |A - B| > K * \mathrm{Max}(|A|,|B|)$
- Orders of magnitude greater than: $A \ll_K B \Leftrightarrow |A| < K * |B|$

In other words, larger values of K make differences less easy to see. Order-of-magnitude representations focus reasoning by highlighting effects that matter. This can help focus modeling by deciding what phenomena are worth thinking about in a situation. For example, the influence of evaporation over short time scales in many everyday situations can reasonably be viewed as almost equal to zero, and hence we can safely ignore it. Similarly, order-of-magnitude information can provide useful ways of resolving qualitative ambiguities. If the water coming into a lake is vastly larger than what is being drained from it, then its level will be rising. One subtlety with order-of-magnitude representations is how absolute the stratification should be. With Dauge's formalism, *K* serves as a knob that can be used to adjust sensitivity, determining when enough small effects can combine to become nonnegligible. On the other hand, in the FOG[4] formalism (Raiman, 1991), the stratification is absolute: no amount of negligible effects can ever become significant. In cognitive modeling, there are situations when the ability to adjust sensitivity based on circumstances is important. (In chapter 18, this representation is used to model the effects of sacred values on moral decision making.)

5.8 Infinitesimals

A key idea in the history of mathematics is the infinitesimal (i.e., a number so small that one cannot get any smaller). Infinitesimals played a key role in the intellectual history of calculus, that is, the notion of a derivative (e.g., dy/dx) being defined as how much y changes for a very tiny change in x. It was used in an intuitive manner by mathematicians for several hundred years, only being successfully formalized in the 1960s (Keisler, 1976). Infinitesimals have been used as the theoretical foundation for several reasoning techniques in qualitative reasoning, including Raiman's (1991) FOG order-of-magnitude representation and Weld's (1990) exaggeration technique for reasoning about changes in a system.

Recall that the real numbers, although originally intended as a formalism of our intuitions about continuous values, ended up having some distinctly counterintuitive properties. Unfortunately, the same is true about infinitesimals (Weld, 1990). Moreover, the additional distinction of a small "halo" around each real number multiplies the number of states in qualitative simulations that have no connection to our intuitions about physical behavior (Davis, 1987). To see this, consider the temperature of a cup of coffee, T(c). In the usual course of events, it might start out being hotter than you would like it to be. As heat flows from it to its surroundings, the temperature will drop to your ideal temperature, Tp(c), for drinking. Alas, it continues to cool and will be below your ideal drinking temperature. Figure 5.2 illustrates how this would look in terms of movement along a number line, in both the reals and the infinitesimals.

Notice that in the reals, this trajectory describes three distinct qualitative states: "too hot," "just right," and "too cold," corresponding to T(c) being greater than, equal to, and less than Tp(c). The distinctions made via these ordinals are just right for capturing these intuitive distinctions (although a more accurate model would define "just right" in terms of a band of temperatures,

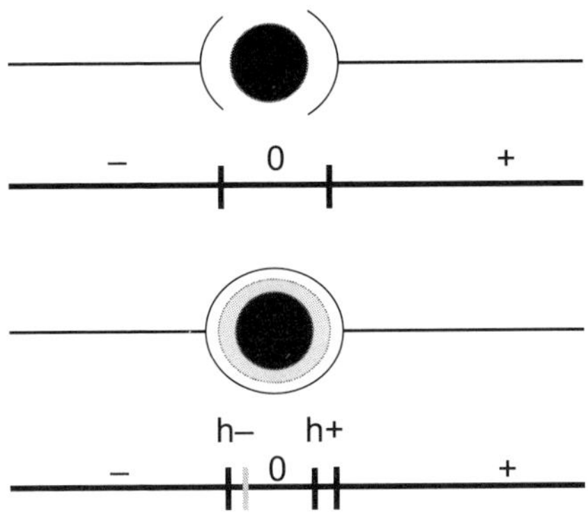

Figure 5.2
Comparing qualitative distinctions imposed by the real number line versus infinitesimals. The real number line imposes three qualitative distinctions, whereas infinitesimals impose five.

using two limit points). On the other hand, introducing infinitesimals forces us to make many more distinctions (i.e., being in the positive and negative halos around the desired temperature). Thus, three states in the real number line have become five when infinitesimals are used. Most of the time, the halo does not correspond to anything intuitively meaningful, and hence this representation is making distinctions that are unnecessary. This is bad enough in our simple example, but when one incorporates this representation into a full-fledged qualitative simulator, the size of the simulation explodes. (Independent ambiguities, as discussed in chapter 10, typically have a multiplicative effect on the size of a simulation.) This makes infinitesimals a poor choice for a qualitative representation, except for very special purposes.

5.9 Status Values

The infinitesimals are the most detailed representation proposed for capturing our intuition for numbers. By contrast, the most abstract representation is to use just two values, essentially "okay" and "not okay." At first, it may seem like a variant of the finite algebra representation described above, but "not okay" includes values outside some nominal range for a parameter, whether higher or lower, and hence there is no ordering assumed between these values. This representation is useful in diagnosis (Abbott, 1988; Bell et al., 1994; Fromherz, Bobrow, & de Kleer, 2003): if the inputs to a component of a system are all okay and the output is not, then that component (or something downstream from it) is where the problem lies.

Table 5.2

Representations for quantity and their trade-offs.

Representation	Obtainable	Relevance	Resolution	Operations	Graceful Extension	Psychological Plausibility
Status values	Thresholds, signal processing	Monitoring, diagnosis	Fixed	Diagnosis	Depends	Yes/yes
Signs	Easy	Reasoning about change	Fixed	Comparison, propagation	Yes	Yes/yes
Fixed symbols	Relatively easy	Reference	Variable	Comparison	No	Yes/yes
Fuzzy logic	Relatively easy	Reference, control	Variable	Comparison, propagation	No	Maybe/yes
Ordinals	Easy	Reasoning about change	Variable	Comparison	Yes	Yes/yes
Order of magnitude	Relatively easy	Ignoring effects, resolving ambiguities	Variable	Comparison, filtering	Yes	Maybe/yes
Floating point	Expensive	Precise answers	Fixed	Numerical processing	No	No/yes
Reals	Mathematics	Precise answers	Fixed	Mathematics	Not applicable	No/yes
Infinitesimals	Mathematics	Precise answers	Fixed	Mathematics	Not applicable	No/yes

Obtaining status values from numerical measurements can be extremely complex. Context matters considerably (e.g., different temperatures are expected in a car engine that is idling versus running at full throttle). Qualitative models of state, as described in later chapters, provide exactly this form of context. Time scale also matters (Doyle, Chien, Fayyad, & Wyatt, 1993): a tire that loses a pound of pressure over a month is safe, whereas one that loses a pound of pressure in a minute will soon be flat.

Status values are not intended to be calculated by using any form of qualitative calculus because their role is in diagnostic algorithms. We include them in our catalog of numerical representations because they are an extreme edge case. They are fixed in resolution but do allow for graceful extension if the criterion for computing them is in a form that fits with more detailed data (e.g., thresholds on a small set of parameters rather than the result of running a signal-processing filtering operation over a stream of time-averaged data).

5.10 Summary

Let us step back and examine our catalog of representations for continuous values. Table 5.2 summarizes the representations, in order from coarser to finer grained, and how well they satisfy the properties of good qualitative representations.

The last column represents an estimation of the psychological plausibility of these representations. We need to distinguish several senses of what we might mean by this, for any representation R:

- That someone, somewhere, can think in terms of R. By this criterion, all must be psychologically plausible, because they were all developed by people, and hence it is uninteresting.
- That neural hardware directly implements R and operations upon it. This is of some interest but less than might be supposed, given the importance of cultural knowledge. Human commonsense reasoning can be subtle and sophisticated, as the examples in chapter 1 illustrate. Cultural knowledge and experience matter, and if we ignore these factors, we ignore much of what is interesting about people as compared to, say, gerbils.
- That some culture uses R in its reasoning about continuous properties. This is perhaps the most interesting, because it can help us understand what we all are likely to share, what specialized groups know, and how people transition from novices to experts. (Here I exclude the culture of

cognitive science researchers, who again arguably use all of these representations as hypotheses, but we are concerned with less self-reflective practices at the moment.)

The entries in the psychological plausibility column are my best estimates based on current knowledge. The first value corresponds to everyday reasoning, the second to expert reasoning. By experts, here I mean scientists, engineers, and mathematicians. The "maybe" entries for everyday reasoning in fuzzy logic and order-of-magnitude reasoning are there because plausible alternatives exist for achieving the same kinds of behavior that they are intended to explain. In the case of fuzzy logic, multiple competing criteria for computing fixed symbols seem to be an equally likely explanation for the feeling of graded membership that fuzzy logic explains via hardwired static distributions. In the case of order of magnitude, simply leaving out phenomena that are viewed as negligible, without any possible consideration of small effects combining to become comparable, may well be what is happening. On the other hand, the case for ordinal relations is quite strong because changes in them correspond to distinctions in qualitative behaviors if the limit points are chosen to be meaningful. Except for the ability to do algebra, signs can be subsumed by ordinals, so the question as to the psychological plausibility of the sign algebra hinges on whether there is evidence for (or against) such reasoning.

Which of these notions of quantity are part of our starting endowment, versus learned, is a fascinating question. Progress has been made on understanding how the natural numbers might be learned (e.g., Carey, 2011), and arguments for grounding mathematics in physical, body metaphors have been made (e.g., Lakoff & Nunez, 2000). I am concerned here with exploring the range of representations that people are using and how they use them for both everyday and expert reasoning. The range of reasoning that people do with continuous properties and numbers is much broader than prior studies have considered, and these additional constraints might shed new light on even the initial mechanisms.

As we have seen, the catalog of representations for quantity is larger than one might first expect, and some of the contenders that might be viewed initially as plausible (i.e., the real numbers) turn out to have properties that rule them out as reasonable representations of everyday commonsense reasoning, even though they are clearly part of the stock in trade of mathematicians and other technical professionals. As we look at causal reasoning, reasoning about change more broadly, and language, we will see the remaining representations play various roles.

6 Relationships between Quantities

As the last chapter illustrated, there are a variety of useful and plausible representations for continuous properties. To reason about such properties requires representing how they can be related. A very simple way to use quantities is in the form of rules, either learned by instruction or via induction (e.g., "To boil a cup of water in the microwave, set the power to 900 watts and the time to 1 minute 30 seconds"). Although such rules are clearly used by people, they are far from the only way that people reason about quantities. We have intuitive notions of how quantities change over time, as our examples of boiling water and freezing ice cubes illustrate. Moreover, we can reason causally about such changes: if the temperature of the stove is higher, the kettle will boil sooner, all else being equal. Qualitative reasoning research has identified forms of *qualitative mathematics* that capture many aspects of such reasoning. This chapter describes two systems of qualitative mathematics, because each captures aspects of human causal reasoning about continuous systems. We begin with a look at traditional mathematics and why qualitative mathematics is needed in the first place. Then we examine the qualitative mathematics of *qualitative process theory* (Forbus, 1984), showing how it provides a representation system that captures partial knowledge about causal relationships between quantities and how it can be used to reason with partial information. Next we examine de Kleer and Brown's (1984) idea of *confluences*, showing how it can be used for causal reasoning in circumstances where QP theory breaks down. Finally, we discuss some of the limitations of these models.

6.1 Why Qualitative Mathematics?

Traditional mathematics has been a thriving enterprise for centuries. It has produced representations and reasoning processes of unprecedented scope and power. Differential equations, for example, provide an exquisitely precise language for describing some aspects of continuous change. A differential equation is an equation that contains one or more terms involving rates (i.e., how a parameter is changing). There are two kinds of differential equations. *Ordinary* differential equations discuss how parameters change with time, ignoring spatial properties except as expressed algebraically in boundary conditions. *Partial* differential equations describe how some parameters change with respect to changes in other variables in addition to time, most often with respect to spatial coordinates. We will be concerned with phenomena modeled by ordinary differential equations in this section, whereas phenomena modeled via partial differential equations will be explored in parts III and IV.

Why can't we just use traditional equations as they are, with qualitative values instead of real numbers (or high-resolution approximations thereof, like floating-point numbers)? There are three fundamental reasons: soundness, minimal knowledge, and causality. Let us examine each in turn.

6.1.1 Soundness

Recall from chapter 3 that soundness refers to whether a logical system always produces correct results, given correct assumptions. It is a useful property to have, because then any errors in results can be attributed solely to the assumptions. Let us consider the kinds of reasoning that can be done with equations to produce values. Broadly, there are two kinds of operations:

1. Substitute values into equations to derive new values. This is often called *propagation*, when the system of equations is viewed as a network of relationships constraining a set of quantities. For example, if $x+y=7$ and $x=4$, then we can substitute the value of x into the equation and derive that $y=3$.
2. Substitute one equation into another. This can reduce the number of variables and hence (perhaps after multiple steps, depending on the size of the system of equations) lead to an equation with only one unknown that can be directly solved. For example, if $x+y=7$ and $x-y=1$, then by solving the second equation for y, we can substitute $x-1$ for y in the first equation. Solving our simplified equation leads again to $x=4$, and then using that value to solve for y in the second equation, we again get $y=3$.

Suppose we use sign values in equations. (Such equations are called *confluences* [de Kleer & Brown, 1984].) Propagation still works; for example, suppose we know that $[x]+[y]=[z]$. If $[x]=+$ and $[z]=-$, we can plug these values into the equation and see that for this to be true, $[y]=-$ must hold. However, suppose we know that $[x]=+$ and $[z]=+$. Then we have no information whatsoever about $[y]$: it could be positive, negative, or zero, depending on what the actual values of x and z are. This is to be expected: we are trying to find out how much we can derive with minimal information, and if we have less information, then we can conclude less. Ambiguity is the price of qualitative representations. This is often a price well worth paying because operations with qualitative values are simpler, and such information can be gleaned from observation and other information sources more easily. Measuring the exact speed of a falling object is much harder than ascertaining that it is falling, for instance.

What about propagating quantity space values (i.e., sets of ordinal relations)? Because there is not a finite set of symbols over which to define an algebra, as we did with signs, propagation of quantity space values per se is not well defined. However, propagation of specific ordinal relationships from quantity spaces can be done, as described below. I will argue that this is in fact essential in understanding human causal reasoning about continuous systems.

Therefore, reasoning via propagation with signs can be done, and it is ambiguous but still sound. What about solving equations via substitution? A key principle of substitution is that one can substitute equals for equals. Unfortunately, the algebraic structure of signs is very different from the reals or even the integers (Williams, 1991). Solving via substitution over signs is not sound. Suppose we know that $[x] = +$ and $[y] = +$. For any x, clearly $[x] - [x] = 0$. If we could substitute equals for equals, then with what we know, we can conclude $[x] - [y] = 0$. But this is incorrect: the actual value of x might be 1 and the actual value of y might be 2, which would result in –, when converted to signs.

6.1.2 Minimal Knowledge

Traditional mathematics provides precise, detailed answers. Precise enough to send people to the moon and back safely, landing within a few meters of their intended splashdown coordinates, for example. Today, many phenomena are so well understood that computational prototyping is being used to design complex artifacts such as airliners, reducing or even eliminating the need for physical models during the design process. This makes sense: mathematics was developed to enable us to go beyond what our native capacities are. And indeed, if one thinks about numerical simulation as a model for what people are doing when they imagine what will happen in a situation, as we did in part I, it has serious problems. First, it imposes unrealistic input requirements. Suppose you are relaxing in front of a crackling fire on a winter evening. Your child, being mischievous, throws a can of hair spray into the fire. Is the decision to run away (taking your child with you) something made on the basis of numerical simulation? Most people do not have the faintest idea what the underlying equations of this newly constructed system might be. And even for the tiny minority who could construct a mathematical model of this situation, they do not have the numerical data needed to provide the initial conditions for a simulation of it, nor do they have enough time to mock up the model and run it with variations of the parameters, Monte Carlo style, to ascertain what might happen. Even when the timeframe is less constrained, as it was for the examples in part I, we generally do not have the mathematical modeling knowledge or the high-resolution data needed to reason with traditional mathematical models.

Qualitative mathematics solves this problem by defining more abstract relationships that are better fits for the kinds of information that we are more likely to be able to extract perceptually, such as ordinals and signs. The conclusions are generally low resolution, in that they specify only what might happen within broad categories of equivalent behaviors. They are also often ambiguous, because the lack of resolution means that we cannot always rule out alternative changes within a situation or derive a unique outcome for a situation. Another property of qualitative representations is that they support partial states of knowledge, in order to be useful in learning. Human learning often occurs piecemeal, with incomplete models being extended through experience (in everyday life) or theoretical analysis and laboratory/field experimentation (in professional life). Being able to represent and reason with partial knowledge is an important constraint on qualitative representations.

6.1.3 Causality

Causality may not be the "cement of the universe" (Mackie, 1980), but it may well be the cement of our conceptual structures. Causality enables us to know what can be changed to bring about desired situations and avoid undesirable situations. Fields using traditional mathematics have rarely attempted to formalize causality because they are building on the intuitions of their practitioners, who already have causal models honed through their everyday lives and their professional work. Causal models are heavily used in scientific research, but they are to be found in the natural-language text surrounding the mathematical equations rather than in the equations themselves. It is formalizing that knowledge that concerns qualitative reasoning. Thus, qualitative mathematics complements traditional mathematics, in modeling expert reasoning, and provides formalisms for describing causality in intuitive models formed from everyday experience.

6.2 Qualitative Mathematics in QP Theory

Qualitative mathematics in QP theory is organized around the notion of *influences*. There are two kinds of influences:

- *Direct influences* are imposed by continuous processes. In QP theory, all changes are stipulated to be ultimately caused by processes. Thus, directly influenced parameters form the start of any causal chain of reasoning involving changes in continuous properties.
- *Indirect influences* propagate changes caused by processes through the other parameters of a situation.

Here we focus on the nature of influences themselves and postpone defining what we mean by continuous processes until chapter 7. For now, think of them as everyday physical processes such as heat flow, liquid flow, boiling, and motion. We discuss direct influences first, then indirect influences.

6.2.1 Direct Influences

Consider a lake with a dam that has a river flowing into it and a spillway that lets water out through the dam, to modulate the flow of the river downstream. Thus, we have two physical processes, a flow in from the river (hereafter `inflow`) and a flow out from the spillway (hereafter `outflow`). Mathematically, we might model the effects of these flows on the amount of water in the lake (call it `WaterL`) as

```
D[WaterL] = inflow – outflow
```

where `D` means the derivative with respect to time[1] (i.e., the amount of water in the lake will increase when the inflow is bigger than the outflow). This seems perfectly fine, but what if we discover, as we look closer at the dam, that there is a second spillway? Let us call its flow rate `outflow2`. Now we have to formulate a new equation:

```
D[WaterL] = inflow – outflow – outflow2
```

What do we know that allows us to do this reformulation? QP theory claims that we express our knowledge about the constituents of equations directly and then compose these constituents as needed to model particular situations. There are two types of *direct influences*, defined as follows:

```
(I+ A b) ≡ D[A] = ... + b + ...
(I– A b) ≡ D[A] = ... – b + ...
```

That is, `(I+ A b)` means that the derivative of `A` is defined as a sum of quantities, one term of which is `b`, and that `b`'s contribution is positive. `I-` is similar, except that `b`'s contribution is negative. Returning to our example,

```
(I+ WaterL inflow)
(I- WaterL outflow)
```

describes the direct influences in our initial state of knowledge, gleaned by inspection of the situation. Looking more closely and discovering the second spillway, our state of knowledge about direct influences becomes

```
(I+ WaterL inflow)
(I- WaterL outflow)
(I- WaterL outflow2)
```

When we wrote down entire equations, we had to start over when we learned something new. When we learn something new with influences, the knowledge accumulates. When we need to do reasoning about a situation, we make *closed-world assumptions* about the set of influences based on what we already know. This enables us to make some surprisingly subtle conclusions. For example, a system is in a dynamic equilibrium when key aspects are constant, even though many properties of the system affecting those aspects are changing. The lake, for example, is at equilibrium when the inflows and outflows are equal. Using + as the operation of combination in direct influences allows relative magnitude information to disambiguate conflicting influences when such information is available. In the simpler model of the lake, if the rate of inflow equals the rate of outflow, the amount of water in the lake should remain constant. If we measure these rates and the change of amount—indirectly, through change of level, as discussed below—and find that amount of water is unchanging, then all is well. But if we find that the level of the lake is still falling, we must reexamine our assumptions—here, the closed-world assumption that we knew all instances of processes that were occurring in the situation. Thus, the compositionality of qualitative mathematics is an aid to diagnosing and repairing problems with models, as well as a means of constructing models for a specific situation in the first place.

These are called direct influences because they represent the direct effects of a process on some quantity. Changes in these quantities indirectly cause changes in other quantities as well. For example, if the outflow from the spillway is larger than the inflow from the river, that causes the level of the lake to fall. The relationships that support causal conclusions like these are called *indirect influences*, discussed next.

6.2.2 Indirect Influences

The quantities of an entity or of a situation are often interconnected by relationships. The level of water in our lake example, for instance, depends on the amount of water in it. Similarly, its temperature depends on both the heat of the water and the amount of water that heat is distributed through. These other parameters are indirectly affected by the processes of liquid flow in this situation. For instance, if the river water flowing into the lake is warmer than the lake water itself, the temperature of the lake will slowly rise, all else being equal. These indirect effects are captured by indirect influences. Like direct influences, each indirect influence will provide one piece of information about the constraints on a quantity and must be combined with others to determine how it might actually be changing. Moreover, the amount of information it specifies about the underlying connection between the quantities is even weaker than with direct influences.

We define a *qualitative proportionality*, `(qprop A B)`, as

`(qprop A B)` ≡∃`f` s.t. `A` = `f(…, B,…)` ∧ `f` is increasing monotonic in `B` and `A` is causally dependent on `B`.

In other words, the function that determines `A` depends at least on `B` (but may depend on other quantities), and all else being equal, if `B` goes up, then `A` goes up, and if `B` goes down, then `A` goes down. `(qprop– A B)` is defined similarly, with `f` being decreasing monotonic. Moreover, the causal interpretation is that the change in `B` is a cause of the change in `A`.

Qualitative proportionalities are the minimal amount one can know and still be able to infer that a change in `B` will cause a change in `A`, all else being equal. The "all else being equal" comes from the partial information about the other causal antecedents in any particular qualitative proportionality. To determine what will actually happen requires making a closed-world assumption about the indirect influences on a quantity. For example, if the shape of the lake bed is constant, then the causal connection between the level of the lake and the amount of water is captured concisely by

```
(qprop LevelOfLake WaterL)
```

This enables us to predict that if the amount of water in the lake is increasing, then we should see that the level of the lake is rising, and conversely, if we see that the level of the lake is falling, then it must be the case that (because this is the only qualitative proportionality constraining `LevelOfLake`) the amount of water in the lake is decreasing.

Similarly, the relationship between the temperature of the lake's water and its amount and heat can be captured thusly:

```
(qprop Twater Heatw)
(qprop- Twater WaterL)
```

If we know that the heat of the lake's water is decreasing while the amount is increasing, these two qualitative proportionalities enable us to conclude that the temperature of the water is falling. However, should the heat of the lake's water be rising as well as the amount, we can infer nothing about the change in the temperature of the water. That is because we only know that the function determining `Twater` is increasing monotonic in one and decreasing monotonic in the other. A wide variety of concrete functions satisfy this constraint: the real relationship might be additive, a quotient (physically true in this example), or some complex nonlinear function that has the appropriate monotonicity properties in the range of interest. Thus, unlike direct influences, conflicting indirect influences cannot simply be resolved by adding information about relative magnitudes of the influences. What one does instead depends on the circumstances. If it is an everyday situation, we might just wait and see. If it is a scientific question (e.g., do clouds have a net positive or negative contribution to global warming?), then this ambiguity tells us that we need to formulate (or discover) more precise models. The partial information provided by the qualitative representation already provides constraints on such models—namely, some of the parameters that they need to include and the overall character of their interdependencies.

These differences in functional roles are central to qualitative representations and reasoning, so it is worth exploring them further. Qualitative representations provide a natural level of detail for many kinds of reasoning, because they allow for partial knowledge and express our intuitions of causality. Consider Newton's second law:

$$F = M \times A$$

This equation tells us all that we need to know if we want to find the value for one of the parameters. By finding the values for the other two, we can solve for any of them. However, it doesn't tell us how to cause motion to occur, so it does not capture causality. It doesn't allow us to express intermediate conclusions we might gather when learning a law. On the other hand, a qualitative version of the law does:

```
(qprop A F)
(qprop- A M)
```

This representation provides causality: it tells us that to affect acceleration, you must change the force or the mass. Moreover, it does some of the work needed in a formal language for expressing hypotheses about continuous systems, namely supporting the incremental accumulation and combination of knowledge. This property is called *compositionality*. Compositionality facilitates learning each of these parts of the law independently, via experience or experimentation (which I view as more organized and controlled experiences). These statements can easily be conveyed by language: chapter 13 argues that sentences such as "The acceleration depends on the force" and "As the mass is increased, the acceleration is decreased" have these two qualitative proportionalities as part of their semantics. This means that qualitative representations support learning via reading and dialogue, powerful forms of cultural transmission. Thus, qualitative representations provide a useful formal language for acquiring, using, and integrating knowledge about continuous systems.

6.2.3 Compositionality and Graceful Extension of Knowledge

Compositionality enables knowledge about continuous phenomena and systems to be accumulated incrementally. This can be valuable in reasoning at different levels of detail. Let us consider the two-container situation in figure 6.1 as a simple example.

Two holding tanks, which contain water, are connected by a pipe that has a valve. We know that if the levels are unequal, once the valve is open, liquid will flow between them until the levels are equal. Let us say that tank `F` has the higher level. What can we say about the rate of liquid flow in such circumstances?

We might know that the rate of liquid flow depends on the pressure difference between the two liquids. We can express this via two qualitative proportionalities:

```
(qprop FlowRate (Pressure WaterF))
(qprop- FlowRate (Pressure WaterG))
```

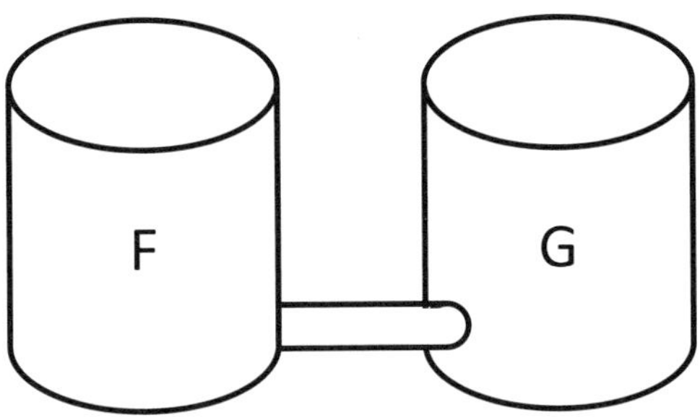

Figure 6.1
Two containers, connected at their bottoms by a pipe.

It also depends on how much fluid resistance the pipe itself has:

```
(qprop- FlowRate (Resistance Pipe))
```

Suppose we needed a more rapid flow, because we want to get the excess water out of `F` quicker. These statements tell us that we have three options:

1. Increase the pressure in `F`
2. Decrease the pressure in `G`
3. Decrease the resistance in the pipe

Which of these makes sense to do? We need to elaborate the models a bit more to figure this out. The pressure on the bottom of each container depends on the mass of the water in it, that is,

> `(qprop (Pressure ?w) (Mass ?w))`, where `?w` is the liquid held in a container.

Because we are trying to empty `F` quickly, adding more water to `F` to increase its pressure and thereby increase `FlowRate` is counterproductive. Reducing the amount of water in `G` might be an option. The third option is to decrease the resistance somehow, because the `qprop-` tells us that will cause the `FlowRate` to be larger. How might we do that?

The resistance of a pipe can be modeled in terms of its area and its roughness:

```
(qprop- (Resistance Pipe) (CrossSectionArea Pipe))
(qprop (Resistance Pipe) (InteriorRoughness Pipe))
(qprop (Resistance Pipe) (Length Pipe))
```

Area, in turn, depends on the geometry of the pipe. Most pipes are round, and knowing from three-dimensional (3D) geometry:

> `(qprop (CrossSectionArea ?obj) (Diameter ?obj))` where `?obj` is a 3D object with a round cross section

This gives us three new options: we can use a pipe that is smoother on the inside, move the tanks closer together and use a shorter pipe, or use a larger-diameter pipe. Which of these options is the best depends on many other factors—the point is that qualitative representations enable us to generate those options via causal reasoning on the relationships between parameters. Importantly, it enables models to be formulated and extended incrementally, which is important cognitively to keep working memory loads manageable. (The representations and one kind of reasoning involved in this are introduced in chapter 11, and another way of doing the reasoning is introduced in chapter 12.)

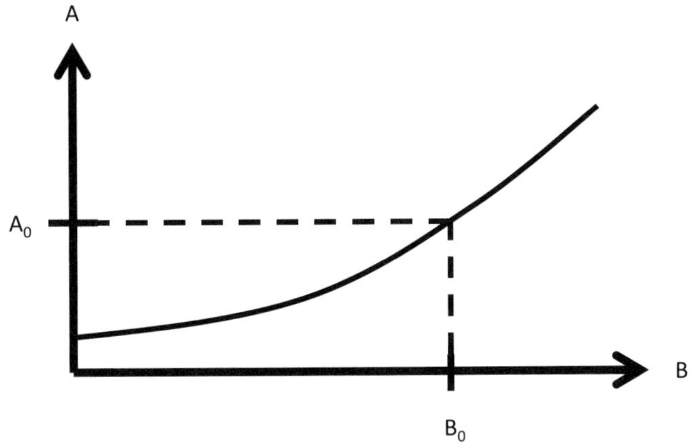

Figure 6.2
By knowing a correspondence involving a single point on the (implicit) curve defining a qualitative proportionality, ordinal information can be propagated.

6.2.4 Specifying Additional Information about Relationships

Qualitative proportionalities are extremely weak. They suffice to propagate signs of derivatives, as long as the influences on a quantity all agree. In some circumstances, though, even a little more information about the implicit function that they specify can provide quite a lot of inferential power. We examine several relationships here that provide more information but still do not completely specify the underlying function, in keeping with the minimality principle.

The first of these are *correspondences*. Intuitively, a correspondence relationship pins down just one point on the graph defining the underlying function, as figure 6.2 illustrates. (Of course, we don't actually know the shape of the curve, only that it is increasing monotonic.) This state of affairs is noted as

```
(qprop A B)
(correspondence (A A0) (B B0))
```

This can be read as "When B = B_0, A = A_0." Correspondences are *n*-ary—that is, if three qualitative proportionalities were constraining A, then the first argument would still be the coordinate for A, whereas the other coordinates would be expressed by other pairs. In reasoning, correspondences enable us to propagate ordinal information across indirect influences. Here, if we know that B is greater than B_0, then we know, because the underlying function is increasing monotonic, that A must be also greater than A_0. Similarly, if B is less than B_0, then A must also be less than A_0.

Correspondences are extremely useful because they enable us to specify constraints involving key values. For example, in thinking of flow rate above, the qualitative proportionalities alone do not tell us the sign of the flow rate or that the flow rate will approach zero as the containers equilibrate. We can add the following correspondence to represent this information:

```
(correspondence  (FlowRate 0)
                 ((Pressure WaterF) (Pressure WaterG))
```

This tells us that, in this situation, the flow rate will be zero when the pressures are equal. Correspondences provide a means of recording experienced linkages between values as well as formulating general laws. (How to use correspondences in logically quantified descriptions, a form of schema, so that we can express insights such as these as general laws is discussed in chapters 7 and 11.)

Sometimes it is useful to propagate ordinal information across objects. For example, if we see that the level in F is higher than the level in G, we can predict that when we open the valve, water will flow from F to G. The difference in levels has told us something about the difference in pressures. In other words, there is an implicit function defining pressure in terms of level that is the same across the two containers. To support such inferences, the relationship explicitFunction provides a name to the implicit function specified by qualitative proportionalities. If the function is the same in two objects, then corresponding values in the antecedent parameters give us information about the relationship between the constrained parameter. For example, suppose we know

```
(explicitFunction PressureLevelFn (Pressure ?cl))
```

where ?cl is the liquid in a container. This says that the function PressureLevelFn designates the particular functional relationship that is partially described by the influences. Suppose this fact were known for contained liquids in general. For any two contained liquids ?l1 and ?l2, knowing that the underlying function, whatever it is, is the same licenses the inference

```
(correspondence  ((Pressure ?l1) (Pressure ?l2))
                 ((Level ?l1) (Level ?l2)))
```

Thus, if we know that the level in F is higher than the level in G, we can infer that its pressure must be higher as well.

Not every causal antecedent of a quantity is a quantity. Consider how the level of liquid in a container depends on the amount of liquid. The shape of the container matters: pouring a cup of water into a coffee cup leads to a higher level than pouring the same cup of water into a saucepan. The relationship `functionDependency` enables such facts to be specified, for example,

```
(functionDependency LevelAmountFn
                    (ShapeFn (Container ?cl)))
```

This does not tell us exactly what the dependency is; it only alerts us that this is a factor. If two holding tanks have the same shape, then we are safe in inferences concerning levels and amounts. Otherwise, we might choose to ignore this factor and continue onward but keep this in mind as an assumption to reconsider if our predictions are off. Moreover, it also serves as an articulation point for extending models: if we deal with cylindrical containers frequently, for example, we might start encoding more aspects of their shape as quantities (e.g., diameter) so that we can use the same representational ideas already introduced to predict at least the relative signs of changes.

For capturing the intuitions of experts, compositional primitives that have somewhat stronger semantics can be useful. The following have proven to be useful in modeling more advanced reasoning (Collins & Forbus, 1989):

`(C+ ?a ?b)` is like `qprop`, except that `?b` is an additive term in the function that determines `?a`.

`(C- ?a ?b)` is like `qprop-`, except that `?b` is a negative term in the function that determines `?a`.

`(C* ?a ?b)` is like `qprop`, except that `?b` is in the numerator of the expression that determines `?a`.

`(C/ ?a ?b)` is like `qprop`, except that `?b` is in the denominator of the expression that determines `?a`.

For example, the impact of heat and volume on the pressure of an ideal gas might be expressed as

```
(C* (Pressure ?g) (Heat ?g))
(C/ (Pressure ?g) (Volume ?g))
```

because, for an ideal gas,

$$P = \frac{nRT}{V}$$

6.3 Naturalness

How intuitive are these representations? One source of evidence for their naturalness is their successful use in educational software aimed at middle school children. Using a concept map interface, Betty's Brain (Biswas et al.,

2001) helped children learn about stream ecosystems by building concept maps whose underlying semantics were simple qualitative relationships. The teachable agents model it was based on had students "teaching" Betty via the concept map, with the goal of having their Betty do well on subsequent quizzes. Northwestern's VModel (Forbus, Carney, Sherin, & Ureel, 2004) also used concept maps to enable students to build qualitative models to test their predictions. Feedback was given by the software, providing a combination of step-by-step causal reasoning, with animation and automatically produced natural language based on their model. VModel was domain independent and also enabled students to use continuous processes in their models (see chapter 7). Qualitative representations have also been used to successfully model learner behavior in intelligent tutoring systems. Kees de Koning, Bredeweg, Breuker, and Wielinga (2000) used a qualitative reasoner to generate dependency structures that indicate, for a particular inference, what knowledge was involved. If a learner gave a different answer than the expected answer, model-based diagnosis techniques were used to construct hypotheses about what mistake was made (e.g., not knowing one of the qualitative proportionalities). All of these systems were successfully used by students, suggesting that the level of causal representations described here is natural for science education.

6.4 Expressiveness

The analogy between qualitative mathematics and traditional mathematics is useful because it gives us a way to explore how complete our qualitative mathematics is. As noted above, ordinary differential equations (i.e., sets of equations that include derivatives with respect to time) have been fantastically successful in high-precision modeling of aspects of continuous change for centuries. Thus, we can use it as a type of gold standard to understand how expressive our qualitative mathematics is. If we can express every ordinary differential equation as a qualitative equation, then our qualitative mathematics should be able to cover the same range of phenomena (albeit at less precision) that they do. It turns out that this can be done. A formal proof is provided by Kuipers (1994); here we summarize the argument to provide the underlying intuitions and also explore some of the connections with other traditional formalisms.

Any system of ordinary differential equations can be divided into two components: a set of purely algebraic equations containing one or more variables and a set of equations constraining derivatives with respect to time of those variables (i.e., $\frac{dx}{dt}$ on the left-hand side and no derivatives on the right-hand side). Let us see, roughly, how to rewrite this set of equations into a set of influences.

The algebraic equations can be modeled via qualitative proportionalities by rewriting every equation so that the left-hand side is a single variable (let's call it X), which does not occur on the right-hand side. Numerical coefficients can be ignored: if they cannot be changed, then there is no reason to include them in the qualitative model. For every variable that occurs on the right-hand side (Y_i), we need to ascertain how X depends on it (i.e., increasing or decreasing monotonic) and add the appropriate qualitative proportionality to the set of influences we are constructing. What if the dependency is non-monotonic, that is, (qprop X Y_i) over some of its range and (qprop– X Y_i) over other parts? In such cases, we must use sets of influences, not just one, providing the appropriate preconditions for when to believe each of them as appropriate. (How to do this is described in chapter 7.)

Appropriately scoped, the extreme generality of qualitative proportionalities pays off handsomely: the original equations might contain polynomials, trigonometric expressions, or complex nonlinear expressions. It doesn't matter because they are all the same at this level of precision. Recall that we have also left out numerical coefficients. Thus, a set of qualitative proportionalities can capture an entire family of algebraic equations.

Now let us consider the equations constraining derivatives. Again, let us say that the parameter being constrained is X, and the other parameters are Y_i. As you might expect, every such equation will be modeled by a set of direct influences (I+ or I–) whose first argument is X. In the simple case where the right-hand side consists of a sum of the Y_is (perhaps with negative signs), for each Y_i, there will be either (I+ X Y_i) or (I– X Y_i), according to whether or not the sign of Y_i is positive or negative. But what about more complex forms on the right-hand side? Here the more concrete means of combination in the definition of direct influences make things a bit more complex. Consider

$$\frac{dX}{dt} = aY_1 - bY_2$$

It would be incorrect to represent this as

(I+ X Y_1)
(I– X Y_2)

because that would imply that when $Y_1 = Y_2$, the derivative of X would be zero. To model these situations requires introducing new parameters that are linked to the originals via qualitative proportionalities. Here,

```
(qprop Y1' Y1)
(qprop Y2' Y2)
(I+ X Y1')
(I- X Y2')
```

The abstract nature of the qualitative proportionalities then expresses our lack of knowledge about what will actually happen while correctly propagating signs of derivatives. For example, if we know that Y_1 is increasing and Y_2 is decreasing, then we can determine that the derivative of X must be increasing. But since we don't know what the relationship between Y_1' and Y_1 really is, we cannot conclude that when $Y_1 = Y_2$, the derivative will be zero.

The argument sketched here suggests that every ordinary differential equation can indeed be expressed by sets of influences. Thus, the potential expressive range of this system of qualitative mathematics is as broad as that of ordinary differential equations and thus capable in principle of capturing a broad range of continuous phenomena. This raises two other questions:

1. Is the range of qualitative mathematics too broad, that is, can it license inferences that are incorrect? This would make qualitative reasoning unsound.
2. Can people's knowledge of the physical world actually be organized in this way?

We postpone consideration of the first question to chapter 10. The answer is, yes, qualitative simulation is unsound, but as discussed there, this is not a problem, given the functional roles that qualitative reasoning plays in human cognition. Let us tackle the second question here.

There are several reasons to believe that human knowledge about continuous systems can be organized this way. First, the distinction between direct and indirect influences is critical for breaking causal loops. A set of traditional equations has no particular causal structure. Influences provide a causal structure: processes impose direct influences, and those changes propagate via indirect influences. Some of the indirectly influenced quantities are rate parameters, which enables causal explanations involving feedback and dynamic equilibriums to be generated. (To ensure coherent causal accounts, QP theory, discussed further in chapter 7, imposes the constraint that no quantity can be both directly and indirectly influenced at the same time.) Second, the distinction between direct and indirect influences reflects a fundamental difference in types of quantities. Quantities can be divided into two categories: *extensive* quantities are those that accumulate over time, such as mass and heat. *Intensive* quantities are those that depend, directly or indirectly, on extensive quantities, such as pressure and temperature. Extensive quantities are always constrained by direct influences, whereas intensive quantities are always constrained by indirect influences. This distinction mirrors widely used, and therefore known to be useful, distinctions drawn in formalisms developed by scientists and engineers. For example, Forrester's (1961) system dynamics formalism divides parameters into stocks (extensive) and flows (intensive), and the state-space model of dynamical systems divides parameters into state parameters (extensive) and dependent parameters (intensive).

QP theory has been used to model a wide variety of domains, including flows, thermodynamics, motion, economics, ecosystems, and social reasoning, as discussed in subsequent chapters. Some of these efforts were intended as cognitive models, whereas others were driven solely by the needs of an application (scientific reasoning, engineering reasoning, instruction). But even those application-oriented efforts provide evidence for the psychological plausibility of QP theory because part of their measure of success is that they can perform humanlike reasoning and generate humanlike explanations. All of this suggests that the representation of causality provided by influences provides a good model of causality in continuous systems. However, there is at least one domain where it clearly fails, and an alternate account is needed. We discuss this next.

6.5 Confluences and Causal Ordering

As it happens, analog electronics is really different. Consider an explanation of how the circuit in figure 6.3 works:

> When the voltage on the input rises, it causes the current flowing from the base to the emitter to increase. This increase causes the current flowing from the collector to the emitter to increase, which then causes the voltage on the output to drop.

You don't need to understand the details of analog electronics to extract the important features of this explanation: first, the explanation is driven by a presumed disturbance to the system ("When the voltage on the input rises"). Second, it contains assertions about causality between voltage and current that go in both directions. In the first sentence, it is a change in voltage that causes a change in current. In the second sentence, it is a change in current that causes a change in voltage. That is not something that is easily modeled in QP theory. QP theory assumes that the direction of causality between properties of particular types is essentially constant. For example,

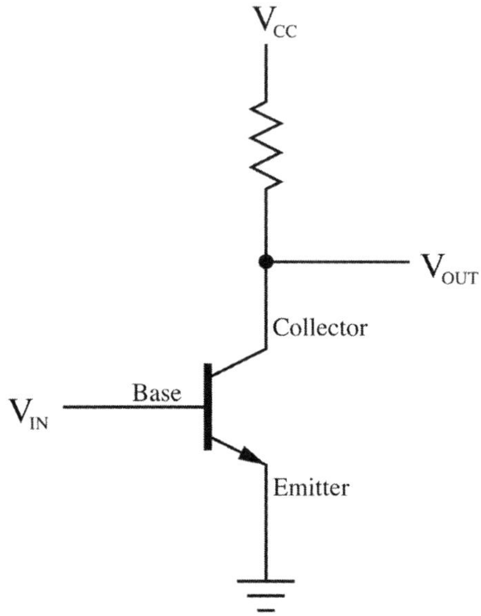

Figure 6.3
A transistor amplifier.

changes in heat cause changes in temperature but never the other way around. What is going on here?

Let us consider Ohm's law, which states that the voltage drop across a resistor is the current flowing through it times the value of the resistor (i.e., $V=I\times R$). This equation was used twice in the explanation above. (Implicit in the above explanation is that the resistance across the collector/emitter junctions of the transistor depends inversely on the current across the base/emitter junctions, but this does not affect the argument.) Notice that none of these parameters is an extensive parameter. Thus, there isn't any particular reason to choose one of them over another. At the level of analysis usually used in analog electronics, the only extensive parameters are those associated with devices that contain history (i.e., charge in capacitors and flux in inductors). Could one introduce more extensive parameters and then use a QP-like model of causality? In electronics, current is defined as the flow of charge, with electrical charge being an extensive quantity. In theory, electrical circuits could be modeled in terms of charges at every node and introducing charge flow processes along every path, but in electrical circuits, the number of nodes and path segments between them is large, and that would cause the complexity of the model to explode. So those who invented analog electronics came up with a different way to think about causality. They grounded causality in some form of input disturbance and followed those disturbances through the laws that governed the devices that constitute the circuit. Formalization of this way of causal thinking for analog electronics was first developed by de Kleer (1984).

Do people use this disturbance/propagation model in other domains? Simon and Iwasaki (1988) argued yes. Simon's (1953) original formulation of *causal ordering* imposed causality onto a set of simultaneous equations by identifying *exogenous variables* (i.e., variables that could be considered "drivers" of the system, external causes like the change in input voltage in the circuit above). Simon's original work was in building qualitative models of economics. If one identifies exogenous parameters of a system with directly influenced parameters, then his causal ordering algorithm can be used as a means of generating candidate sets of indirect influences for a specific system. However, Simon and Iwasaki's (1988) account requires formulating a new causal model for each specific system. QP theory, as described in chapter 7, proposes that people's knowledge of these causal laws is more general and that models for specific situations are assembled from more fragmentary knowledge.

6.6 Summary

Qualitative mathematics attempts to formalize the intuitive notions people have about relationships between quantities. It provides ways to reason about change with minimal information (ordinal relationships between values, signs of derivatives), providing a good fit for what is actually easily available perceptually in many everyday situations. It provides representations of causality that, between QP theory and confluences, may suffice for expressing human mental models of continuous phenomena and situations. Qualitative representations are compositional, making them useful in communicating via natural language and learning (this is explored in more depth in chapter 13 and chapter 17, respectively). The level of resolution that qualitative mathematics provides is coarser than traditional mathematics, which means that qualitative reasoning is often ambiguous. This is actually desirable, because it is what enables qualitative representations to serve as a means of identifying possibilities. Such alternate predictions (and explanations) can be used to identify where more detailed information, such as observation or experience or traditional mathematical models, is needed.

In illustrating these ideas, relevant knowledge has been introduced as needed. What is missing from the discussion so far is how these statements are organized in our conceptual structure. This is a complex question. The next chapter (chapter 7) describes part of the answer by introducing the notion of continuous process and formalisms that provide schemas for organizing such knowledge. Chapter 11 explains what I believe is happening when someone's knowledge about an area has been reasonably well systematized. Our theory of what else is going on, as well as how people reach these more systematic levels of knowledge, is discussed in part IV.

7 Qualitative Process Theory

The world is full of a variety of things. And these things are not static: they move, flow, heat up, cool down, mix, separate, get created, grow, decline, and die. Our conceptual structures for understanding the world must support the ability to characterize, explain, and predict changes in it. Thus, we need representations that capture the broad regularities that occur, so that we can appropriately contextualize what we learn. Representations of the types of things in the world and the categories of behaviors that they manifest provide our mental resources for expressing context. In philosophy, ontology is the study of what exists. Thus, formalizations of the broad categories of what exists are ontological hypotheses. Ontology plays two major roles in qualitative reasoning:

1. Ontology solves the *applicability* problem. The causal, qualitative relationships we have seen need to be appropriately scoped and contextualized so that they are only applied when appropriate. Ontologies provide the conceptual structure needed for this.
2. Human causal reasoning appears to be grounded in hypotheses about the existence of mechanisms (Bechtel & Abrahamsen, 2005; Chi, Slotta, & de Leeuw, 1994; Forbus, 1984). A central idea of QP theory is that a particular ontological category, *continuous processes*, provides this notion of mechanism for a broad range of human causal reasoning.

This chapter focuses on the ontology provided by qualitative process theory. It begins by describing the standard account of how modeling (of systems and the world) is treated in qualitative reasoning. This provides a useful idealization of the modeling process, elaborated more fully in chapters 11 and 13. The ontological hypotheses made by QP theory are discussed next. This includes formal representations for *model fragments* and *continuous processes*, which provide a language for describing knowledge about the continuous world. Many kinds of commonsense and expert models have been formulated in QP theory, but in a few domains, different ontologies seem to be required, as discussed in chapter 6. We will not see those other ontologies again until chapter 9.

7.1 Modeling the Modeling Process

In reasoning about the world, scientists and engineers make models. They apply general laws about the nature of the universe, combined with assumptions and approximations honed through practice and professional norms, to construct models of the situation or system under study. We call this reasoning *model formulation*. Once a model has been formulated, there are various ways to reason with it, depending on what kind of model it is. The results of reasoning might directly answer the question(s) that the model was designed to answer, or it might indicate that the model itself is inadequate, and hence a new model is required. I call this account *first-principles modeling*, because it relies on applying general laws to specific situations. A reasonable hypothesis, held by many, is that modeling in everyday reasoning is carried out via first-principles modeling. That is, in figuring out what might happen in a situation, we apply our general knowledge of the world to build a model, and then we reason with that model to make predictions, generate explanations, diagnose problems, and so on. In psychology, these are often referred to as *mental models* (Gentner & Stevens, 1983; Johnson-Laird, 1983).[1]

It is worth being more precise about this reasoning, so we can examine it more closely. As later chapters describe, I agree with a slightly different version of this account: professional reasoning of scientists and engineers works the same in most ways as everyday reasoning, but everyday reasoning actually works differently than most qualitative reasoning researchers believe (chapter 12). Nevertheless, the first-principles modeling account is intuitively simple and likely to be part of what people do, so it is a helpful starting point. Moreover, this account has proven quite useful in creating systems that do human-level professional reasoning, so it is worth understanding for its own sake.

The first-principles account of modeling relies on the following ideas:

- A *domain theory* is a collection of knowledge about some class of phenomena or type of system. Examples of domains include aspects of thermodynamics, biology, mechanics, and chemistry. It is general-purpose knowledge (i.e., logically quantified laws and schemas that can be applied to particular situations via instantiation). Given their intended purpose for reasoning, they are called *model fragments*, because models are built from them.

- A *scenario* is the problem, system, or situation under study. Scenarios are described in naturalistic terms. For some technical domains, this may include specialized entities (e.g., pumps, flywheels, or electronic components). There are always one or more questions that models must strive to answer, which can range from the very general (e.g., "What might happen?") to the very specific (e.g., "Will we introduce problems if we shorten this wire by an inch?").
- A *scenario model* is automatically constructed via model formulation by applying the knowledge in the domain theory to the scenario description. Model formulation can involve substantial reasoning, both in finding ways to cast the entities in the scenario into the idealizations assumed in the domain theory and in selecting what subset(s) of the domain theory are appropriate, given the intended purpose.

This account was in part a response to the practices of the expert systems community (Hayes-Roth, Waterman, & Lenat, 1983), which would build systems of rules that intertwined general knowledge of a domain with knowledge about particular artifacts and knowledge about specific tasks. These conflations meant that knowledge could not be reused: the rules used to diagnose one type of printer could not be used to diagnose a different type of printer, nor could they be used to do design modifications of the original type of machine. This is not only expensive and inelegant but also clearly not the way that human engineers work. An expert troubleshooter's knowledge generalizes, to some degree, and scientists don't have to get a new PhD to study a new phenomenon in their field.

The general-purpose nature of knowledge in domain theories helps ensure coverage. Given any new scenario for a domain, with a complete and correct domain theory, a sufficiently powerful model formulation algorithm can construct a model that, by reasoning, can be used to answer that scenario's questions. Possessing and using such general knowledge is surely part of the explanation for the flexibility of human expertise, as evidenced by how science and engineering are taught.[2] As chapter 12 describes, another part of the explanation is experience and generalizations constructed from experience. But here we shall focus only on general domain knowledge for clarity.

7.2 Model Fragments

A variety of representations for model fragments have been developed and used in the qualitative reasoning literature (Bobrow et al., 1996; Falkenhainer & Forbus, 1991). Here are the conventions we use:

- A *model fragment type* is a logically quantified schema describing a type of entity, relationship, or concept in a domain theory.
- The schema variables are called its *participants*. Each variable has one or more *constraints* that must be satisfied by any binding for that variable. Constraints can involve multiple participants. No free variables are allowed in constraints.
- The instantiation of a model fragment type is allowed for each set of bindings for its participants, such that the constraints are satisfied.
- Model fragments can have *conditions* that indicate when an instance of them is *active*. Conditions can mention participants, but no free variables are allowed. If there is no condition for a model fragment type, then instances of it are always active.
- Model fragments can also have *consequences* that hold when an instance of them is active. Like conditions, consequences can mention participants, but no free variables are allowed.

Note the distinction between the conditions for instantiation (i.e., the existence of a set of bindings for participants that satisfies its constraints) and the conditions for an instance being active (i.e., the instance's conditions holding). This helps determine when a potential phenomenon is worth thinking about in a situation (i.e., when a model fragment of that type should be instantiated) versus whether or not it actually occurs or is something to bring about/avoid (i.e., that model fragment instance is active). The possibility of our kettle melting on the stove is something we want to think about when making water for coffee. On the other hand, all of the substances that might appear in my coffee cup (including water, whisky, arsenic, plutonium) are not even worth thinking about.

Figure 7.1 shows an example of a model fragment, using a syntax close to that commonly used in the literature.

The participants, conditions, and consequences are indicated by keywords. Its constraints are the union of the statements in the participant descriptions (e.g., the `:type` and `:constraints` keyword specifications). For this example, the constraints are the conjunction of

```
(defModelFragment ContainedGasProperties
  :participants ((?stuff :type ContainedStuff
                         :constraints (phaseOf ?stuff Gas))
                 (?sub :type Substance
                       :constraints (substanceOf ?stuff ?sub))
                 (?can :type Container
                       :constraints (containerOf ?stuff ?can)))
  :conditions ((active ?stuff))
  :consequences ((>= (Temperature ?stuff) (TBoil ?sub ?can))
                 (qprop (Pressure ?stuff) (Mass ?stuff))
                 (qprop- (Pressure ?stuff) (Volume ?stuff))
                 (qprop (Pressure ?stuff) (Heat ?stuff))))
```

Figure 7.1
A model fragment defining a causal model for gas in a container.

```
(ContainedStuff ?stuff)
(phaseOf ?stuff Gas)
(Substance ?sub)
(substanceOf ?stuff ?sub)
(Container ?can)
(containerOf ?stuff ?can)
```

This model fragment illustrates a causal model for the properties of gas inside a container. The gas itself is represented by the variable `?stuff`, which will be bound to an instance of `ContainedStuff`. `ContainedStuff` is itself a model fragment, introduced below. The other participants bind relevant aspects of the stuff (i.e., its substance and container). The condition for this type of model fragment is when the instance of stuff it is describing is actually there, denoted by that model fragment instance being active.

The consequences of this model fragment provide two types of information. The first is a constraint on the temperature of the gas: it has to be at or above the boiling point for that kind of material in that container. Including the container is a way of leaving room for more or less detailed models. A simple model might be that there is a uniform boiling point for each substance. A more sophisticated model might include a dependency on the pressure inside the container, which can be discussed using the `?can` variable in that term. The second type of information it provides is a simple causal model for the thermodynamic properties of a gas. Recall that for ideal gases, the equation $PV = nRT$ describes the relationship between pressure (P), volume (V), mass (n), and temperature (T). For nonideal gases, these parameters are still related, but no simple analytic form accurately captures this relationship, so extensive tables are used to calculate one parameter based on others. In either case, the causal model is the same.

Let us step through the reasoning underlying how this model fragment was constructed. To figure out what indirect influences are needed, it is useful to first determine what parameters can be directly influenced. As discussed in chapter 6, a quantity cannot be both directly and indirectly influenced, so directly influenced parameters will be used to constrain the others. Recall that directly influenced parameters are always extensive parameters. Here there are three extensive quantities—namely, mass, volume, and heat. (Heat is not mentioned explicitly, but it is always in mind when talking about thermodynamics.[3]) For every extensive parameter, we can identify continuous processes that will affect them: we can add gas to a container, we can heat it up, and (unless it is rigid) we can stretch or compress the container. Hence, these are indeed going to be directly influenced quantities. That leaves pressure and temperature to constrain. A general property of thermodynamic entities (expressed elsewhere in another model fragment) is that temperature depends on heat. Therefore, we do not need to state that here. But how could we come up with the rest of the constraints?

There are two ways to do this. The first is to use physical examples. If we think about how pressure changes as we perform individual manipulations on the extensive parameters, we can quickly arrive at a set of constraints on pressure. Adding gas increases its pressure, as we know from filling and emptying tires. This justifies a `qprop` between pressure and mass. Similarly, heating a gas causes its pressure to increase, a problem that bedeviled early designers of steam engines. This justifies a `qprop` between pressure and heat. Finally, squeezing a balloon causes its pressure to increase, because the same amount of it is now confined to a smaller volume and hence a smaller surface area. This method is very heuristic but, if the examples are chosen carefully, quite productive.

The second way is to analyze an equation governing the phenomenon, when available. Think again about the ideal gas law:

$$PV = nRT$$

We can imagine perturbations in each extensive parameter at a time and look at how pressure must change accordingly. If amount or heat is increased, this equation tells us that P must go up if V is constant. On the other hand, if amount and heat are held constant, then an increase in V must be accompanied by a decrease in P, and a decrease in V must lead to an increase in P. Thus, we end up deriving the same causal model.

7.3 The Ontology of QP Theory

Processes are ubiquitous in human explanations of continuous change. Everyday solid objects move, crumple, melt, stretch, crack, and break. Liquids flow, freeze, and boil. QP theory concerns models of such processes, where the direct effects on the world can mainly be expressed via continuous changes in quantities.

In QP theory, processes are first-class entities in the ontology of naive physics. Some philosophers and other cognitive scientists have identified processes with patterns of behavior over time (e.g., Hayes, 1985b; Mackie, 1980). The problem with identifying processes with patterns of behavior is that it is conflating the generator (processes) with their output (behavior). Consider modeling pouring and leaking as patterns of events, as in Hayes's (1985b) naive physics for liquids. If I pour water into a container, the level of water in that container rises while the pouring is happening and stops rising when I stop pouring. In leaking, a container starts with a lot of water in it, and its level drops as the water comes out of the hole in the container. That seems fine so far. In fact, we can even use qualitative proportionalities to construct partial causal models on top of these patterns (e.g., the amount of water leaking depends on the size of the hole, the number of holes, etc.). But what happens when we pour water via a hose into a leaky bucket? The pouring behavior predicts that the level is rising, and the leaking behavior predicts that the level is falling. These static descriptions of behavior are noncompositional, yet people routinely reason about systems that are composed out of multiple parts and in which multiple processes are operating. By including processes in our ontology, we gain the ability to reason about continuous systems compositionally, because we describe the causal relationships between quantities in terms of influences that are part of the processes that are generating the behavior. A sophisticated naive physics might model both pouring and leaking in terms of liquid flow, thereby reducing the problem to resolving competing direct influences on the amount of water in the bucket. A less sophisticated naive physics might model pouring and leaking as distinct types of processes but still be able to compose them via influences to reason about combined effects.

These two different ways that someone might model a leaky bucket raises an important point. Qualitative process theory concerns the form of dynamical theories, not their specific content. It does not assume conservation of matter or conservation of energy, for example. Heat flow can be described in ways that adhere to energy conservation or involve flow of a "caloric" substance and violate energy conservation. This enables it to represent a wide range of human mental models. Moreover, it supports representing higher-order constraints such as energy conservation, because they can be framed as laws restricting what patterns of influence are allowed in processes.

The central assumption of qualitative process theory is the *sole mechanism assumption*, namely,

> All changes in continuous systems are caused directly or indirectly by processes.

This is a psychological assumption about how we tend to structure our knowledge about the world. I believe it holds for a broad range of human mental models, although this level of understanding is not the first level of understanding people achieve, as discussed in part IV. Similarly, in at least one domain, analog electronics, a different form of causality is required, as discussed in chapter 6. Nevertheless, processes and influences are remarkably effective representations for supporting a broad range of human reasoning about continuous systems.

The sole mechanism assumption has some interesting consequences. First, our models of a continuous domain must include a set of types of processes that might occur. This *process vocabulary* can be viewed as the dynamics for that domain. Second, explanations of causal changes in continuous systems must be grounded in processes. Agents must work through processes to cause changes: to boil water, for example, one must arrange for enough heat to flow into it that its temperature rises to its boiling point. Third, it allows us to reason by exclusion. If we see a quantity changing, it must be explainable in terms of the types of processes we know about, so if we can rule out all but one type, instances of that type must be the culprit. Fourth, it supports learning. If we cannot explain a behavior, then we know that our set of processes, or our understanding of them, must be incomplete or incorrect. In other words, QP theory imposes an inductive bias on causal theories of continuous phenomena.

Processes are represented as a special kind of model fragment. (Model fragments that are not processes are called *views*.[4]) A model fragment representing a type of continuous process must have at least one direct influence in its consequences. Moreover, only processes may include direct influences in their consequences. This constraint is so important that the syntax for modeling languages using QP theory typically includes a separate field for direct influences to highlight them visually. Figure 7.2 illustrates a simple model of heat flow.

In this model, an instance of heat flow will exist whenever one thinks of two entities as thermal objects (`ThermalPhysob`) connected by a path that can support heat flow (`HeatConnection`). When the path can support

```
(defModelFragment HeatFlow
 :participants ((?src :type ThermalPhysob)
                (?dst :type ThermalPhysob)
                (?path :type HeatPath
                   :constraints (heatConnection ?path ?src ?dst)))
 :conditions ((heatAligned ?path)
              (> (Temperature ?src) (Temperature ?dst)))
 :consequences ((Quantity (HeatFlowRate ?self))
                (qprop (HeatFlowRate ?self) (Temperature ?src))
                (qprop- (HeatFlowRate ?self) (Temperature ?dst))
                (I- (Heat ?src) (HeatFlowRate ?self))
                (I+ (Heat ?dst) (HeatFlowRate ?self))))
```

Figure 7.2
A description of the process of heat flow.

heat flow given its particular conditions (`heatAligned`) and the temperature of the source is greater than the temperature of the destination, then that instance of heat flow will be active. When it is active, there will be a quantity representing its flow rate (i.e., `(HeatFlowRate ?self)`, where `?self` always refers to the current instance) that is causally constrained by the temperature difference (i.e., the qualitative proportionalities) and causes changes in the heat in the source and destination (i.e., the direct influences).

When should one think of particular objects as thermal objects, and what physical configurations correspond to heat paths? That is a separate issue from describing the nature of heat flow itself. In scientific reasoning, heat flow arises in circumstances ranging from microscopic effects in semiconductors to the raging inferno of stars. A premium is put on explaining new phenomena and effects in terms of existing processes, whereas introducing new processes is relatively rare. In engineering practice, multiple types of heat flow are identified: heat can flow via conduction, convection, and radiation. These can be expressed via additional model fragments that add information to this basic concept of heat flow. (The example of modeling fluid conductance of a pipe, in chapter 6, provided an illustration.)

For any continuous domain, there is a *process vocabulary* that is the set of types of processes that exist. Similarly, the *view vocabulary* consists of the nonprocess model fragments that someone uses in reasoning about that domain. Given a scenario, a scenario model includes the original entities, attributes, and relationships, plus instances of views and processes. Prediction proceeds by finding what views and processes are active, combining influences to ascertain their immediate effects, and determining what changes might occur over time in the set of active process and view instances. Explanation occurs by searching for combinations of active process and view instances that, when their influences are combined appropriately, could lead to the observed behavior. When planning or designing, the types of processes available are scrutinized with respect to the current context, and entities are selected (in the case of planning) or postulated (in the case of design) that will give rise to instances of processes and views that will achieve the desired outcome or behavior. These different types of reasoning all rest on a set of *basic inferences* that QP theory supports. I describe these next.

7.4 Basic Inferences of QP Theory

There are four basic inferences of QP theory:

1. *Model formulation*. This is the answer to the intuitive question "What phenomena might be relevant?"
2. *Determining activity*. This answers the question "What's happening?"
3. *Influence resolution*. This answers the question "What's changing?"
4. *Limit analysis*. This answers the question "What might happen next?"

We discuss each in turn.

7.4.1 Model Formulation

The purpose of this inference is to use the process and view vocabularies for the domain to construct a scenario model. In its simplest form, it involves finding matches for the participants and constraints of the model fragments of the domain theory among the entities, attributes, and relationships of the scenario, as well as constructing an instance of each type of process or view. For example, if we think about a kettle of water on a stove, we might consider the water and the stove as `ThermalPhysobs` and the kettle as constituting a `ThermalPath` that connects them. If our process vocabulary consists of the definition of heat flow in figure 7.2, then we would have two instances of heat flow in our scenario model, one representing the possibility of heat flow from the stove to the water and one representing the possibility of heat flow from the water to the stove. Figure 7.3 illustrates.

As domain knowledge grows, model formulation can become quite complex. People often understand phenomena at multiple levels: a physicist can choose between using classical, relativistic, and quantum mechanics, depending on the situation, for example. Human professionals have a wealth of intuitive qualitative models as well: even well-trained scientists do not use equations to reason about when to add milk to their coffee on a daily basis. Chapter 11 is devoted to these issues.

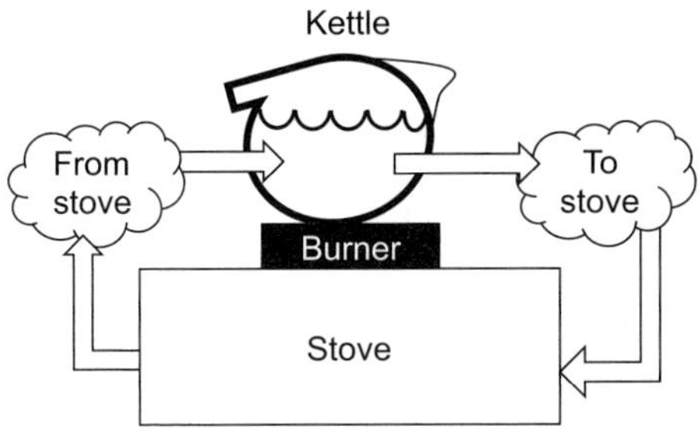

Figure 7.3
Two instances of heat flow between the stove and the water in the kettle. Because the instances have opposite conditions, at most one of them can be active at a time.

Model formulation conceptually involves multiple closed-world assumptions. The process and view vocabularies are, often tacitly, sometimes explicitly, assumed to be closed. Our knowledge about the entities, attributes, and relationships is also assumed to be closed. This enables the subsequent inferences to be much simpler and provides several points for backtracking that support diagnosis when reasoning goes awry and learning when one's models prove to be insufficient.

7.4.2 Determining Activity

Once the scenario model has been constructed, finding out which processes and views are active provides information about the current state of the situation or system. By knowing the active processes, we know what quantities are being directly influenced, and by knowing the active views plus the active processes, we know how those direct effects might be propagated. Thus, this information provides the overall answer to the question of what is happening in the situation. Returning to our kettle example, suppose that the stove is hotter than the water:

```
(> (Temperature Stove) (Temperature Water))
```

There isn't anything that can be changed about the kettle itself to block heat, and hence we assume that it is always `heatAligned`. Figure 7.4 illustrates that the heat flow from the stove to the water will be active, whereas the heat flow from the water to the stove will be inactive.

Where does information about conditions come from? It depends. If we are thinking about a situation in front of us, we can use a combination of perception (e.g., look if the stove is on, hold our hand near the burner and kettle to see if how warm they are) and knowledge (e.g., that the heat path

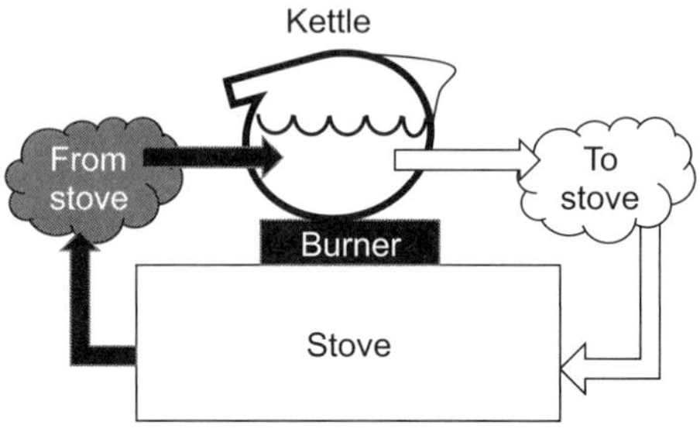

Figure 7.4
If the stove is hotter than the water, the heat flow from the stove will be active, and the heat flow to the stove will not be.

here is unchangeable). If we are trying to explain a set of observations, we might reason backward from the observations to find patterns of activity that could explain them.

In thinking about conditions, it is useful to divide them into two types:

- *Dynamic conditions* consist of statements concerning ordinal relationships and the status of other model fragments. Examples of dynamic conditions introduced so far include the temperature difference that causes heat flow and the `ContainedGasProperties` model fragment depending on that gas actually existing.
- *Preconditions* consist of all other types of statements. They express how dynamics depends on factors outside of itself. One use of preconditions is to express boundary conditions, which hold throughout a dynamical analysis. Preconditions also provide a conceptual interface between dynamics and agency. Turning off the stove, for example, is a way of preventing overcooking.

This distinction helps identify the information requirements for reasoning about change. It becomes crucial when reasoning about changes in a qualitative state: changes in dynamic conditions can be predicted solely within QP theory, assuming preconditions remain constant. How this works is described below. But to get there, we must first talk about how changes within a state are derived.

7.4.3 Resolving Influences

Recall that influences provide partial information about how quantities are related. Knowing the activity of a scenario model means we can make the necessary closed-world assumptions to gather all of the constraints on each quantity and use those constraints to determine the possible signs of derivatives (i.e., `Ds` values) for each of them. This is the essence of influence resolution. It is called *influence resolution* by analogy with resolving multiple forces in classical mechanics.

Every quantity in a scenario model is either

- *Directly influenced* (i.e., there is one or more process instances that constrain it via an `I+` or an `I-` relationship)
- *Indirectly influenced* (i.e., there is one or more model fragments that constrain it via a `qprop` or a `qprop-` relationship)
- *Uninfluenced.* By the sole mechanism assumption, if a quantity is uninfluenced, its `Ds` value is 0.

Influence resolution proceeds by determining the `Ds` values for the directly influenced quantities, then propagating that information through the indirect influences to determine the `Ds` values for the rest. Let us consider each case in turn.

Resolving a directly influenced quantity requires adding up the influences on it. If the signs of the influences are all the same, then the `Ds` value is simply that sign. For example, in our kettle example, the only influence on the heat of the water is the heat flow to it, which is `I+`, and so the temperature of the water is increasing. We represent this via the binary predicate `Ds`, whose first argument is a quantity and whose second argument is its sign value—here,

```
(Ds (Temperature Water) 1)
```

There is a subtlety here: this tacitly assumes that the heat flow rate is positive. For simplicity, in this book, it is assumed that all rate quantities (i.e., those introduced by processes) are positive. I believe this is psychologically realistic and is a common assumption in practice (e.g., Bredeweg, Linnebank, Bouwer, & Liem, 2009). If there are ambiguous direct influences, then information about relative magnitudes of the rates can determine which direction of influence dominates. If we are filling a bucket that has a tiny leak with a hose, for instance, the rate of flow out the leak is smaller than the flow in from the hose, so while this is occurring, the amount of water in the bucket will be increasing.

Resolving indirectly influenced quantities requires gathering up the qualitative proportionalities that constrain it and determining their net contributions. Suppose we have a quantity `Q` such that

```
(qprop Q A)
(qprop- Q B)
```

If `(Ds A 1)` and `(Ds B -1)`, then both are making a positive net contribution, and we infer that `(Ds Q 1)`. On the other hand, if `(Ds B 1)`, then we cannot determine the `Ds` of `Q`, because we don't know the relative effects of the two contributions, given how little information qualitative proportionalities provide. There are three ways to respond to such ambiguities:

1. *Wait and see*. If the quantities are observable or inferable from observable quantities, and the goal isn't prediction, the conclusion that it is ambiguous has given us a signal about where to focus our attention to find out more.
2. *Search alternatives*. If our goal is prediction, then reasoning about the potential consequences of each possible value becomes a way of constructing multiple possible outcomes. These possible outcomes might be scrutinized for potential danger (in prediction, planning, or design) or their ability to produce predictions that match observations (in explanation).
3. *Apply more knowledge*. Through experience, including professional knowledge, we may already know how influences will resolve for particular classes of situations. We might know that a particular influence is negligible with regard to another (e.g., that heat lost from a kettle to the atmosphere is negligible compared to the heat flow into it from the stove). Or we might have more detailed knowledge of the function underlying the qualitative proportionalities. For instance, Black's law tells us that when there is thermal mixing during a liquid flow, the temperature of the source remains unchanged, whereas the temperature of the destination will rise or fall depending on whether the source is warmer or colder than the destination.

Note that the influence resolution algorithm honors the centrality of processes in causality: by starting with the direct influences of processes, we first calculate the changes they are causing. Then we propagate this information through the qualitative proportionalities (hence their synonym, *indirect influences*) to ascertain what this means for the rest of the system. To keep this causal story clear, QP theory stipulates that no quantity can be both directly and indirectly influenced and that the graph of qualitative proportionalities is loop-free. At first glance, it might seem that banning loops in qualitative proportionalities would make it impossible to model systems with interdependent parameters, like feedback systems. This is not the case. The loops in feedback systems always contain a derivative relationship, which is modeled via a direct influence, not a qualitative proportionality. Even our simple kettle example includes loops, as figure 7.5 illustrates.

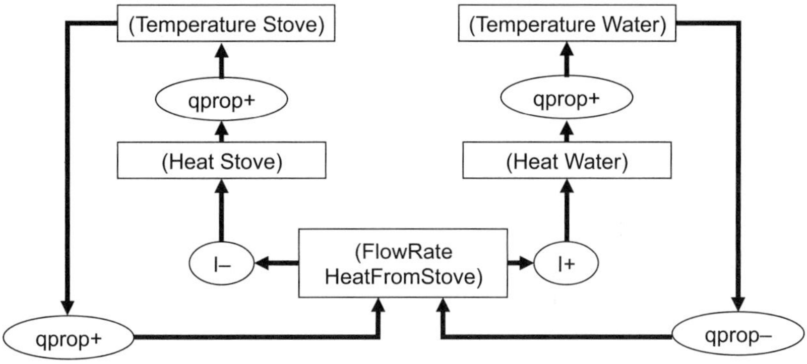

Figure 7.5
Influences when water in a kettle is heating on a stove. Notice that the causal loops are broken by direct influences (`I+`, `I–`).

Although the flow rate determines the change in heat, which in turn determines the change in temperature and hence the flow rate, the `I+` and `I-` relationships ground the causal explanation, thereby breaking the loop.

The operations to this point provide a means of deriving the *qualitative state* of a system. What is a qualitative state in QP theory? It is a pattern of activity, grounded in processes, that describes what is happening in a system and how it changes. More specifically, qualitative state consists of the following:

- What entities, attributes, and relationships from the scenario model hold.
- The status of model fragment instances (i.e., which are active versus inactive). This includes instances of processes.
- The ordinal relationships that hold. This includes `Ds` values, which are defined by ordinal relationships between the derivative of a quantity and zero.

The `Ds` values describe what is changing within a state: what quantities are increasing, decreasing, or steady. Next we see how this definition of state leads to the prediction of changes of state.

7.4.4 Limit Analysis

Processes can start and stop, entities can be created and destroyed, and stable patterns of behavior can be established and disrupted. Being able to reason about such changes in qualitative state is a crucial aspect of reasoning about change. QP theory takes a cue from mathematical modeling by holding boundary conditions and background assumptions constant throughout an analysis. However, because it is also representing modeling knowledge and how to integrate reasoning about continuous change with the rest of human reasoning, QP theory makes these dependencies explicit, through the participants, constraints, and preconditions of model fragments. The operation of *limit analysis* derives how a qualitative state might change, based solely on consideration of its dynamics.

Limit analysis starts by finding the neighboring points within the quantity spaces of each changing quantity, that is, the limit points that are closest to the value of the quantity. If there is no neighboring point in a quantity's current direction of change, that means there is no limit point there, and thus no potential changes in the process structure can occur due to that quantity changing. If there is a neighbor, then their current relationship and how the neighbor is changing must be combined to determine if the relationship between them can change. Figure 7.6 summarizes this inference. Every such potential change in an ordinal relationship constitutes a *limit hypothesis* (i.e., a hypothesis about how that state might end).

Returning to our kettle of water on the stove, the temperature of the water is increasing (i.e., `(Ds (Temperature Water) 1)`, and the only limit point in its quantity space, given that heat flow is the only type of process that we know about, is the temperature of the stove, because they are related by two instances of heat flow. Currently, the temperature of the stove is higher than the temperature of the water. Because the stove's temperature is constant,[5] if this goes on, eventually the temperature of the water will reach the temperature of the stove. That change in the ordinal

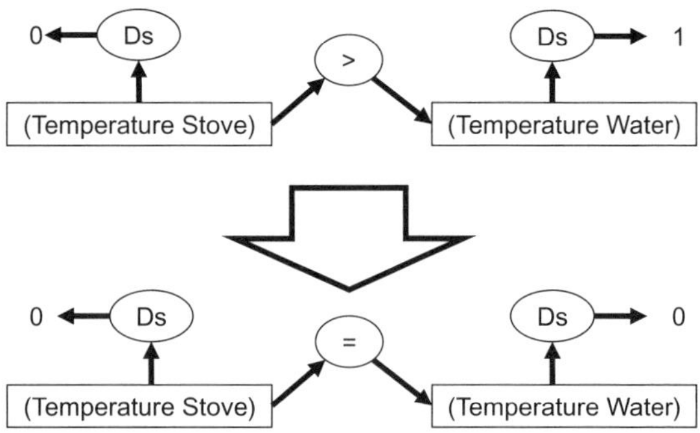

Figure 7.6
Example of a limit hypothesis. When the water is heating on the stove, one possible outcome is that the temperature of the water reaches the temperature of the stove.

relationship of their temperatures marks the end of the heat flow between them and thus constitutes one way for the current qualitative state to end. It is the only way, in fact, given that this is the only process occurring in this situation. Thus, this state gives rise to only one limit hypothesis:

```
(> (Temperature Water) (Temperature Stove))
  → (= (Temperature Water)   (Temperature Stove))
```

Note that this limit hypothesis assumes that the entities and preconditions remain constant: if one pours the water out of the pot, all bets are off. QP theory further assumes that such changes are not what mathematicians call asymptotic (i.e., something that would take an infinite amount of time to occur). This assumption rules out a simple version of Zeno's dichotomy paradox.[6]

Now let us complicate our example slightly. Suppose we extend our domain theory with a process representing boiling. One of its conditions will be that the temperature of the water will be greater than or equal to its boiling point. That means the quantity space for the temperature of the water will now have a second limit point, let's call it `TBoil`. Suppose we know, in our current state, only the following ordinal about `TBoil`:

```
(> TBoil (Temperature Water))
```

This tells us that boiling is not occurring, and if `TBoil` is steady, then another possible transition is

```
(> TBoil (Temperature Water))
  → (= TBoil  (Temperature Water))
```

That is, another possible outcome of our current state is that the temperature of the water reaches its boiling point, which causes it to start boiling. Do we know that this must occur? No, because we do not know the relative magnitude of `TBoil` versus `(Temperature Stove)`, so this limit hypothesis is a distinct possibility. If the stove happened to be set so that its temperature is `TBoil`, then we could also have both hypotheses being true (i.e., the water reaches its boiling temperature and that of the stove at the same time).

In general, there can be multiple limit hypotheses. There are three reasons for this. First, if the ordering of limit points within a quantity space is not a total order but only a partial order, then that quantity might have multiple neighbors in a direction. For example, one might not know the ordinal relationship between the temperature of a stove and the melting point of a kettle placed on it, so whether the kettle reaches thermal equilibrium safely versus melts cannot be determined for certain in that case. Second, a process can influence more than one quantity. Third, multiple processes can be occurring at the same time. Thus, limit analysis often involves evaluating multiple limit hypotheses. Moreover, conjunctions of limit hypotheses must be considered. If the changes are truly independent, one might be tempted to ignore this possibility because such coincidences are rare. (How often can you set your stove to be exactly 100°C?) But the changes aren't always independent: because quantities can be functionally related, a change in one ordinal relationship might indeed imply a change in another. (Consider two quantities linked by an indirect influence, which also has an associated correspondence.)

QP theory does not in general allow the determination of which change occurs first when multiple changes are hypothesized. If we use the traditional calculus as a model theory, it is easy to see why: the alternative that occurs next is the one for which the time to integrate the quantities involved to their limit points is minimal. Because we know so little about the functions involved, we cannot determine what will actually happen. Conversely, this is why qualitative reasoning is so valuable: by determining what might happen, it provides the information we need to figure out when more knowledge is needed. In other words, QP theory cannot always guarantee a unique answer for what might happen next, but it guarantees that what will happen will be among those alternatives it generates (chapter 10 elaborates).

Even though limit analysis cannot always provide a unique answer, the rest of the operation filters out changes that are clearly not possible. This is done in two steps. The first is to use continuity to rule out transitions that violate our intuitions about change. The second is to use the nature of equality to distinguish changes that happen in an instant from those that take an interval of time—clearly the former will happen before the latter. We discuss each in turn.

We assume that changes in quantities do not violate continuity. That is, if

```
(> A B)
```

holds in a qualitative state, then it is inconsistent for

```
(< A B)
```

to hold in the next state—there must be some state in which A and B are equal in between them. This provides a surprisingly powerful filter. To see how this filter operates, we must first consider how possible next states are found for a limit hypothesis or a conjunction of them. (For convenience, we hereafter just say "limit hypothesis" to mean both the single-change case and a conjunction of changes assumed to all happen at the same time.) In other words, we are finding triples of the form

<qualitative state>, <limit hypothesis> $\rightarrow$ *<next qualitative state1>*

...

<qualitative state>, <Limit hypothesis> $\rightarrow$ *<next qualitative stateN>*

This part of the operation is done by constructing qualitative states that might result from a limit hypothesis, initially generating a broad set of candidates and then filtering them.

How can candidate next states be generated? Recall that the dynamical constituents of qualitative states are as follows:

- Ordinal relationships between quantities
- Status assignments to model fragment instances (i.e., processes and views)

The other constituents of state can be considered stable conditions, statements that must be true for any state in the current analysis. They provide the starting point for our construction.

Next, consider the union of the pairs of quantities mentioned in the set of limit hypotheses for a state. Any specific limit hypothesis (single change or conjunctive) represents the possibility that the pair(s) of quantities it mentions change and no others, because those prospects are represented by yet other limit hypotheses. Thus, every next state must include the change(s) represented by the limit hypothesis being explored, as well as all of the current ordinal relationships for the pairs of quantities mentioned in other limit hypotheses, but not the one under consideration.

Given this starting point, the initial candidates for next states are found by searching for consistent assignments of truth values to the constituents of the dynamical state that have not already been constrained by the assumptions made so far. The logical interconnections between these statements implied by the domain theory significantly constrain the number of candidates to be considered (e.g., in our water on the stove example, knowing that the temperature of the water is the same as the temperature of the stove means that neither heat flow instance can be active, so we do not have to think about those possibilities).

Formally, this can be modeled as a *constraint satisfaction problem* (Mackworth, 1977), where the constraints are the instantiated laws of the domain theory plus the laws of QP theory. Continuity is one such law: if there is a pair of quantities whose change across the state transition would violate continuity, then that transition can be ruled out. Another intuition that is expressed via filtering is that the causal effects of change are minimized: those implied by the change must occur, but other random changes do not occur. This is done by sorting the candidate next states into equivalence classes based on the number of changed statements in them and eliminating all but those with the minimum number of changes.[7]

As mentioned above, the equality change law provides another powerful filter. Suppose we have four quantities in a qualitative state, two of which are equal and two of which are unequal:

```
(= A B)
(> C D)
```

Suppose further that these quantities are changing in ways that will cause these ordinal relationships to change:

```
(Ds A 1)
(Ds B 0)
(Ds C -1)
(Ds D 0)
```

The finite difference between `C` and `D` means that the change to equality will take a finite interval of time. On the other hand, the change from equality of `A` and `B` will occur instantaneously, because QP theory does not assume fuzzy values for numbers. Therefore, the change involving `A` and `B` will happen first, because instants are shorter than intervals. QP theory further assumes that the change that can occur in an instant is infinitesimal, thereby less than any finite value. Thus, if the influences in the new state were such that A were to decrease, it would transition back to equality in an instant as well. The *equality change law* expresses the consequences of these arguments:

> *Equality change law:* With two exceptions, a process structure lasts over an interval of time. It lasts for an instant only when either
>
> 1. a change from equality occurs, or
> 2. a change to equality occurs between quantities that were influenced away from equality for only an instant.

(Readers with certain backgrounds will recognize that this rules out impulses in direct influences, that is, instantaneous finite changes in quantities. QP theory has been extended to include impulses [Kim, 1993], but we will not concern ourselves with that here.)

The equality change law provides an important filter on limit hypotheses. Recall that the set of limit hypotheses consists of single changes and conjunctions of single changes. Consider the set of hypotheses that contains only changes that occur in an instant. The largest such set, in the sense of set inclusion, must represent what occurs next, because instants are shorter in duration than intervals. Thus, the equality change law can rule out limit hypotheses, sometimes leading to a unique predicted next state.

What happens if filtering is so extreme that no candidate transitions remain? Because QP theory does not model the asymptotic approach, the only conclusion that can be reached is that the state itself must be inconsistent! This may seem unintuitive at first, but this conclusion seems to be broadly compatible with expert human reasoning about continuous change. Consider, for example, a ball connected to a support by a string. Assume the string, ball, and support are completely inelastic—they cannot stretch or compress at all. (No real material is completely inelastic; this is an ideal model that is often used to ignore such effects when they are assumed to be small.) Suppose the ball is released. What happens? Assuming Newtonian dynamics and gravity, during the fall, the velocity of the ball continues to rise due to the acceleration by gravity. But in the final state, when the string reaches its maximum extension, the fall must end. This situation is inconsistent with the previous state because the acceleration is downward, thus increasing the velocity, but to stop, the velocity has to be decreasing, which violates continuity. Thus, the transition to this state must be ruled out. This leaves us with a falling object that is always approaching but never reaching the length of the string. Intuitively, this is impossible (unless you are a fan of Zeno's). Hence, we consider the qualitative state itself as an inconsistent description of the world, relative to our knowledge of it. The outcome of limit analysis on a qualitative state is a set of transitions from that state to one or more other states. If that state lasts indefinitely, then there are no state transitions.

This subtlety, along with phenomena like dividing time into subsequent instances in some situations (a consequence of the equality change law), is something that most people do not think about. Professionals who model complex systems do and, depending on their task, often make similar conclusions. Other models of qualitative dynamics (e.g., Kuipers, 1994) allow asymptotic approach and enforce a strict alternation between instances and intervals in decomposing time. These are models of human thinking about continuous phenomena, and given the differences in how people think about the world, there isn't a single correct model. However, it appears that only a handful of models suffice to capture the broad generalities in people's reasoning about qualitative dynamics.

Limit analysis can provide some surprisingly subtle conclusions. Consider, for example, the three water holding tanks shown in figure 7.7.

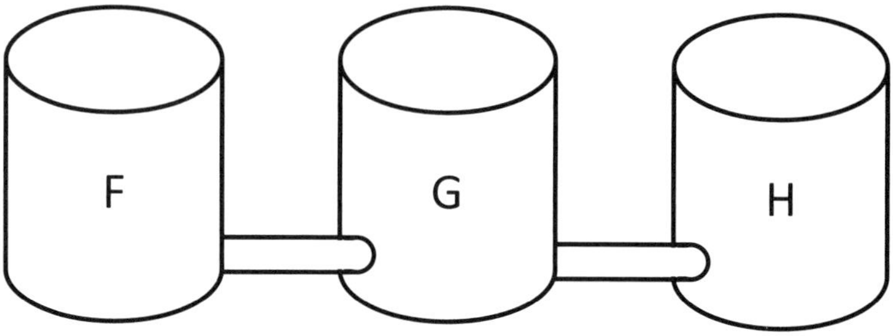

Figure 7.7
Three containers connected by pipes. Initially, the level of water in F is higher than that in G, which in turn is higher than that in H.

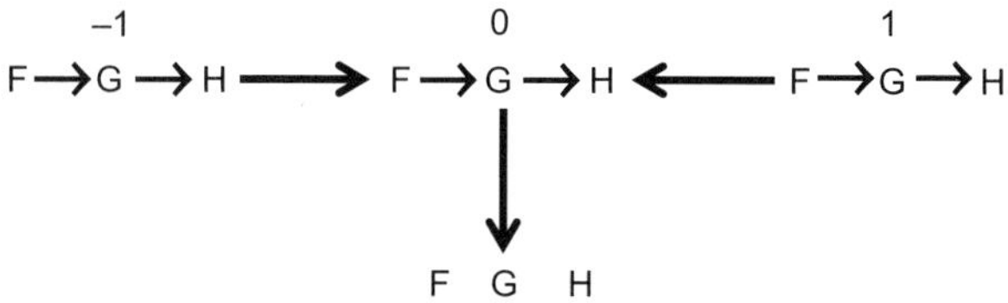

Figure 7.8
The possibility for dynamic equilibria can be detected by looking for changes in relative rates.

Given the cross-flow that is occurring, we know immediately that the level and pressure of the water in F are decreasing, and the level and pressure in H are increasing, because the only influences on their amounts are positive and negative, respectively. But what about G? Depending on the relative rates between the flow from F to G and the flow from G to H, it could be increasing, decreasing, or constant. Suppose we generate a qualitative state for each possibility and perform limit analysis on each state. The network of transitions we get between these states is shown in figure 7.8. This shows that one outcome of this situation is that the level of G can equilibrate, something that is not obvious from the initial description of the situation.

7.5 Encapsulated Histories

The temporal semantics of model fragments, including processes, is that the statements in the consequences hold throughout any instant or interval of time over which that model fragment is active. Given that the participants in these model fragments typically have a spatial extent, the set of model fragments active over time defines a *history*, a piece of space-time bristling with properties that describe what attributes and relationships hold there (Forbus, 1984; Hayes, 1979). It is useful to be able to reify such histories as an abstract pattern for two reasons. First, as discussed in chapter 17, such patterns of behavior are a useful intermediate representation for learning and conceptual change because they can be constructed through observation of specific behaviors via analogical generalization. Second, such abstract patterns provide the appropriate context for stating laws and constraints that explicitly refer to portions of intervals (e.g., equations that describe the motion at the end of an interval based on properties of its start and what is happening during that interval). QP theory includes *encapsulated histories*, again a form of schema, to represent such patterns. The details of their representation are postponed until chapter 17.

7.6 Summary

The basic inferences of QP theory provide a model for reasoning at two levels. First, the decomposition of tasks involved in reasoning about continuous systems is, I believe, very plausible psychologically. The ability to identify the kinds of processes that are occurring, to understand what is actually happening in a situation, including how it is changing and how those changes lead to events that change the nature of what is occurring, seem essential constituents of any account of reasoning about continuous change. These are fundamental types of questions that people ask about the everyday world and, as later examples will illustrate, are involved in professional reasoning: identifying what is possible and thereby framing the important questions for subsequent work with more detailed representations.

The second level concerns the degree to which the specific methods proposed here for solving these tasks can be taken as psychological models. Here the story is, I believe, more complicated. The reasoning described here proceeds entirely from first principles, completely ignoring experience. It may be a reasonable model for how people who have achieved some level of expertise do at least some of their reasoning. But the paradox of the falling ball fools a surprising number of people, who see nothing wrong with it until they work through it, step by step. What are they relying on initially? Experience, it seems, applied to the current situation—incorrectly—via analogy. Chapter 12 makes this argument in detail.

Chapter 8 explores QP theory further by delving into a variety of examples to show how it can be used to represent and reason about intuitive physical models in several domains.

8 Examples Using QP Theory

One of the key hypotheses of QP theory is that the representations it provides are expressive enough to formally describe a wide variety of human mental models of continuous systems. To argue for this hypothesis, this chapter uses QP theory to model a variety of phenomena. This includes two of the motivating examples from chapter 1 (the kettle on the stove and speed of freezing examples) and other examples from the literature. These include the following:

- One-dimensional motion, including two common misconceptions found in physics students (i.e., Aristotelian and Impetus models)
- A simple model of materials sufficient to state Minsky's conundrum that one can pull with a string but not push with it
- A spring-block oscillator and the impact of dynamic and static friction on its possible behaviors
- How dynamic equilibria can be detected qualitatively

Any theory intended to capture the range of human mental models ought to be able to handle at least these phenomena.

8.1 Modeling Fluids

Many objects, like chairs and cats, are easy to carve up into individuals. Carving up liquids and gases into individuals that can be discussed and reasoned about is a bit more complicated.[1] Hayes (1985b) points out that we have two basic ontologies for liquids:

- The *contained stuff ontology* individuates fluids via the space that they are in. When we talk about the Atlantic Ocean, the Rhine River, or the coffee in our cup, we are using this way of carving up the world.

- The *piece of stuff* ontology individuates fluids via a particular collection of molecules. When we talk about the gasoline consumed by a country each year, the water involved in a flood, or pouring our coffee into the sink, we are using this way of carving up the world.

We flexibly invoke each of these ontologies as needed in our daily lives: we understand what it means to not be able to walk in the same river twice, even while planning a return trip to snorkel in the Pacific Ocean. This chapter examines how to formalize a simple version of the contained stuff ontology using QP theory, to show how it can perform some of the reasoning discussed in chapter 1. It is intended as one model of intuitive knowledge about liquids and gases. No single model can capture the broad range of individual differences in people's models, and matching a particular person's model involves laborious effort to first uncover what their model is, as discussed in chapter 17. The psychological plausibility of this model is based on ability (i.e., models built upon it can be used to draw conclusions that are similar in nature to those people are able to do).

We start by examining how dynamic changes in existence affect reasoning, then construct a simple domain theory of liquids and gases using QP theory that suffices to handle these problems. This is by no means the most advanced domain theory that has been or could be built using these ideas—such extensions will be pointed out along the way.

8.2 Existence and Why It Matters

Nothing lasts forever, although we like to pretend that certain things do. Some objects in our conceptual environment are very stable, such as buildings and rocks, whereas some are ephemeral, such as the furniture arrangement for a party or the coffee in our cup. Individuation can be quite complex. A very simple, but surprisingly powerful, individuation method offered by QP theory is *quantity-conditioned existence*. That is, suppose we can define a quantity such that an individual only exists when that quantity is positive. Consider contained liquids. The coffee in my cup has a particular amount in it at the moment. I take a sip, and it is less. Enough sips, the amount becomes zero, and the coffee is gone. That is an example of quantity-conditioned existence. Another example is the existence of a biological species, which is conditioned on its population being larger than zero. While the organisms that make up the individuals of a species are discrete, for many purposes, we approximate population as a continuous parameter.

Quantity-conditioned existence is a useful conceptual tool because it enables us to reason about dynamic changes in existence. If we pour coffee into a cup, we have, in a sense, created "the coffee in the cup." Once it has been consumed, that individual is gone. By including processes that directly influence the parameter that existence is conditioned upon, we can infer the creation and destruction of quantity-conditioned individuals via the basic inferences of QP theory.

Let us construct a representation for contained fluids. First we need to formalize the notion of amount. Since we are talking about the "amount of coffee in the cup," we need to talk about substance (e.g., "coffee") and the container (e.g., "cup"). We also need to talk about the phase the substance is in. Coffee in a cup is conventionally a liquid. Air in a balloon, on the other hand, is conventionally a gas. It would be more elegant to have one representation that covers both cases, especially given that phase transitions such as boiling and freezing convert from one to the other, so we will define `AmountOf` as a function of three parameters. That is,

> `(AmountOf ?sub ?phase ?can)` denotes the amount of substance `?sub` in phase `?phase` inside container `?can`. It is always nonnegative. When positive, there is an individual of that substance and phase inside that container.

Some readers may be wondering what the units of `AmountOf` are. Basic QP theory ignores units, by design. Otherwise, its scope would be limited to modeling people with at least a modicum of expertise. Units, like conservation laws, are one of those grand conceptual innovations that I believe are built on top of our intuitive models and serve to provide layers of precision that improve reasoning. (When education works correctly, our intuitive models get reorganized via these ideas, but that happens less than often supposed—see models of motion later in this chapter and in chapter 17.)

Our domain theory must include model fragments that introduce appropriate `AmountOf` quantities and the individuals that exist as a consequence. The only phases we consider here are `Liquid`, `Gas`, and `Solid`. (Although plasmas are relevant to everyday life, most of us are unaware of that, and condensates that exist only at extreme laboratory conditions are out of bounds.) The only substances we need for our examples are `Water` and `Air`, although again, the model should be constructed so that it can be gracefully extended to new substances as someone learns about them. Because there are a lot of substances in the world, one thing we want to avoid is thinking about them when they are not relevant. (Someone dwelling on the possibility of arsenic in tea, for example, is either paranoid or reading a murder mystery.) The domain theory uses the relationship `canContainSubstance` to indicate interest in a combination:

(canContainSubstance ?can ?sub ?phase) is true exactly when container ?can is capable of containing substance ?sub in phase ?phase, and this is relevant to the current analysis.

Note that this relation covers both the physical possibility of existence and its relevance. Conflating these two is a simplification, and techniques have been developed that allow them to be cleanly separated and reasoned about (see chapter 11). However, it is not clear to me that people's untrained intuitions are always as nicely separated, so this simplification may not be that egregious.

Figure 8.1 describes two model fragments. The first, ContainedStuff-Possibility, describes the circumstances under which it makes sense to consider a potential contained stuff, introducing an AmountOf quantity. Recall that a model fragment instance is not even instantiated if the constraints do not hold, so the placement of the canContainSubstance pattern achieves our goal of not considering possible entities for which this statement does not hold. The second model fragment, ContainedStuff, represents the contained stuff itself. (This is denoted by the :subclassOf field, which is taken to be the type of entity that the model fragment instance is. defProcess is equivalent to defModelFragment with :subclassOf being ContinuousProcess.) Because it is conditioned on the AmountOf being positive, it implements the constraint for quantity-conditioned existence that we intended. It also introduces a quantity to represent the

```
(defModelFragment ContainedStuffPossibility
 :participants ((?can :type Container)
                (?phase :type Phase)
                (?sub :type Substance
                      :constraints (canContainSubstance ?can ?sub ?phase)))
 :consequences ((Quantity (AmountOf ?sub ?phase ?can))
                (>= (AmountOf ?sub ?phase ?can) 0)))

(defModelFragment ContainedStuff
 :subclassOf (PhysicalObject)
 :participants ((?can :type Container)
                (?phase :type Phase)
                (?sub :type Substance
                      :constraints (canContainSubstance ?can ?sub ?phase)))
 :conditions ((> (AmountOf ?sub ?phase ?can) 0))
 :consequences ((Quantity (Mass ?self))
                (qprop (Mass ?self) (AmountOf ?sub ?phase ?can))
                (correspondence ((Mass ?self) 0)
                                ((AmountOf ?sub ?phase ?can) 0))))
```

Figure 8.1

The model fragment ContainedStuffPossibility encodes the conditions under which a contained stuff might exist. The model fragment ContainedStuff encodes the bare minimum properties—namely, that it has mass.

```
(genls VolumetricPhysob PhysicalObject)

(implies (VolumetricPhysob ?o)
         (and (Quantity (Pressure ?o))
              (Quantity (Volume ?o)))
              (> (Volume ?o) 0))

(genls ThermalPhysob PhysicalObject)
(implies (ThermalPhysob ?o)
         (and (Quantity (Heat ?o))
              (Quantity (Temperature ?o))
              (> (Heat ?o) 0)))

(genls TemperatureSource ThermalPhysob)
(genls FiniteThermalPhysob ThermalPhysob)
(implies (FiniteThermalPhysob ?o)
         (qprop (Temperature ?o) (Heat ?o)))
```

Figure 8.2
We can define several specializations of objects based on whether one needs to reason about volumetric properties and/or thermal properties.

mass of our new individual, which is causally dependent on the amount of stuff that there is. It does not specify exactly how mass depends on amount, just that (via the correspondence), when the `AmountOf` is positive, the mass is, too.

So far, our contained stuff has only one property. One way to model contained stuffs is to add to the `ContainedStuff` model fragment all of the properties that we might wish to reason about (e.g., pressure, volume, level, heat, temperature, etc.). This doesn't quite work for two reasons. First, not all properties apply to all phases: the idea of level makes sense for liquids but not for solids or gases. (We are ignoring sand and other amorphous solids here.) Second, even when a quantity makes sense, we may not want to be thinking about it. Consequently, we will define several subcategories of physical objects, each containing a bundle of properties that make sense for some analyses. Figure 8.2 defines four classes of objects.

`VolumetricPhysob` represents things for which pressure and volume matter. `ThermalPhysobs` involve heat and temperature, with `FiniteThermalPhysobs` being the usual case. Temperature sources are modeled as `ThermalPhysobs` with no particular connection between their heat and temperature, and thus an arbitrary amount of heat can be added or removed without affecting their temperature.

Now that we have some basic object properties, we can extend our domain theory to incorporate phase-specific properties of contained substances.

```
(defModelFragment ContainedLiquidProperties
 :participants ((?cl :type ContainedStuff
                     :constraints (phaseOf ?cl Liquid))
                (?sub :type Substance
                     :constraints (substanceOf ?cl ?sub))
                (?can :type Container
                     :constraints (containerOf ?cl ?can)))
 :conditions ((active ?cl))
:consequences ((Quantity (Level ?cl))
               (explicitFunction LevelMassFn (Level ?cl))
               (qprop (Level ?cl) (Mass ?cl))
               (correspondence ((Level ?cl) 0)((Mass ?cl) 0))
               (explicitFunction PressureLevelFn (Pressure ?cl))
               (qprop (Pressure ?cl) (Level ?cl))
               (explicitFunction PressureContainerFn (Pressure ?can))
               (qprop (Pressure ?can) (Pressure ?cl))
               (≤ (Temperature ?cl) (Tboil ?sub ?can)))
```

Figure 8.3
Contained liquids have levels whose changes causally affect their pressure.

8.3 Representing Contained Liquids

One of the salient properties of contained liquids is that they have a level. If we think about the pressure on the bottom of a container, the more liquid there is above it, the higher that pressure is. Because we don't have the representational machinery yet to describe shapes and space, we will only consider containers with uniform, flat bottoms here—no deep ends to a swimming pool or tidal pools on a beach. We further assume we are living somewhere in the Midwestern United States, where it is absolutely flat (or at least seems that way), and all of our containers are on the ground, and hence we can leave the container geometry implicit.[2]

We can express these relationships via the model fragment in figure 8.3. This model fragment introduces the idea of level and that it depends on the mass of the liquid. By using a named function (see chapter 6), we are stating that the function that determines it is always the same, which enables us to propagate ordinal information across objects. Given that we have done the same for pressure here, it means we can infer that, if the level of the liquid in one container is higher than another, its pressure must also be higher, too, if these are the only influences on it.

Pressure differences matter, of course, because they drive liquid flow. Figure 8.4 describes a simple model of liquid flow. It requires a liquid path between the source and destination, which must not be blocked or turned off (expressed via the predicate `aligned`, which is a traditional nautical term

```
(defProcess LiquidFlow
 :participants ((?src :type ContainedLiquid)
                (?sub :type Substance
                      :constraints (substanceOf ?src ?sub))
                (?src-can :type Container
                          :constraints (containerOf ?src ?src-can))
                (?dst-can :type Container
                          :constraints (canContainSubstance
                                          ?sub Liquid ?dst-can))
                (?path :type LiquidPath
                       :constraints (liquidConnection ?path ?src-can ?dst))
 :conditions ((aligned ?path)
              (> (Pressure ?src-can) (Pressure ?dst-can)))
 :consequences ((Quantity (LiquidFlowRate ?self))
                (qprop (LiquidFlowRate ?self)(Pressure ?src-can))
                (qprop- (LiquidFlowRate ?self) (Pressure ?dst-can))
                (correspondence ((LiquidFlowRate ?self) 0)
                                ((Pressure ?src-can) (Pressure ?dst-can)))
                (I+ (AmountOf ?sub Liquid ?dst-can) (LiquidFlowRate ?self))
                (I- (AmountOf ?sub Liquid ?src-can) (LiquidFlowRate ?self))))
```

Figure 8.4
A representation for liquid flow.

for enabling flow in a piping system). Notice that it specifies the source liquid but not a destination liquid. If a destination liquid were required, then liquid could not flow into an empty container. Notice also that the direct influence is on `AmountOf`. This makes sense compared to using mass for two reasons. First, it enables liquid to come into existence even if a container is empty. Second, it maintains a consistent causal story: `AmountOf` influences mass.

We now have enough model fragments to support the reasoning about the situation involving two holding tanks that we saw in chapter 6. The two holding tanks, `F` and `G`, are modeled as containers. The piping system `P` is modeled as a liquid path that connects the two tanks. Given these choices, there will be two potential occurrences of liquid flow, one from the water in `F` to tank `G` and one from the water in `G` to tank `F`. Given that `F` initially has a higher level (see figure 6.1), the liquid flow from `F` to `G` will be active. The results of resolving influences and limit analysis follow our intuitions: the amount of water in `F` is decreasing, which causes its mass and level to decrease, which in turn causes the pressure in `F` to decrease. Simultaneously, the liquid flow is causing the amount of water in `G` to increase, which causes its mass and level to increase, which in turn causes the pressure in `G` to increase. Eventually, the two pressures become equal, which causes the liquid flow to stop (i.e., the status of that instance of `LiquidFlow` changes from active to inactive). The structure of the system plus the initial conditions caused liquid flow to occur, which caused the system to change in such a way that eventually ended the flow itself.

Notice the simplifications we made in modeling this system: the volume of the liquid in the pipe is considered negligible, the pipe is connected to the bottom of the containers, and the pipe is horizontal. Such assumptions are common in many engineering analyses, but of course some analyses do require more detailed models. Chapter 11 focuses on this issue.

8.4 Representing Gases

Gases provide an abundance of interesting examples of physical phenomena. They are often the bane of thermodynamics students because modeling them quantitatively is quite subtle. Unlike the simple idealizations of Newtonian motion, most calculations with real substances require extensive numerical tables. Nevertheless, the essence of the causal structure involving gases can be stated succinctly and, as chapter 19 outlines, provides a robust plank in the construction of expertise for thermodynamics. Here we lay out that causal structure, developing a simple causal model by analyzing the kinds of processes that affect gases. Then we ground that causal structure with sufficient model fragments so that we can make some everyday conclusions about them, as well as some more subtle ones.

The everyday properties of gases that we all know about are pressure, temperature, volume, and mass. Some gases, called ideal gases, can be described by a simple equation:

$$PV=nRT$$

That is, the product of pressure and volume equals the product of the number of molecules in it (n) times a constant (R, which is the ideal gas constant) and its temperature. Here we do not care about whether it is accurate quantitatively but what it says about the parameters that are relevant to each other. It tells us (as do the more complex methods of calculation that involve property tables) what set of parameters we must consider together. As with $F=MA$, our qualitative model must make explicit our causal intuitions involving these parameters. Moreover, if we want to model the intuition of a human expert, our qualitative model must be broadly consistent with the quantitative model. (We may not always want to do that: the misconception literature in education and learning sciences is rife with opportunities for modeling what learners construct in intermediate states of grappling with the world.) Thus, this equation, combined with observations from everyday experience, will lead to a causal model for gases.

We start by considering the processes that might be applied to a gas `G`, recapitulating the argument in chapter 6 but in more detail. Heat flow can certainly occur. That will directly influence heat, and if we assume that a gas is a thermal entity, we can assume the usual constraint on their relationship:

```
(qprop (Temperature G) (Heat G))
```

Suppose `G` is in some container. We can also in principle add more gas to that container via a flow, for instance. We can take the *n* in the equation to be the amount—that is, `(AmountOf S Gas C)`, where `S` is the substance of `G` and `C` is the (hypothetical) container it is in. S does not matter if we assume the same causal structure holds for all gases. The container `C` does matter: think about a pneumatic piston versus a balloon. In a rigid container, the volume is fixed. In that case, adding more gas will lead to the pressure increasing:

```
(qprop (Pressure G) (AmountOf S Gas C))
```

In a flexible container, the volume can be directly influenced by processes such as stretching. This will cause the pressure to decrease, so we have

```
(qprop- (Pressure G) (Volume G))
```

So far we have ignored the temperature. What happens if we heat a balloon? It expands. One way to capture this is that the pressure of `G` also changes as a consequence of its temperature:

```
(qprop (Pressure G) (Temperature G))
```

Have we covered everything? We can consider different combinations of processes and see if we get causal explanations that make sense for them. If we think about a mass of air rising in the atmosphere, it cools as it expands. The expansion itself is doing work against the atmosphere. Doing work is essentially a form of flow, increasing the internal energy of the thing that work is done against and decreasing the internal energy of the thing doing the work. Therefore, it makes sense for the air mass to cool as it expands because it is doing work against the rest of the atmosphere. Conversely, what happens when one fills up a bicycle tire with air? It gets hot. Here, too, a work flow is in effect, with the destination being the air in the tire that is being compressed. Because its internal energy rises, its temperature will rise. Because the amount of air is increasing, the pressure is increasing. Therefore, we can cover these two new phenomena (assuming a reasonable model of work flow) with our same causal model.

Figure 8.5 puts it all together, adding the conditions under which we should think about this (i.e., we are dealing with a contained gas) and the

```
(defModelFragment ContainedGasProperties
 :participants ((?g :type ContainedGas)
                (?sub :type Substance
                      :constraints (substanceOf ?g ?sub))
               (?can :type Container
                     :constraints (containerOf ?g ?can)))
 :conditions ((active ?g))
 :consequences ((≥ (Temperature ?g) (TBoil ?sub ?can))
                (qprop (Pressure ?g) (AmountOf ?sub Gas ?can))
                (qprop- (Pressure ?g) (Volume ?g))
                (qprop (Temperature ?g) (Pressure ?g)))
                (qprop (Pressure ?can)(Pressure ?g))
```

Figure 8.5

Describing contained gases.

appropriate constraint on its temperature. Gases exert pressure on their containers, of course. This is captured by the last indirect influence in the description.

Recall that gases are individuated in the same two ways as liquids, either as pieces of stuff or as contained stuffs. Here we will stick with a simple version of the contained stuff ontology. Thinking about gases means we have to think of where they will be. One canonical place is the atmosphere, which we can consider as the place taken up by the air surrounding our planet. (This example is a bit circular, because we are using the piece of stuff perspective to carve out this region of space, but it is a useful example of how reasoning often depends on multiple ontologies.) The atmosphere is often approximated as infinite,[3] which we could model via a variant of `ContainedGasProperties` that introduced the relevant quantities but did not connect them, so that in reasoning about everyday systems, we ignore the fact that cooking raises the Earth's temperature ever so slightly.

We also need to consider gases inside containers, such as balloons, tires, bottles, rooms, and pneumatic cylinders. The relationship `canContainSubstance` that we introduced earlier handles expressing the idea that a particular real-world entity can serve as a container for individuating gases. We also need the idea of a gas path, analogous to that of a liquid path, with a similar relationship for expressing connectivity between two things that can be viewed as containers. We assume the concepts `GasPath` and `gasConnectionBetween` accordingly.

Suppose there are multiple contained stuffs in a container. What sorts of interactions might we consider? One is the effect of their pressures on the pressure of the container. The indirect influence of the pressure of the gas on the pressure of the container captures this nicely. The compositionality of qualitative proportionalities provides a reasonable causal model if there is only a single gas, multiple gases, or liquids and gases in the container. More detailed models can be expressed by elaborating this idea (i.e., the concept of partial pressures, with more precise relationships). Another interaction comes from the fact that liquids are essentially incompressible. If we start filling one of our holding tanks with water and there is air in it, the rising level of the water will reduce the volume available for the air if the tank is sealed. To express this fact, we need to add this influence:

```
(qprop- (Volume Air-In-Tank) (Level Water-in-Tank))
```

when there are both water and air in a container. Given this extra law, we can already predict one possible consequence: in the two tanks example, equilibrium might be achieved before the levels are equal due to the additional pressure applied by the air in the destination tank. Similarly, the pressure in the source tank will drop both as a consequence of the change in the amount of liquid and due to the increased volume available to the air inside it. (This is one reason why there are vent valves in many real liquid piping systems.)

A third class of interaction that might be considered is thermal interactions between multiple substances in the same container. If we want to model this, all we need to do is add a model fragment that stipulates that, between any two contained stuffs in the same container, a heat path exists between them. Similarly, we will want to thermally connect stuffs inside a container to thermal entities that are in contact with that container (see figure 8.6).

8.5 Phase Changes

Now we are in a position to think about phase changes. These come in pairs, linked to transitions involving the boiling point and freezing point of the fluid. Because these points themselves depend on the pressure of the environment, recall that we have used the term `(TBoil` *<substance>* *<container>*) to denote the boiling point, so that the dependence on *<container>* can be expressed by laws that refer to *<container>*. We will do the same thing with `(TFreeze` *<substance><container>*).

Let us begin with boiling and condensation. There are a couple of ways to think about boiling. In everyday life, boiling occurs when we are heating water, so in the simplest way of thinking about it, boiling relies on a heat flow into the water to make it happen. (In reality, boiling can also occur by dropping the pressure so that the boiling point drops below the current temperature, but we will ignore that here.) When boiling occurs, it causes the amount of water to decrease and the amount of steam to increase. We

```
(defModelFragment WithinContainerHeatPath
 :subclass (HeatPath)
 :participants ((?s1 :type ContainedStuff)
                (?can :type Container
                      :constraints (containerOf ?s1 ?can))
                (?s2 :type ContainedStuff
                     :constraints (and (containerOf ?s2 ?can)
                                       (different ?s1 ?s2))))
 :conditions ((active ?s1)(active ?s2))
 :consequences ((heatConnection ?self ?s1 ?s2)
                (heatConnection ?self ?s2 ?s1)))

(defModelFragment ContainerToStuffHeatPath
 :subclass (HeatPath)
 :participants ((?stuff :type ContainedStuff)
                (?can :type Container
                      :constraints (containerOf ?stuff ?can))
                (?to :type ThermalObject
                     :constraints (thermalContact ?to ?can)))
 :conditions ((active ?stuff))
 :consequences ((heatConnection ?self ?to ?stuff)
                (heatConnection ?self ?stuff ?to)))
```

Figure 8.6
Introducing heat paths between different contained stuffs.

can model this via two direct influences of the boiling process. Figure 8.7 illustrates.

This version of boiling is grossly simplified: it does not capture, for example, that heat associated with the water is transferred as part of the boiling process. People can paper over this problem by simply stipulating constraints such that the appropriate outcomes occur (i.e., that the temperature of the water does not change during boiling and that the temperature of the steam in the container is the same as the water it was produced from). This could give many correct predictions about everyday behavior but fails as more complex situations are considered. Another route is to explicitly model latent heat transfer, so that the causal account of boiling is more complete. We will stick with the simple route here because this is a book about qualitative modeling, not engineering thermodynamics (but see Collins & Forbus, 1989). Boiling occurs faster when one turns up the heat on the stove, which we capture by the qualitative proportionality between the generation rate of boiling and the heat flow rate to the liquid.

Another way in which this model of boiling is simplified is that it requires an explicit heat flow into the liquid to drive the process. The actual conditions under which boiling occurs are more complex. For example, having

```
(defModelFragment Boiling
 :subclassOf (ContinuousProcess)
 :participants ((?liquid :type ContainedLiquid)
                (?sub :type Substance
                      :constraints (substanceOf ?liquid ?sub))
                (?can :type Container
                      :constraints (containerOf ?liquid ?can))
                (?heating :type HeatFlow
                          :constraints (destinationOf ?heating ?liquid)))
 :conditions ((active ?liquid)(active ?heating)
              (≥ (Temperature ?liquid) (TBoil ?sub ?can)))
 :consequences ((Quantity (GenerationRate ?self))
                (qprop (GenerationRate ?self) (HeatFlowRate ?heating))
                (I+ (AmountOf ?sub Gas ?can) (GenerationRate ?self))
                (I- (AmountOf ?sub Liquid ?can) (GenerationRate ?self))))

(defModelFragment Condensation
 :subclassOf (ContinuousProcess)
 :participants ((?gas :type ContainedGas)
                (?sub :type Substance
                      :constraints (substanceOf ?gas ?sub))
                (?can :type Container
                      :constraints (containerOf ?gas ?can))
                (?cooling :type HeatFlow
                          :constraints (sourceOf ?heating ?gas)))
 :conditions ((active ?gas)(active ?cooling)
              (≤ (Temperature ?gas) (TBoil ?sub ?can)))
 :consequences ((Quantity (CondensationRate ?self))
                (qprop (CondensationRate ?self) (HeatFlowRate ?cooling))
                (I- (AmountOf ?sub Gas ?can) (CondensationRate ?self))
                (I+ (AmountOf ?sub Liquid ?can) (CondensationRate ?self))))
```

Figure 8.7
A simple model of boiling and condensation.

nucleation sites for bubbles inside the liquid to form around helps facilitate boiling. (This is why small inert rocks, called boiling chips, are often introduced into the water when being boiled in a chemistry experiment.) Liquid in a very smooth container, undisturbed, can become superheated, eventually leading to explosive boiling. (This can happen when boiling water in microwave ovens, for example.) But these are complexities beyond the simple, everyday model of boiling that most of us have.

One way that condensation can occur is essentially the inverse of boiling: by coming in contact with something cooler, the steam loses heat and changes back into its liquid phase. That is captured by the condensation process description of figure 8.7. Notice that, for compositionality, this description does not try to incorporate where the heat is going—all it needs to know is that heat is being drawn from the contained gas.

A gentler form of liquid/gas phase change is evaporation and condensation. Evaporation happens at the liquid-gas interface. It can occur at any temperature, unlike boiling, but it tends to happen very slowly. In professional models of fluids, both evaporation and boiling are considered forms of *vaporization* but involving different mechanisms, and hence they have identities as distinct processes. (One can build model fragments that capture this hierarchical relationship, but that is an additional depth of modeling we will skip here.) The complementary process is called condensation, which can be a bit confusing because the same term is used for the mechanism described above but also for the return of water vapor molecules from the air back into the liquid, which does not require an external heat source. (Evaporation occurs faster when a fluid is being heated, but that is because the rate of evaporation depends on the temperature of the fluid.) Figure 8.8 provides a simple model for this pair of processes.

Notice that, in keeping with our intuitions, there are no external heat flows required to drive either process. (Hence, we call this case of condensation `CondensationUnaided` to distinguish it from the version of figure 8.7.) The direct influences in these processes represent a common

```
(forAll ?s (implies (ContainedStuff ?s)
                    (Quantity (InterfaceSurfaceArea ?s))))

(defModelFragment Evaporation
 :subclassOf (ContinuousProcess)
 :participants ((?liquid :type ContainedLiquid)
                (?sub :type Substance
                      :constraints (substanceOf ?liquid ?sub))
                (?can :type Container
                      :constraints (containerOf ?liquid ?can)))
 :conditions ((active ?liquid))
 :consequences ((Quantity (EvaporationRate ?self))
                (qprop (EvaporationRate ?self) (Temperature ?liquid))
                (qprop (EvaporationRate ?self)
                       (InferfaceSurfaceArea ?liquid))
                (I+ (AmountOf ?sub Gas ?can) (EvaporationRate ?self))
                (I- (AmountOf ?sub Liquid ?can) (EvaporationRate ?self))))

(defModelFragment CondensationUnaided
 :subclassOf (ContinuousProcess)
 :participants ((?gas :type ContainedGas)
                (?sub :type Substance
                      :constraints (substanceOf ?gas ?sub))
                (?can :type Container
                      :constraints (containerOf ?gas ?can)))
 :conditions ((active ?gas))
 :consequences ((Quantity (CondensationRate ?self))
                (qprop- (CondensationRate ?self) (Temperature ?gas))
                (qprop (CondensationRate ?self)
                       (InferfaceSurfaceArea ?gas))
                (I- (AmountOf ?sub Gas ?can) (GenerationRate ?self))
                (I+ (AmountOf ?sub Liquid ?can) (GenerationRate ?self))))
```

Figure 8.8
Evaporation and condensation.

pattern, a *mass transfer*. When people learn about conservation of matter, they sometimes realize that their models have been incomplete, requiring additional elaboration to make influences consistent with this new higher-level physical principle. Both of these processes have rates governed by temperature but also by the surface area of the liquid-gas interface. Here we have simply stated the existence of this quantity for contained stuffs, without specifying any particular causal model of it. Such incremental elaborations occur frequently during learning by reading or by conversation, and capturing such intermediate states of knowledge is why compositionality in representations is so important.

Two aspects of our everyday models of evaporation and condensation are missing from the formalization in figure 8.8. The first is that evaporation happens very slowly compared to boiling and liquid flow. This can easily be expressed via axioms stating ordinal relationships between rates for instances of these types of processes or, better still, using one of the order-of-magnitude formalisms described in chapter 6. The other aspect is that evaporation provides cooling, which is why we sweat to shed excess heat. To capture this aspect of evaporation and condensation, we will use separate processes to describe their thermal effects. That way, we can start by thinking about the volumetric effects and only add the additional complexity of thermal effects when the thinking we are doing requires it.

Figure 8.9 provides one model for the thermal effects of evaporation and condensation. Arguably, many people's intuitive models of evaporation contain only the `I-` influence on liquid, without being concerned about where the heat is going, and they may not think of the thermal consequences of condensation at all. It is easy to see what these models might look like by simply ablating most of the contents of figure 8.9.

Finally, to finish our simple domain theory of phase transitions, we create simple models of freezing and melting (figure 8.10), which provide transitions between liquids and solids. (Because most people do not think much about deposition and sublimation, which are phase changes directly between gases and solids, we will not include them in this domain theory.) In terms of their direct influences and conditions, they are analogous to boiling and condensation. Although there often is a driving heat flow, it is captured indirectly by the dependence of their rates on the temperature difference between the freezing point and the actual temperature. This captures the idea that, however an ice cube is warmed up, it will melt.

Now that we have a domain theory for phase changes, let us put it to work describing some everyday reasoning, including one of the motivating examples of chapter 1.

```
(defModelFragment EvaporationThermalEffects
 :subclassOf (ContinuousProcess)
 :participants ((?liquid :type ContainedLiquid)
                (?sub :type Substance
                      :constraints (substanceOf ?liquid ?sub))
                (?can :type Container
                      :constraints (containerOf ?liquid ?can))
                (?gas :type ContainedGas
                      :constraints ((substanceOf ?gas ?sub)
                                    (containerOf ?liquid ?can)))
 :conditions ((active ?liquid)(active ?gas))
 :consequences ((Quantity (EvaporationRate ?self))
                (qprop (EvaporationThermalTransferRate ?self)
                       (Temperature ?liquid))
                (qprop (EvaporationThermalTransferRate ?self)
                       (InferfaceSurfaceArea ?liquid))
                (I+ (Heat ?gas) (EvaporationThermalTransferRate ?self))
                (I- (Heat ?liquid) (EvaporationThermalTransferRate ?self))))

(defModelFragment CondensationThermalEffects
 :subclassOf (ContinuousProcess)
 :participants ((?gas :type ContainedGas)
                (?sub :type Substance
                      :constraints (substanceOf ?gas ?sub))
                (?can :type Container
                      :constraints (containerOf ?gas ?can))
                (?liquid :type ContainedLiquid
                         :constraints ((substanceOf ?liquid ?sub)
                                       (containerOf ?liquid ?can))))
 :conditions ((active ?gas)(active ?liquid))
 :consequences ((Quantity (CondensationThermalTransferRate ?self))
                (qprop- (CondensationThermalTransferRate ?self)
                        (Temperature ?gas))
                (qprop (CondensationThermalTransferRate ?self)
                       (InferfaceSurfaceArea ?gas))
                (I- (Heat ?gas) (CondensationThermalTransferRate ?self))
                (I+ (Heat ?liquid) (CondensationThermalTransferRate ?self))))
```

Figure 8.9
Thermal effects of evaporation and condensation.

8.6 Boiling Water and Its Consequences

Returning to the example of water heating on the stove, we can now generate a concise explanation of what happens and why. Figure 8.11 shows the results of limit analysis repeatedly applied to our initial situation, after suppressing evaporation temporarily from our process vocabulary.

What happens if we boil water in a sealed container instead of an open kettle? As boiling begins, the pressure and temperature in the container begin to rise. Why? Because boiling is the only influence on the amounts of water and steam, we can deduce that the amount of water is decreasing and the amount of steam is increasing. What consequences does this have? Because liquid water is, to the first order, incompressible, the volume

```
(defModelFragment Melting
 :subclassOf (ContinuousProcess)
 :participants ((?solid :type ContainedSolid)
                (?sub :type Substance
                      :constraints (substanceOf ?solid ?sub))
                (?can :type Container
                      :constraints (containerOf ?solid ?can)))
 :conditions ((active ?solid)
              (≥ (Temperature ?solid) (TFreeze ?sub ?can)))
 :consequences ((Quantity (MeltingRate ?self))
                (qprop (MeltingRate ?self)
                       (- (Temperature ?solid) (TFreeze ?sub ?can)))
                (I+ (AmountOf ?sub Liquid ?can) (MeltingRate ?self))
                (I- (AmountOf ?sub Solid ?can) (MeltingRate ?self))))

(defModelFragment Freezing
 :subclassOf (ContinuousProcess)
 :participants ((?liquid :type ContainedLiquid)
                (?sub :type Substance
                      :constraints (substanceOf ?liquid ?sub))
                (?can :type Container
                      :constraints (containerOf ?liquid ?can)))
 :conditions ((active ?liquid)
              (≤ (Temperature ?gas) (TFreeze ?sub ?can)))
 :consequences ((Quantity (FreezingRate ?self))
                (qprop (FreezingRate ?self)
                       (- (TFreeze ?sub ?can) (Temperature ?liquid)))
                (I- (AmountOf ?sub Liquid ?can) (FreezingRate ?self))
                (I+ (AmountOf ?sub Solid ?can) (FreezingRate ?self))))
```

Figure 8.10
A simple model of melting and freezing.

available for steam is the difference between the volume of the container and the volume of the liquid:

```
(qprop (Volume Steam) (Volume Boiler))
(qprop– (Volume Steam) (Volume Water))
```

The decrease in amount of water will decrease its volume, which in turn increases the volume available for the steam. If we look at the indirect influences on the properties of a contained gas (figure 8.5), we have two influences on pressure:

```
(qprop (Pressure Steam) (Mass Water))
(qprop– (Pressure Steam) (Volume Steam))
```

Because both causal antecedents are increasing, we have an ambiguity. How might we resolve it? Here we can use a fact about steam: at any particular temperature and pressure, the volume of steam is very much greater than the volume of water it was produced from (at standard temperature and pressure,

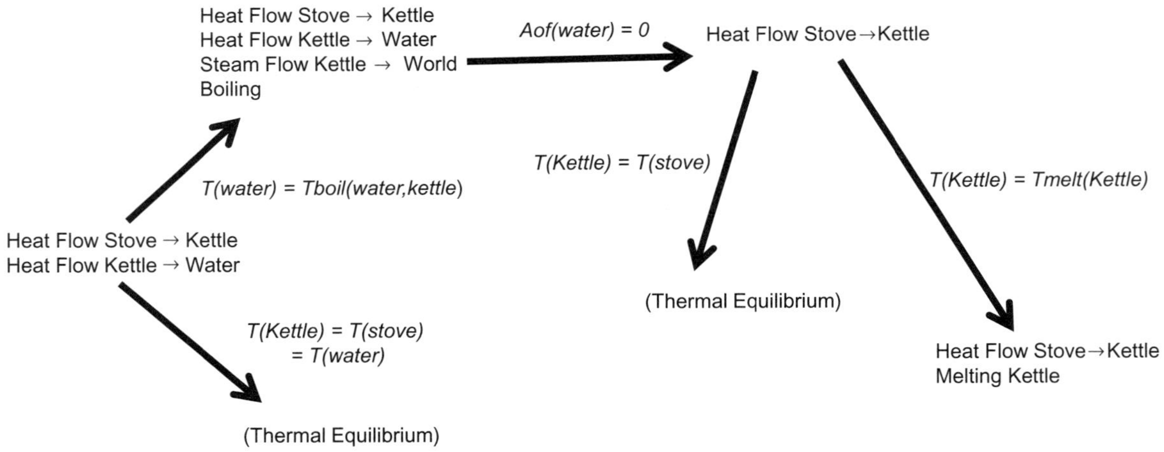

Figure 8.11
What can happen when boiling water on a stove.

about 220 times greater, for example). This means that the influence of the amount should dominate over the effect of volume, leading to an increase in pressure. This, in turn, leads to an increase in temperature (via another qualitative proportionality in `ContainedGasProperties`). To model the effect of an explosion, we add a bursting pressure to the quantity space for pressure of the container and thereby get as a possible outcome of boiling in a sealed container that the container could explode. Figure 8.12 illustrates.

As usual, we could go into much more detail in exploring this phenomenon, even qualitatively (e.g., Forbus, 1984). For example, we have ignored here the fact that the boiling temperature of a substance actually depends on its pressure, so that the boiling temperature will also be continuously rising. We have not considered exactly how heat is transferred across phase changes. But that is part of the point. We can get a surprising number of intuitive conclusions from qualitative models that, compared to traditional mathematical models, are far less detailed. These models serve an important, complementary purpose: they help us identify broad categories of possible behaviors, rather than trying to predict exactly what will happen in a situation.

8.7 Ice Cubes in Freezers, Revisited

We can use this phase change domain theory and QP theory to express the reasoning involved in arguing why ice cubes made with warm water might freeze faster than ice cubes made with cold water. The causal structure for this situation, derived from the domain theory, is shown in figure 8.13. What processes are at work in the freezer? There are three:

1. Heat flows from the water in the ice cube tray to its surroundings.
2. The water in the ice cube tray is freezing.
3. The water in the ice cube tray is evaporating.

The third process is the factor that most people considering this problem overlook. How long will it take to freeze the water? That depends on the difference in temperatures and how much water there is. Less water means it will take less time to freeze. Also, evaporation is taking away heat from the water, which also helps reduce the temperature difference. This provides a reasonable and satisfying explanation for why this occurs.

Prediction and explanation both depend on the modeling assumptions we make in looking at the world. If we fail to consider a relevant process, here evaporation, we will make mistakes. But at least our framework for modeling provides a constrained set of places to look in order to diagnose the reasons for those mistakes.

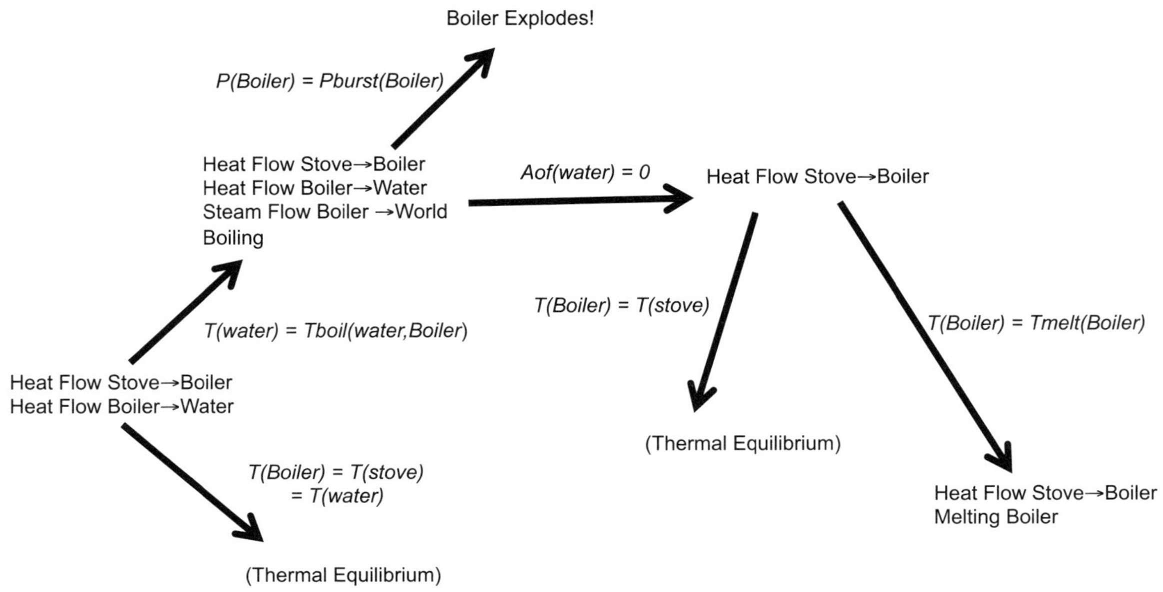

Figure 8.12
Explosion as a possible consequence of boiling water in a sealed container.

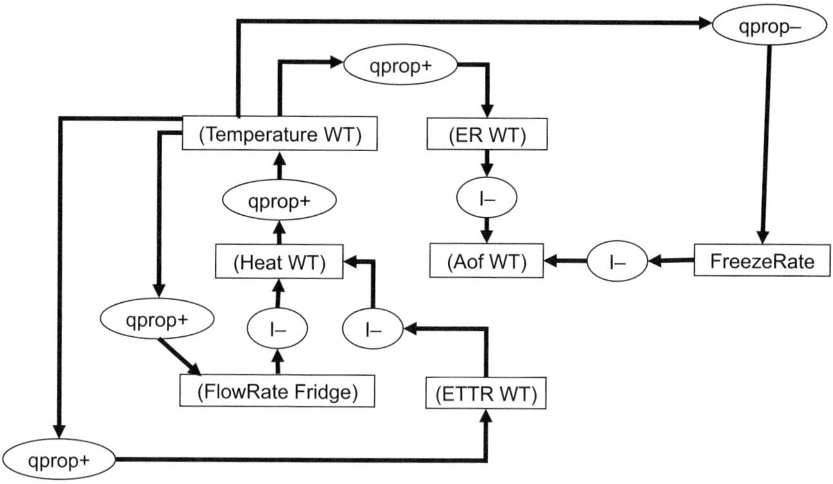

Figure 8.13
Causal relationships for freezing water in a refrigerator. Influences where one of the parameters is constant are not shown.

8.8 Modeling Motion

Motion is one of the first processes we experience and is something we reason about every day. Motion also is intimately intertwined with our concepts of shape and space. We only consider dynamical aspects of motion here, returning to it more fully in part III. Prior research (Clement, 1983; McCloskey, 1983) has shown that students often have pre-Newtonian models of motion. Sadly, these models often persist, even after successfully taking a physics course (Hestenes, Wells, & Swackhamer, 1992). To illustrate the expressive power of QP theory, this section describes how Aristotelian, Impetus, and Newtonian dynamics for motion can be represented.

Because we are factoring out space for now, we limit consideration to a single object moving in one dimension. Moreover, because we are not reasoning about shape or surface contact—again, those require the spatial representations discussed in part III—we ignore for now the differences between flying, sliding, swinging, and rolling. Instead, we develop a very abstract vocabulary for describing motion. Even this abstract vocabulary has its uses: we can deduce that if we kick something, it can move, and if we rule out the most abstract version of motion, we have ruled out all of the more specific kinds.

```
(defProcess Motion
 :participants ((?obj :type Object
                      :constraint (Mobile ?obj))
                (?a :type Axis
                    :constraint (onAxis ?obj ?a))
                (?dir :type Direction))
 :conditions ((freeDirection ?obj ?a ?dir)
              (directionOf (velocity ?obj) ?a ?dir)
              (> (magnitude (velocity ?obj))0))
 :consequences ((I+ (position ?obj) (velocity ?obj))))
```

Figure 8.14
Newtonian motion.

We assume a concept of *axis*, a one-dimensional line, upon which motion is occurring. Directions along an axis will be described by signs (i.e., –1 and 1). We need a few predicates to describe spatial relationships in this abstract one-dimensional (1D) world:

- `(directionOf` *<Q><a><dir>*) is true exactly when the sign of *<Q>* along axis *<a>* equals *<dir>*.
- `(onAxis` *<Obj><a>*) is true exactly when object *<Obj>*'s position is always somewhere on axis *<a>*.
- `(freeDirection` *<Obj><a><dir>*) is true exactly when the position of object *<Obj>* on axis *<a>* is free to move in direction *<dir>*. That is, there is no immobile object directly in contact with *<Obj>* in that direction.
- `(directionToward` *<Obj1><Obj2><a><dir>*) is true exactly when object *<Obj2>* lies along direction *<dir>* on axis *<a>* from *<Obj1>*.
- `(contact` *<Obj1><Obj2><a><dir>*) is true exactly when object *<Obj2>* lies along direction *<dir>* on axis *<a>* from *<Obj1>* and is in contact with *<Obj1>*.
- `(mobile` *<Obj>*) is true exactly when object *<Obj>* is free to move.

With these conventions in hand, we can define a 1D process of motion, as shown in figure 8.14.

This description of motion expresses some basic ideas about it: that it occurs when a mobile object has a nonzero velocity in a direction that it is free to move in. Given the two directions we have introduced, there will be two instances of motion for any mobile object on an axis. To reason about whether an object is free in a direction, we must use a rule like

```
(<== (freeDirection ?obj ?a ?dir)
     (uninferredSentence
```

```
        (and (onAxis ?blocker ?a)
             (not (Mobile ?blocker))
             (contact ?obj ?blocker ?dir))))
```

that is, there is no immobile object on the axis that is positioned to be in contact with the object in question in the direction of motion. (As introduced in chapter 3, the predicate `uninferredSentence` conveys a form of nonmonotonic inference: if we fail to prove its argument, it is taken as being true.)

Notice that this model does not quite capture the fact that velocity is defined to be the derivative of position. What it says is that velocity is a direct influence on position. The compositionality of influences means that, potentially, some other influence might be added as well. We could add another statement that precludes such influences to rule out any consideration of other potential processes having an influence on position.

To complete our model of Newtonian motion, we must also model the process of acceleration. Figure 8.15 describes such a process. This model of acceleration incorporates a version of Newton's second law, $F=MA$, but written in a causal form: acceleration depends causally upon net force and mass. If the net force is zero, acceleration will no longer occur, but if the velocity is nonzero, an object will continue in motion forever. This captures Newton's first law.

Newton's first law is counterintuitive, because everyday experience tells us that moving things stop unless we keep pushing on them. This is the heart of Aristotle's dynamics. Figure 8.16 shows a process description of

```
(defProcess Acceleration
 :participants ((?obj :type Object
                      :constraint (Mobile ?obj))
                (?a :type Axis
                    :constraint (onAxis ?obj ?a))
                (?dir :type Direction))
 :conditions ((freeDirection ?obj ?a ?dir)
              (directionOf (netForce ?obj) ?a ?dir)
              (> (magnitude (netForce ?obj)) 0))
 :consequences ((Quantity (acceleration ?obj))
                (qprop (acceleration ?obj) (netForce ?obj))
                (qprop- (acceleration ?obj) (mass ?obj))
                (correspondence ((acceleration ?obj) 0)
                                ((netForce ?obj) 0))
                (I+ (velocity ?obj) (acceleration ?obj))))
```

Figure 8.15
Newtonian acceleration.

```
(defProcess Motion
 :participants ((?obj :type Object
                      :constraint (Mobile ?obj))
                (?a :type Axis
                    :constraint (onAxis ?obj ?a))
                (?dir :type Direction))
 :conditions ((freeDirection ?obj ?a ?dir)
              (directionOf (netForce ?obj) ?a ?dir)
              (> (magnitude (netForce ?obj)) 0))
 :consequences ((Quantity (velocity ?obj))
                (qprop (velocity ?obj) (netForce ?obj))
                (qprop- (velocity ?obj) (mass ?obj))
                (correspondence ((velocity ?obj) 0)
                                ((netForce ?obj) 0))
                (I+ (position ?obj) (velocity ?obj))))
```

Figure 8.16
An Aristotelian model of motion: objects stay in motion only when pushed.

this idea. Note that velocity, because it is only defined within motion, has no independent existence from it. Thus, if force stops being applied to an object, its position is no longer influenced, and its motion stops.

Aristotle's model of motion has the problem of explaining what keeps an object moving once it doesn't touch anything. Medieval scholars attempted to explain this by postulating that pushing something gives the object what we might think of as a kind of internal force, called *impetus*. Impetus theory is very attractive, intuitively: there is evidence that students today have misconceptions that are very much like it (McCloskey, 1983). Although superficially like momentum, impetus is different because it spontaneously dissipates. Figure 8.17 describes a QP model of impetus theory. Pushing an object imparts impetus to it.[4] Motion occurs when the impetus is nonzero. But having a nonzero impetus also causes dissipation, so that over time, the impetus will become zero. Note that the decaying impetus will lead to the object slowing down, which fits with our observations of everyday motion.

8.9 Modeling Materials

Think about what happens when we pull on something. If it doesn't move, then its internal structure is taking up the force somehow. Depending on what it is made of, three things can happen. First, it might do nothing—this is what we expect from rigid objects that are fixed in place. Second, it might stretch if it is elastic. Third, it might break. Pushing can be thought of in similar terms, with *compress* and *crushed* substituted for *stretch* and *break*. These everyday intuitions about materials can be expressed using

```
(defProcess Impart
 :participants ((?obj :type Object
                      :constraint (Mobile ?obj))
                (?a :type Axis
                    :constraint (onAxis ?obj ?a))
                (?dir :type Direction))
 :conditions ((freeDirection ?obj ?a ?dir)
              (directionOf (netForce ?obj) ?a ?dir)
              (> (magnitude (netForce ?obj)) 0))
 :consequences ((Quantity (imp ?self))
                (qprop (imp ?self) (netForce ?obj))
                (qprop- (imp ?self) (mass ?obj))
                (correspondence ((imp ?self) 0) ((netForce ?obj) 0))
                (I+ (position ?obj) (impetus ?obj))))

(defProcess Motion
 :participants ((?obj :type Object
                      :constraint (Mobile ?obj))
                (?a :type Axis
                    :constraint (onAxis ?obj ?a))
                (?dir :type Direction))
 :conditions ((freeDirection ?obj ?a ?dir)
              (directionOf (impetus ?obj) ?a ?dir)
              (> (magnitude (impetus ?obj)) 0))
 :consequences ((I+ (position ?obj) (impetus ?obj))))

(defProcess Dissipate
 :participants ((?obj :type Object
                      :constraint (Mobile ?obj))
                (?a :type Axis
                    :constraint (onAxis ?obj ?a))
                (?dir :type Direction))
 :conditions ((directionOf (impetus ?obj) ?a ?dir)
              (> (magnitude (impetus ?obj)) 0))
 :consequences ((Quantity (diss ?self))
                (qprop (diss ?self) (impetus ?obj))
                (correspondence ((diss ?self) 0) ((impetus ?obj) 0))
                (< (magnitude (diss ?self)) (magnitude (impetus ?obj)))
                (I- (impetus ?obj) (diss ?self))
```

Figure 8.17
An impetus model of motion: impetus is imparted to a body but spontaneously dissipates.

QP theory by defining model fragments that introduce these concepts and what they are conditioned on. Readers steeped in material science will recognize that this analysis leaves out a number of interesting cases, such as sheer, but our goal here is not a comprehensive qualitative theory of materials. As with the models of motion in the previous section, we focus here on 1D objects and furthermore constrain them to be fixed at one end. By convention, forces into an object (pushes) will be negative and applied forces directed outward (pulls) will be positive.

```
(defModelFragment ElasticObjectProperties
 :participants ((?o :type 1D-Object)
                (?sub :type Substance
                      :constraints ((madeOf ?o ?sub)
                                    (hasElasticRange ?sub)))
 :conditions ((elasticRange ?o))
 :consequences ((Biconditional (ElasticObject ?o))
                (Quantity (InternalForce ?o))
                (Quantity (RestLength ?o))
                (qprop (InternalForce ?o) (Length ?o))
                (correspondence ((InternalForce ?o) 0)
                                ((Length ?o) (RestLength ?o))))

(defModelFragment RelaxedDefinition
 :participants ((?o :type ElasticObject))
 :conditions ((= (Length ?o) (RestLength ?o))
 :consequences ((Biconditional (Relaxed ?o))))

(defModelFragment StretchedDefinition
 :participants ((?o :type ElasticObject))
 :conditions ((> (Length ?o) (RestLength ?o))
 :consequences ((Biconditional (Stretched ?o))))

(defModelFragment CompressedDefinition
 :participants ((?o :type ElasticObject))
 :conditions ((< (Length ?o) (RestLength ?o))
 :consequences ((Biconditional (Compressed ?o))))
```

Figure 8.18
Some states of an elastic object.

Let us start with elastic objects. Elasticity can be thought of as a relationship between applied force and internal force. If the magnitude of the applied force is greater than that of the internal force, the length of the object will change. That change in length causes, in turn, a change in the internal force that counterbalances the applied force. We can describe these intuitions about elastic objects via four model fragments (figure 8.18): one provides the basic properties of elastic objects, and the other three describe the possible states of an elastic object, which is relaxed, stretched, or compressed.

The `ElasticObjectProperties` model fragment applies to 1D objects made out of a substance that could be elastic (i.e., the relation `hasElasticRange`). Thus, we wouldn't think about elastic behavior of chairs, eggs, or rocks in everyday life, for example. If we didn't know that elasticity could change with environmental conditions, then the conditions might be empty (i.e., instances of it would always be true). If we know that elasticity can vary with the environment—frozen rubber bands don't stretch well, for example, and to a geoscientist, rocks can be plastic—then the condition `(elasticRange ?o)` expresses our state of knowledge that there's some dependence there. People with a sophisticated model of materials would have an elaborated theory of when this condition holds, defined via other model fragments, whereas those who know less might just leave it as a placeholder, an acknowledgment that there could be some reason why the conclusions drawn here could be wrong, even if they don't know the details at their current state of knowledge. (This ability to state a dependence without having a fully articulated model behind it is perhaps one of the causes of Keil's *illusion of explanatory depth* effect [Rozenblit & Keil, 2002; Weisberg, Keil, Goodstein, Rawson, & Gray, 2008].) In a learner, such placeholders are scaffolds, promissory notes to be filled in as they learn more.

The statement `Biconditional`, which appears in the consequences field, indicates that the statement that is its argument is true if and only if that model fragment instance is active. The usual semantics of consequences is conditional, which means there could be more than one model fragment that would support a consequence. `Biconditional` indicates that a particular model fragment is a necessary, not just sufficient, condition for that statement to hold.

The qualitative proportionality and correspondence in `ElasticObjectProperties` express the nature of the restoring force within an elastic object. If the object is relaxed (i.e., its length is the same as its rest length), then the internal force is zero. If the object has been pushed, it will be compressed, and the internal force will be positive—in other words, the object starts pushing back. Its response to a pull will be similar, in that the internal force will act in the opposite direction.

In terms of predicting the dynamics of the system, the three model fragments describing state are not strictly necessary. Being compared in a correspondence statement places a value into a parameter's quantity space. Hence, the correspondence in `ElasticObjectProperties` is sufficient to infer that the `Length` and `RestLength` of an elastic object are in each other's quantity spaces (and similarly, for `0` to be in the quantity space for `InternalForce`). Therefore, in terms of causing a difference in qualitative state when we would intuitively think there was one, the correspondence is sufficient. The intent of the additional model fragments describing the state of the elastic object is to model the grounding of language about the system in physical terms: when we talk about an elastic object being stretched, the model we construct of that situation will include an instance of the `StretchedDefinition` model fragment, whose condition or negation will be believed depending on whether the statement was positive or negative (e.g., "the bungee cord was stretched" versus "the bungee cord was not stretched").

Pushing an elastic object can cause it to compress, and pulling it can cause it to stretch. Moreover, if we stop pushing or pulling, the internal force of a relaxing object will cause its length to change. We can represent this via processes for stretching, compressing, relaxing, and decompressing (figure 8.19).

There are several things to notice about these process representations. First, they use predicates describing state (i.e., compressed, stretched) defined by other model fragments as part of their condition. Considering collections of object states and processes on them together creates a higher-level perspective on interrelated phenomena, analogous to how the Periodic Table organizes chemistry. Object states and processes are intertwined in a robust domain theory. Higher-level knowledge, like conservation laws, provides guidance for building one's knowledge of the world: if there is a change in one direction and it is reversible, then there must be some process that carries it out in the other direction (it can be the same type of process, as in flows, or different, as with this elasticity model). Second, notice that this model predicts that, for an object at rest, if the applied force and the internal force are perfectly balanced, nothing happens. Here is why: suppose these forces are perfectly balanced. Then no instance of the types of processes in figure 8.19 can be active. If these are the only types of processes that can influence length, then by the sole mechanism assumption, the length cannot change. We did not have to explicitly "wire in" this prediction into the model; it comes directly from the domain theory and the laws of QP theory. This kind of generativity is an important phenomenon of human reasoning about continuous systems that QP theory captures well.

So far, we have focused on perfectly elastic substances. Of course, real materials have their limits. (Group bungee jumps do not involve just one cord in simultaneous use by multiple people.) If too much force is applied, an object can break or crush. If the applied force is very small, objects can behave rigidly. Thus, these other kinds of behavior can be conditioned on new limit points on applied force. However, crushing and breaking involve irreversible changes. Thus, they are not well represented by model fragments and are better described in terms of encapsulated histories. The differences in quantity spaces for materials and what processes they undergo give rise to a taxonomy for classifying them, as figure 8.20 illustrates.

A classic artificial intelligence conundrum is to be able to express the fact that one can pull with a string but not push with it (Minsky, 1974). Reasoning about the interactions of shape, forces, and motion requires rich spatial

```
(defProcess Stretching
 :participants ((?o :type ElasticObject))
 :conditions ((> (AppliedForce ?o) 0)
              (> (magnitude (AppliedForce ?o))
                 (magnitude (InternalForce ?o)))
              (not (Compressed ?o)))
 :consequences ((Quantity (StretchRate ?self))
                (qprop (StretchRate ?self) (AppliedForce ?o))
                (qprop- (StretchRate ?self) (InternalForce ?o))
                (correspondence ((StretchRate ?self) 0)
                                ((AppliedForce ?self)
                                 (InternalForce ?self)))
                (I+ (Length ?o) (StretchRate ?self)))

(defProcess Compressing
 :participants ((?o :type ElasticObject))
 :conditions ((< (AppliedForce 0)
              (> (magnitude (AppliedForce ?o))
                 (magnitude (InternalForce ?o)))
              (not (Stretched ?o))))

(defProcess Relaxing
 :participants ((?o :type ElasticObject))
 :conditions ((Stretched ?o)
              (< (magnitude (AppliedForce ?o))
                 (magnitude (InternalForce ?o))))
 :consequences ((Quantity (RelaxRate ?self))
                (qprop (RelaxRate ?self) (InternalForce ?o))
                (qprop- (RelaxRate ?self) (AppliedForce ?o))
                (I- (Length ?o)(RelaxRate ?self))))

(defProcess Decompressing
 :participants ((?o :type ElasticObject))
 :conditions ((Compressed ?o)
              (< (magnitude (AppliedForce ?o))
                (magnitude (InternalForce ?o))))
 :consequences ((Quantity (DecompRate ?self))
                (qprop (DecompRate ?self) (InternalForce ?o))
                (qprop- (DecompRate ?self) (AppliedForce ?o))
                (I+ (Length ?o) (DecompRate ?self))))
```

Figure 8.19

These processes describe intuitive notions of stretching, compressing, expanding, and decompressing for elastic materials.

Rigid: No processes affect length
Elastic: Stretching and compressing can occur
Breakable: Limit points: `0, BreakingForce;`
`(> BreakingForce 0)`
Crushable: Limit points: `0, CrushingForce;`
`(< CrushingForce 0)`
Partially stretchable: Limit points: `0, StretchThreshold;`
`(> StretchThreshold 0)`
Partially compressible: Limit points: `0, CompressThreshold;`
`(< CompressThreshold 0)`
Brittle: Limit points: `0, CrushingForce, BreakingForce;`
`(< CrushingForce 0) (> BreakingForce 0)`
Partially Elastic: Limit points: `0, CompressThreshold, StretchThreshold;`
`(< CompressThreshold 0), (> StretchThreshold 0)`
Normal: Limit points: `0, CrushingForce, CompressThreshold, StretchThreshold, BreakingForce;`
`(< CrushingForce CompressThreshold) (< CompressThreshold 0)`
`(> StretchThreshold 0) (> BreakingForce StretchThreshold)`

Figure 8.20
Different types of materials give rise to different quantity spaces because they can participate in different collections of processes. This taxonomy should allow a material to be classified by applying forces and observing what kinds of things actually occur.

representations (see part III), but the dynamical aspects of this fact can be expressed in QP theory. Think of pushes and pulls in terms of transmitting forces via one object to another. We can define conditions under which particular kinds of objects can be transmitters of pushes and pulls via model fragments. Using the conventions above, if the force is negative, its direction is toward an object (push), and if positive, its direction is away from an object (pull). Suppose we define a string as a form of nonrigid object with two parameters:

1. `EndsLength` denotes the current distance between the ends of the string in the physical configuration it is in.
2. `Length` denotes the length of the string.

Note that, unless we allow elastic strings, EndsLength is never greater than `Length` for a string. Figure 8.21 shows two model fragments that suffice to convey the desired intuition. The first, `RigidForceTransmission`, indicates that a rigid object connected to a 1D object can transmit pushes. Because strings are, by assumption, not rigid, this definition does not apply to them. The second, `StringPullTransmitter`, indicates that a string can be a pull transmitter for an object when it is at its full length and not otherwise. This captures the essence of the intuition.

```
(defModelFragment RigidForceTransmission
 :participants ((?s :type RigidObject)
                (?o :type 1D-Object
                    :constraints (connectedTo ?s ?o)))
 :consequences ((PushTransmitter ?s ?o)
                (PullTransmitter ?s ?o)))

 (defModelFragment StringPullTransmitter
 :participants ((?s :type String)
                (?o :type 1D-Object
                    :constraints (connectedTo ?s ?o)))
 :conditions ((= (EndsLength ?s)(Length ?s))
 :consequences ((PullTransmitter ?s ?o)))
```

Strong constraint:

```
(forAll (?s ?o) (implies (PushTransmitter ?s ?o) (RigidObject ?s)))
```

Weak constraint:

```
(forAll (?s ?o) (implies (PushTransmitter ?s ?o) (not (String ?s))))
```

Figure 8.21
A way of saying that one can pull with a string but not push with it.

The alert reader might observe that this does not, by itself, rule out the existence of some other model fragment that allows strings to transmit pulls. However, making a closed-world assumption over the domain theory does rule that out. Suppose, however, we want to model the knowledge of someone with lots of experience in the world who has more direct knowledge about the impossibility of strings transmitting pushes. The weak and strong constraints also shown in figure 8.21, combined with the model fragments, provide a way to express this. The strong version is too strong if the learner's domain theory includes partially elastic materials, for example. The weak version provides a better model of what one might take away from the frustrating experience of string pushing. Both versions only make sense in the context provided by the model fragment definitions of positive cases for these concepts.

This example, like the others involving motion, elides many complex representational issues concerning shape and space. For example, evaluating the validity of an ordinal relationship involving `EndsLength` and `Length` of a string can be quite complex when looking, for example, at a clockwork mechanism. Nevertheless, the dynamical implications of such visual processing are nicely represented by these model fragments. They can also be used for explanation: if something is moving, one way that can happen is that it is being pushed or pulled. And if it is being pushed or pulled, one can look for objects capable of transmitting those pushes or pulls. These model fragments suggest that, in such cases, looking for rigid objects and strings to explain what is happening.

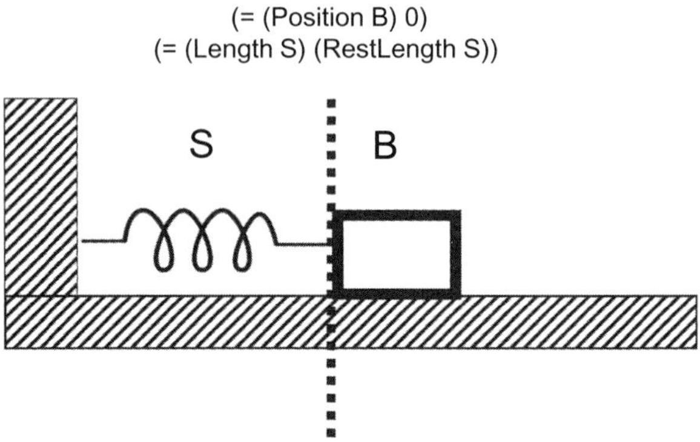

Figure 8.22
A spring-block oscillator.

8.10 Modeling an Oscillator

Let us combine the models of motion and materials described so far to analyze a spring-block oscillator, to show how qualitative representations can capture harmonic motion. Figure 8.22 shows how the geometry of the situation can be translated into quantities and the abstract entities introduced so far. We can treat the spring `S` as a kind of 1D object, made of some elastic material `M`, such that `(ElasticMode ?o)` holds in this situation. The block `B` will also be viewed as a kind of 1D object, albeit a rigid one that is free to move, that is, `(Mobile B)` holds throughout our analysis.

To model the connection between the spring and the block, we must do two things. First, we need to represent the connection between the length of the spring and the position of the block. Which should we view as causing changes in the other? We can move the block, which causes the length of the spring to change, which suggests

```
(qprop (Length S) (Position B))
```

Note that this qualitative proportionality means that we cannot use the processes of figure 8.19 because then `(Length S)` would be both directly and indirectly influenced, which is not allowed. How to model the process of reasoning about what subsets of a domain theory should be considered during model formulation is examined in chapter 11. To express the geometric condition that the block's position is taken to be zero when the spring is at its rest length, we need to use a correspondence:

```
(correspondence ((Length S) (RestLength S))
                ((Position B) 0))
```

The second consequence of connection that we must represent is that the force of the spring is applied to the block. Again, we can do this via a qualitative proportionality plus a correspondence:

```
(qprop (AppliedForce B) (InternalForce S))
(correspondence ((AppliedForce B) 0)
               ((InternalForce S) 0))
```

For our initial state, suppose the block is pulled back, so that the spring is extended. In this analysis, let us assume further that the contact between the block and the floor is frictionless. What will happen?

Because the spring is stretched initially, it will exert a force. This, in turn, will exert a force on the block that, because the block is free to move, will lead to acceleration being active. Thus, in our initial state,

```
VS: {(Stretched S)}
PS: {(Acceleration B -1)}
(Ds (Velocity B) -1)
(= (Velocity B) 0)
(> (Length S) (RestLength S))
```

Let us call this state S0. S0 will only last an instant, because there is a change from equality (equality change law) involving the block's velocity and zero. The next state, S1, looks like this:

```
VS: {(Stretched S)}
PS:{(Acceleration B -1)(Motion B -1)}
(Ds (Velocity B) -1)
(Ds (Position B) -1)
(Ds (Length S) -1)
(< (Velocity B) 0)
(> (Length S) (RestLength S))
```

This state lasts for an interval of time. The only quantity space with a limit point in the direction of change is `(Length S)`, where it can change from being larger than the rest length to being equal to it. At that point, the spring becomes relaxed. This also means (via our correspondence modeling connectivity) that the acceleration stops, because the applied force becomes zero. The next view and process structures (call it S3) are as follows:

```
VS:{(Relaxed S)}
PS:{(Motion B -1)}
(Ds (Velocity B) 0)
(Ds (Position B) -1)
(< (Velocity B) 0)
(= (Length S) (RestLength S))
```

This state of affairs will last only an instant, because the position of B is going to transition away from zero, leading to S4:

```
VS:{(Compressed S)}
PS:{(Motion B -1) (Acceleration B 1)}
(Ds (Velocity B) 1)
(Ds (Position B) -1)
(< (Velocity B) 0)
(< (Length S) (RestLength S))
```

Now the only quantity space with a neighbor in a direction of change is velocity: at some point, the velocity of `B` goes from negative to zero. This will take an interval of time. The new situation, S5, looks like this:

```
VS:{(Compressed S)}
PS:{(Acceleration B 1)}
(Ds (Velocity B) 1)
(Ds (Position B) 0)
(= (Velocity B) 0)
(< (Length S) (RestLength S))
```

Because the change in velocity from zero to positive is a change from equality, S5 will only last for an instant. The new state, S6, will be

```
VS:{(Compressed S)}
PS:{(Motion B 1) (Acceleration B 1)}
(Ds (Velocity B) 1)
(Ds (Position B) 1)
(> (Velocity B) 0)
(< (Length S) (RestLength S))
```

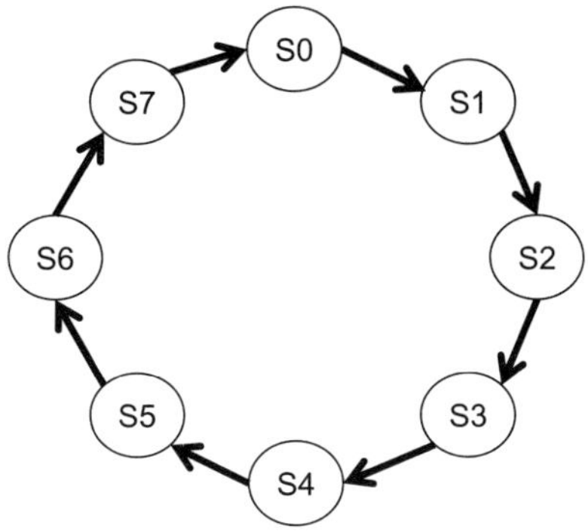

Figure 8.23
The oscillation of the spring-block system of figure 8.22 can be seen in the cycle of states here.

Continuing this process for several more states, we find that we get back to a state that is precisely the same as S0, as figure 8.23 illustrates. This cycle in qualitative states indicates that the system is oscillating. Thus, the behavior we expect from a spring-block oscillator can be derived from these models.

Now, what if we extended the domain theory to include friction? There are two types of friction. *Dynamic friction* occurs when an object is moving, resulting in a force that opposes motion. *Static friction* occurs when an object isn't moving: if the sum of the other applied forces is smaller than the friction force, the object does not move; otherwise, it does. If we reanalyze the spring-block oscillator model fragments representing these concepts added to our domain theory, we get the results shown in figure 8.24.

Notice that it includes the same behaviors as before but also three new states. The state with two transitions into it represents the possible outcome of the oscillator stopping due to dynamic friction: if the block decelerates at the right rate, its velocity could reach zero exactly when the spring is at its rest length. In that state, there is no applied force on the block and hence no acceleration. There are no transitions out of that state, so once it has been reached, the system will stay there forever. The two new states with single transitions represent the possibility that the excursions of the block are (or become) so small that the applied force by the spring is smaller than static friction.

Thus, we have derived some subtle properties of a harmonic system, oscillation and how it might end, from a purely qualitative model of motion and materials.

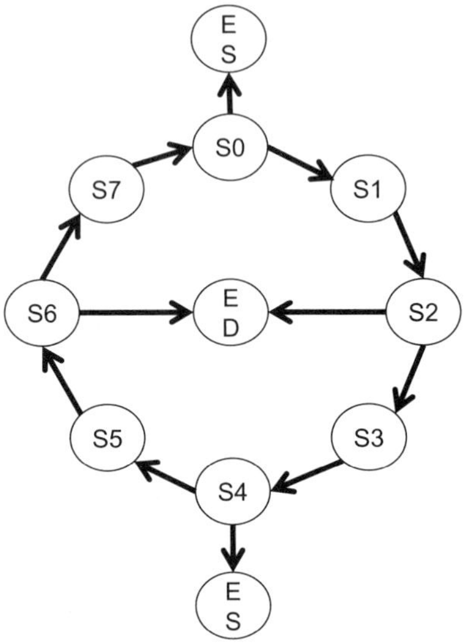

Figure 8.24
Additional states predicted from the effects of friction. The states ES represent how the behavior can end due to static friction, and the state ED indicates how it can end due to dynamic friction.

8.11 Analyzing Stability

The ability to qualitatively reason about derivatives supports the identification of more complex causal patterns. Consider again the dam example from chapter 6, where there is an open spillway and water coming in from a river at a constant rate. This is idealized as a finite container (the lake) connected to a source and sink of water, modeled as infinite-volume contained stuffs. How does the lake's level change over time?

The first thing to notice is that the initial situation as we have described it so far is slightly ambiguous. What we know so far is shown in figure 8.25.

Because there are two flows affecting the `AmountOf` in opposite directions, unless we know their relative rates, we cannot determine how the level is changing. Let us explore all three possibilities:

1. `(> (LiquidFlowRate Outflow) (LiquidFlowRate Inflow))` implies that the level will be dropping.
2. `(= (LiquidFlowRate Outflow) (LiquidFlowRate Inflow))` implies that the level will be steady.
3. `(< (LiquidFlowrate Outflow) (LiquidFlowRate Inflow))` implies that the level will be increasing.

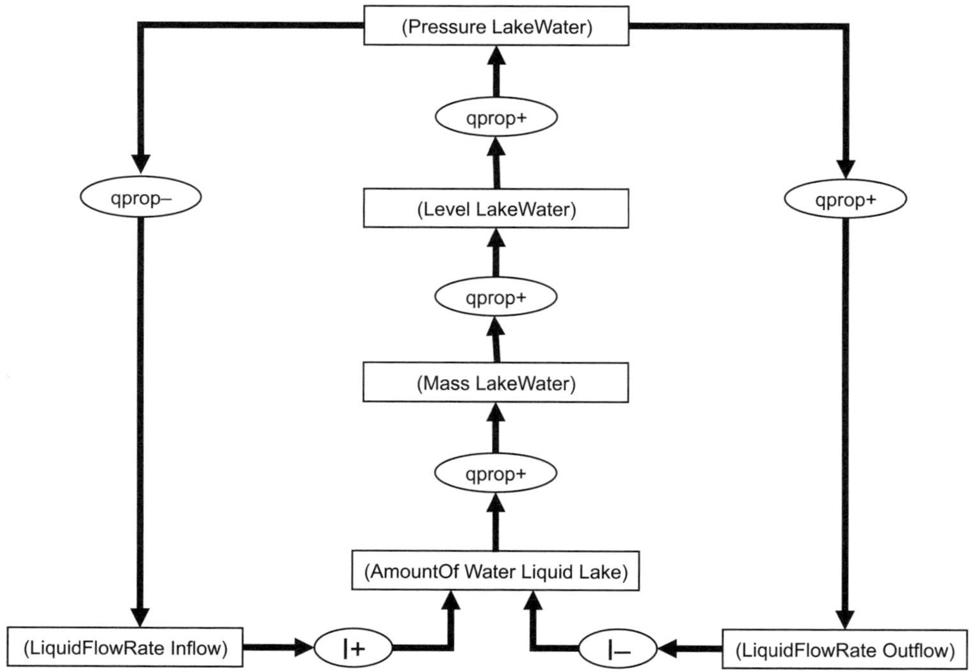

Figure 8.25
Influences for the level of the lake.

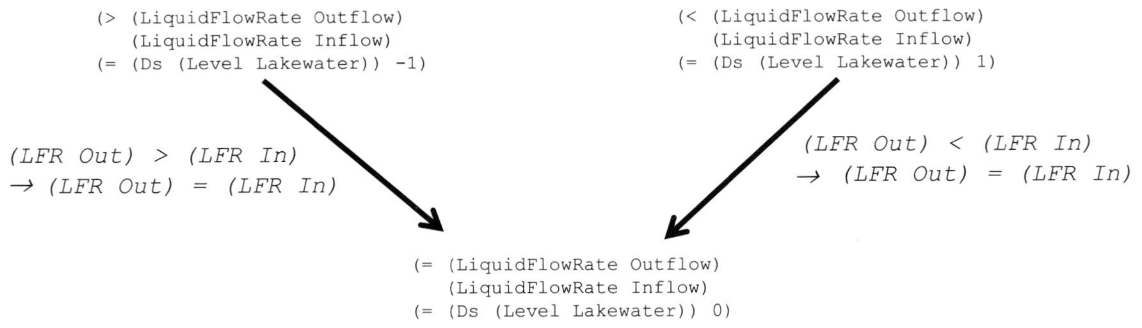

Figure 8.26
Derivation of a dynamic equilibrium.

Ordinals between rates, like other relevant ordinals, have quantity spaces and help determine state. This gives us three initial conditions we must consider. Beginning with the first, the dropping level means that the flow rate out is decreasing, because it is qualitatively proportional to the pressure of the dam, which is decreasing under this assumption. The flow rate in is increasing, because there is a negative qualitative proportionality to the pressure of the dam. This reasoning leads to the conclusion that there is a possible transition in the comparison of the flow rates, from > to =. The third condition is the same but with signs reversed: the increasing level increases the flow rate out and decreases the flow rate in, leading to a transition between the rates from < to =. The final case, where the level is unchanging, does not lead to any transitions, because no ordinal relationships can change. This is illustrated in figure 8.26.

This is a remarkable conclusion. We have derived the existence of a dynamic equilibrium, as indicated by the transitions from states with changing levels to a state where the level is constant. Being able to recognize such behaviors is important for understanding complex changes in the world. This example shows that it can be done via purely qualitative reasoning.

We can do even more if we are willing to construct abstract models based on the results of qualitative reasoning about behaviors, such as reasoning about conservation of mass and energy (Forbus, 1984).

8.12 Discussion

Some of the conclusions that QP theory is capable of drawing are quite subtle. It may be surprising that purely qualitative reasoning is sufficient to obtain them. But these are conclusions that human experts can and do draw, and the hypothesis is that the representations described here are used in doing so. I believe that human novices share similar representations, using qualitative representations for quantities and relationships, and this notion of continuous process. However, in novices, the representations of processes may be less fully articulated. That is, they contain a combination of concrete representations of experience and partial generalizations constructed via analogical processing from that experience, combined with clues from their culture. This account is explored in depth in chapter 12 and part IV.

Qualitative representations and reasoning have implications for understanding causality and change that go beyond what we have seen so far. Chapters 9 and 10 explore them.

9 Causality

The model of causality presented so far has interesting similarities and differences with other models in cognitive science. This chapter first summarizes the theory of causality introduced by QP theory and then compares it with other accounts. I argue that the causal theories developed in qualitative reasoning research have significant advantages over other theories in terms of handling continuous phenomena.

9.1 What Is Causality?

Causality is sometimes treated as a metaphysical phenomenon, mostly by philosophers. Many have argued against its utility: Minsky (2007) describes it as a "suitcase word," containing a variety of concepts that are not very compatible with each other but described using the same word. Hayes (1979) argues that there is very little deep content in a theory of causality per se and that most of the depth lies in the laws of specific domains. I think both Minsky and Hayes are correct in some ways, but from progress in qualitative reasoning, a clear, albeit somewhat more complex, picture is emerging. What is encouraging is that only a handful of types of causal models seem to suffice to explain most of human reasoning about continuous systems.

Our starting point is that we treat causality as a psychological, not a physical (or metaphysical), phenomenon. This means we take as our evidence what people say and do in terms of causal reasoning and learning. From this perspective, the first thing to notice about causality is that it is more complicated than one might expect. Many philosophical analyses of causation, for example, focus on discrete events, a kind of "billiard ball" causality, where one event causes another, and the cause always precedes the effect. When thinking about continuous changes, subtleties arise. For example, change requires defining a frame of reference against which change is measured. There seem to be four distinct types of measurements used in thinking about causality (Forbus & Gentner, 1986a):

1. *Incremental measurements* describe changes in the same situation, distributed sequentially in time, over an interval of interest. This is the billiard ball case.
2. *Continuous measurements* describe what is happening during some qualitative state simultaneously.
3. *Differential measurements* describe how a change in the situation would lead to other changes in it, essentially comparing two possible worlds.
4. *Discrete measurements* describe changes in the same situation between two distinct times, with no description of what is happening in between.

To illustrate these, let us return to our kettle on the stove (figure 9.1).

Suppose at time t_0 the kettle is on the stove, half full. Sometime later, at t_1, someone turns the stove on. At that point, a heat flow begins, from the stove to the water in the kettle. At some later point, call it t_2, the water begins to boil. A causal account using incremental measurements would be as follows: "At t_1, the stove being turned on causes a temperature difference between the water and the burner, which causes heat to flow. This then causes the heat of the water to rise, which then causes the temperature of the water to rise."

Incremental measurements require that one change occurs before another. It extends into the realm of continuous changes the kind of causality we use for macroscopic discrete events, such as a row of dominos falling in succession after one is pushed. It preserves an ancient philosophical claim about causality (i.e., that a cause must precede its effect). Thus, it has some intuitive appeal. Unfortunately, this extension to the continuous realm leads to problems. Although it is reasonable to treat the act of

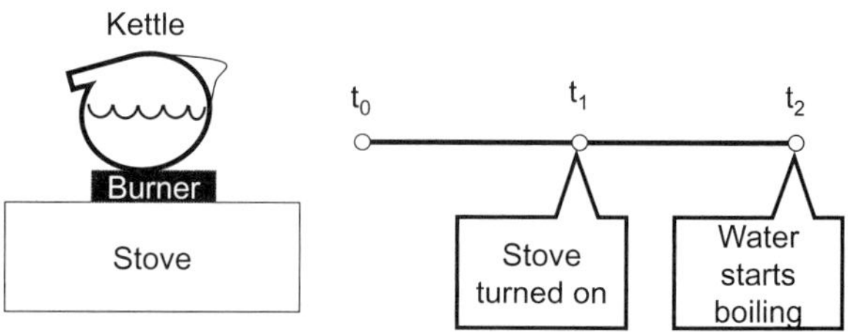

Figure 9.1
Timeline of events for boiling water. Casual explanations hold both within states and across states.

turning on the stove as a discrete event, it is not so clear that it makes sense to talk about the rises of heat and temperature as sequential. Temperature is defined as an algebraic equation involving the heat of a body, multiplied by coefficients to scale units appropriately. Physically, it is nonsense to talk about a rise in heat without a rise in temperature. The existence of other options illustrates that it is unnecessary to assume a conflation of sequence in explanation with a sequence in physical change.

Attempts to generalize the sequential notion of causality illustrate just how counterintuitive it can become. For example, to preserve sequentiality, de Kleer and Brown (1984) introduced the notion of *mythical causality.* The idea is that, even within a continuous system, there is some extremely short period of time within which changes are incremental. To make this notion of causality work correctly required the introduction of *mythical time*, where events are partially ordered and no real time passes between mythical time instants. This extra layer of temporal structure seems like a high price to pay.

Continuous measurements avoid this difficulty. Here is a continuous measurement description of the same piece of behavior: "At t_1, the stove being turned on causes a temperature difference between the water and the burner, which causes heat to flow. This causes the heat of the water to rise, which causes the temperature of the water to rise."

Here the immediate effects of turning on the stove are all happening at once: the temperature difference, the heat flowing, and the rising heat and temperature of the water. The cause and its effects start simultaneously. Nevertheless, we can still identify *dependency* without sequentiality. And dependency is, arguably, the core of causality. Why? To explain why something happened or to understand how to make something happen, dependency is crucial. Sequentiality is a form of dependency but not the only form. Sequentiality is useful in causality when one can actually intervene to change outcomes (i.e., when removing a few dominos can prevent a long line of them from falling). But imposing sequentiality where none exists is a needless complication. That is not to say that incremental measurements are not psychologically plausible: they are when dealing with discrete events. But people seem quite willing to sacrifice sequentiality when thinking about continuous change, as the example above illustrates.

An example of an explanation using differential measurements in the kettle scenario is as follows:

> If the temperature of the burner were higher, the water would boil sooner.

Notice that this describes a change in parameter values (here, burner temperature) across two different worlds, alike in every other way. The parameter affected is an indirect property of the history (i.e., how long a qualitative state would last). Of course, the very structure of the behavior itself can change as a consequence of such counterfactuals: If the temperature of the burner were lower, the water might not boil at all.

This kind of reasoning is called *comparative analysis* (Weld, 1990) and provides an account of counterfactual reasoning involving continuous causality.

Discrete measurements are useful when there is no model of the internals of how a change occurs or the internals do not matter—for example, the process of interpreting geological strata in terms of the processes that give rise to them (Simmons, 1983) and the fact that a process has occurred (i.e., uplift, deposition, erosion) matters, as well as how it changed the situation in which it occurred. But reasoning about the internals of the change doesn't matter for this task; it is enough that it occurred.

9.2 Causality in QP Theory

Recall that the central assumption of qualitative process theory is the sole mechanism assumption (i.e., that all changes in continuous systems are caused directly or indirectly by processes).[1] It is compatible with continuous, differential, and discrete causal explanations. It is not compatible with incremental measurements: if there were *n* quantities in a causal chain, incremental measurements would require creating *n* qualitative states, which for most everyday explanations does not occur.

Let us continue to use the kettle example to see some further implications of this account of causality. Figure 9.2 shows the set of active processes and influences for the qualitative state when the water is heating up.

The cause of the heat flow is the temperature difference. The heat flow process is causing the heat of the water to increase (via the `I+` relationship and a closed-world assumption), which causes the temperature of the water to rise (via the `qprop+` relationship and a closed-world assumption). The change in the water's temperature might eventually change the ordinal relationship upon which heat flow is conditioned (or, as an engineer might say, the temperature difference driving the heat flow). Thus, changes within a state (signs of derivatives) can cause changes in the constituents of state, expressed via limit hypotheses about how ordinals in that state can change. Thus, this account naturally covers causal explanations that extend over time. And as we have seen from chapter 8, qualitative simulation can be used to infer oscillation, dynamic equilibrium, and other subtle properties of behavior, so the range of causal accounts that can be supported is quite broad.

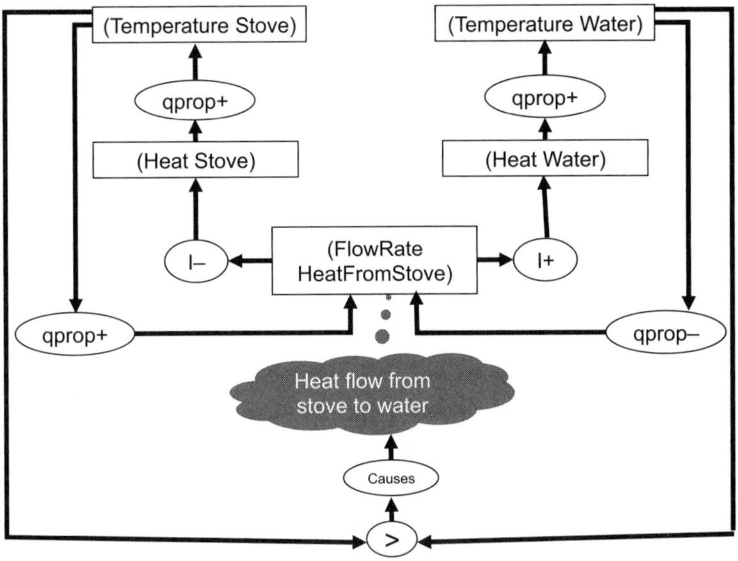

Figure 9.2
Within-state causal relationships for heating water in the kettle.

Not everything in the world is continuous, of course. How does this account interact with agency and discrete events? A complete model of human causal thinking must incorporate them as well. QP theory provides conceptual interfaces that enable its models to be linked with other models. Recall that the conditions of processes and views include both ordinals and arbitrary propositions. Those arbitrary propositions provide a means for discrete actions in the world to affect continuous change. Turning on a switch or opening a valve, for example, might establish a precondition for a process. Similarly, propositions in the consequences of a process might describe perceptual properties that hold while it is active (e.g., some tea kettles whistle when the water in them is boiling), thereby providing signals that can be used by models of action and in planning, as discussed below.

There can be advantages in modeling actions at multiple levels. Consider, for example, pushing a moving object. The act of pushing can be translated into QP theory terms in at least two ways. The simplest is to use a model fragment to describe an applied force. This force would then be added to the set of forces on an object used to determine net force. A more complex model might add to that the idea that this force is a consequence of exertion, modeled as a continuous process, which can occur only until the muscles involved are depleted. We might model this as a "fuel" parameter for that muscle. A muscle that is not exerting itself slowly rebuilds via a process providing a positive influence on its fuel level. Some deep relationships between actions and processes are still being explored. For example, Weld (1986) showed that a process of aggregation over discrete actions could be used to generate continuous process representations, which then could be used for causal simulation in domains such as molecular genetics and digital circuits. Going in the other direction, in causal simulation of military operations, it has proven useful to use QP-style continuous processes to ground discrete action representations (Hinrichs et al., 2011). Alternation between discrete and continuous representations between level shifts might be a general property of human conceptual structures for causal reasoning.

Any account of causality needs to support the following types of reasoning:

1. *Prediction*: What will happen and why?
2. *Explanation*: Why did this happen?
3. *Planning*: How can I make this happen?
4. *Diagnosis*: What went wrong with my prediction/explanation/plan?
5. *Learning*: How do I model what kinds of things can happen?

We have already seen how the causal account of QP theory handles prediction. Explanation involves using the same concepts abductively. That is, given a set of observations, construct a sequence of qualitative states that is compatible with that behavior. The vocabulary of processes and views defines a space of hypotheses within which, assuming that they are complete and correct, the appropriate answer must be. If the structure of the system is well understood, this can be done via envisionment (i.e., generating all possible qualitative behaviors, considered in detail in chapter 10) and filtering (deCoste, 1991). If the structure of the system is only partially understood, then hypotheses about structure must be made, as well as hypotheses about the underlying causes of behavior (Friedman et al., 2011b). We will return to these topics in more detail later, in chapter 17.

Planning can be viewed as setting up a situation (or sequence of situations) in which the appropriate processes will happen. Cooking a meal, for example, usually involves heating, cooling, flows, mixing, and others, in a mixture of parallel and sequential occurrences. Several methods for planning have been developed that use QP models, including compiling process descriptions into temporal plan operators (Hogge, 1987), extending envisionments with STRIPS-style operators[2] (Forbus, 1989), and integrating qualitative reasoning with a traditional planner (Drabble, 1993). These techniques have complementary trade-offs: integrating processes with other plan operators (albeit ones that nature will execute, unless we arrange the world to prevent it) does the least reasoning about the potential additional other consequences, whereas integrating actions into an envisionment as another kind of instantaneous transition focuses on the dynamical consequences of actions.

Diagnosis is facilitated by QP theory because of the sole mechanism assumption and the closed-world assumptions used to drive reasoning. That is, if a prediction (or plan) is incorrect, the only possible types of hypotheses to consider are the following:

- An alternative prediction derivable by qualitative reasoning over the model is correct. (Recall that qualitative models tend to make multiple predictions, given the abstract nature of the information they are operating over.)
- The information about the structure of the system is incomplete or incorrect. For example, a fluid path not originally in the system (i.e., a leak) or a valve that should be closed is open may explain why the level of brake fluid is dropping in a car.
- The information about the initial conditions of the system is incomplete or incorrect. For example, the mechanic might not have completely refilled the brake fluid system after servicing the car.
- Our model of what can happen in the system is incomplete or incorrect. One typically does not consider boiling in a circulating-water hot water heating system, for example, but if it does occur, there can be surprising outcomes.
- Our model of what kinds of things can happen in general is incomplete or incorrect. That is, our process and/or view vocabularies are incomplete or incorrect.

Each of these types of hypotheses can give rise to many specific hypotheses, of course: diagnosis in general is not easy. But this structure does make it more tractable. It has the additional advantage of being able to construct models of failure mechanisms, for example, a leak being an unintended flow (Collins, 1993). And it also shows how diagnosis can lead naturally to learning (i.e., when the last type of hypothesis turns out to contain the correct explanation for our failure).

Experience in using QP theory indicates that this account produces natural explanations for many domains of human reasoning. Everyday models of liquids and gases, engineering thermodynamics (Collins & Forbus, 1989; Skorstad, 1992), cooking (Tenorth & Beetz, 2012), motion, aspects of chemistry (Mustapha, Jen-Sen, & Zain, 2002) and chemical engineering (Catino, Grantham, & Ungar, 1991), biology (Noy & Hafner, 1998; Rousu & Aarts, 2001), ecology (Salles & Bredeweg, 2003), and even conceptual metaphors (Forbus & Kuehne, 2005) have been modeled this way. The systems that use these models produce explanations that are viewed by people as reasonable. We take this as evidence for QP theory as a model for human causal reasoning about continuous systems.

9.3 Causality via Propagation

Two other models of causality in continuous systems have been proposed in qualitative reasoning research. Both are based on propagation, albeit in different ways. We discuss each in turn.

9.3.1 Causality in Confluence Models

As discussed in chapter 6, analog electronics appears to operate under a very different causal model. In QP theory, the direction of causal relationship between two parameters, if there is one, is determined once and for all by the collection of model fragments that constrain them. Some of the constraints of the theory (i.e., that there are no loops involving only qualitative proportionalities and that no quantity can be both directly and indirectly influenced) maintain coherence in causal arguments. Directly influenced parameters are extensive parameters, and the causal relationships with intensive parameters are expressed via qualitative proportionalities. In systems for which the traditional state-space model of engineering works, this causal account can be directly applied.

This model can, in principle, be applied to analog electronics. People can and do talk about charges and their flows in very simple circumstances. Charge at a location is an extensive parameter, so if we were willing to reify charge at every node in a circuit and introduce charge flow as a process to cause charges to equilibrate when there are voltage differences across two connected nodes, we could use QP theory to model electronics. The problem with this approach is that electronic circuits have a lot of nodes and a lot of paths between them. For example, a schematic of a transistor radio at the level of discrete components has many more nodes and connections than an equivalent-level schematic of a power plant. This makes the conceptual overhead using QP theory in electronics quite high. Moreover, the amounts of charge that can be stored at particular nodes are negligible. The dynamics of charge flow in electronics is so uninteresting in many cases that analysis techniques were developed to simplify the structure of the circuit into something that is equivalent but much simpler.[3] So instead of thinking of charge, electronics experts instead just work directly in terms of current and voltage, a different idealization (granularity assumption, as per chapter 11) that enables them to reason more effectively about circuits.

In the confluences model of causality, changes are caused due to a perturbation or disturbance. This means that an input has to be identified as part of the analysis of the circuit.[4] Changes in the input, expressed qualitatively (e.g., + or −) are propagated through the model of the circuit via the constraint laws associated with the models for each component and for the nodes. The order of changes in this propagation of the disturbance through the circuit is identified with the order of causal events in the system. Although this model works well for electronics, it doesn't work that well for reasoning about natural systems. What is the input to the water cycle? Or the solar system? The idea of inputs does not always make sense even when there is human agency involved. In boiling water on a stove, for example, is the input just the turning on of the burner? Or is adding water to the kettle an input as well? When entities can come and go (unlike electronics), propagation within a fixed-component structure makes less sense as a way to think about the world. What about expert reasoning in, say, chemical engineering, where complex plants are set up to process materials to create products? To my knowledge, no one has done a systematic survey of explanations in chemical engineering to determine if this kind of shift in idealization occurs there. Informally, I have not seen such explanations. I suspect the reason is that the stuffs being thought about at each node of a chemical plant are real and manipulable in ways that charge isn't, at least to those of us who aren't electrical engineers.[5]

9.3.2 Causal Ordering

The third model that has been proposed is also the oldest. As mentioned briefly in chapter 6, it was first proposed by Herb Simon (1953) as a way of thinking about economics. The idea is that, for any system of algebraic equations, some set of parameters can be identified as *exogenous* (i.e., constrained or driven by something outside the system). These parameters can be thought of as inputs, and an ordering of computation found between these inputs and the rest of the parameters is then taken to be the causality implied by that system of equations.

More specifically, there are two inputs to a causal ordering algorithm:

1. A set of equations. In Simon's original work, these were quantitative algebraic equations, but later work extended this to qualitative equations (Iwasaki & Simon, 1994).

2. A subset of parameters mentioned in the equations that are identified as exogenous. These are viewed as not requiring explanation within the causal account being constructed.

A causal ordering algorithm produces a directed graph of causal relationships, grounded in the exogenous parameters and linking every parameter in the set of equations. Informally, causal ordering algorithms work like this:

1. Find all equations that have exactly one parameter not yet explained. For each such equation,
   a. Add causal link from explained parameters to the unexplained parameter.
   b. Add the unexplained parameter to the set of explained parameters, and remove that equation from the set of equations being considered.
2. Continue until there are no more equations.

As long as there are enough exogenous parameters, this algorithm will construct a directed graph of causal links. If there aren't enough exogenous parameters, there will be equations that have not been consumed that have more than one unexplained parameter. Such cases are handled by picking more unexplained parameters as exogenous.

This method has the attraction of simplicity: given any set of algebraic parameters and enough exogenous parameters, it will construct a causal story explaining the rest of the parameters. Moreover, if the set of exogenous parameters were to change, it could construct a new causal story explaining everything in terms of that set. Such flexibility is prized in algorithms.

This flexibility does lead to the question of whether or not the causal explanations it produces are always psychologically plausible. The answer is no. To see this, consider the following equation that defines the concentration of sodium in the blood of an organism:

```
Concentration(Na, Blood) =
      AmountOfIn(Na,Blood)/Volume(Blood)
```

We need to pick two exogenous parameters to construct a causal story from this equation. If we pick `AmountOfIn` and `Volume`, all is well: changes in one of those parameters will cause a change in concentration. But if we pick `Concentration` and `AmountOfIn`, we get a causal story that is far from correct: "If the blood sodium goes up, the volume of blood goes up." It is true that, within ranges, if salt intake goes up, a well-functioning animal will, over time, retain fluid—but that follows from another set of mechanisms in the body, not the definition of concentration!

QP theory suggests a constraint on exogenous parameters for constructing psychologically plausible explanations: the set of exogenous parameters should be the set of extensive parameters in the system of equations. In that case, the causal links can be interpreted as qualitative proportionalities, albeit with some further analysis to ascertain sign of effect:

```
(qprop Concentration AmountOf)
(qprop- Concentration Volume)
```

A reasonable hypothesis, albeit one yet unproven, is that if extensive parameters are always chosen as exogenous, the causal order generated for a set of algebraic equations would be equivalent to what a well-constructed QP domain theory would instantiate for the specific scenario that the equations were written to model. The problem, of course, is that by the time one has generated a set of equations for a scenario, a lot of modeling work has already been done. Nevertheless, one could imagine creating a useful technique for reverse-engineering a domain theory from a set of legacy equation models for some domain.

9.4 Other Notions of Causality in Cognitive Science

There has been an increased interest in causality in cognitive psychology and artificial intelligence (Gopnik & Schulz, 2007; Halpern, 2016; Pearl, 2009; Sloman, 2009). These models tend to focus on binary variables (e.g., occurrence of an event or not), which qualitative reasoning accounts ignore (with the exception of limit points being reached as an event, e.g., starting/ending a process or an individual's existence). The ability of qualitative models to handle causality in continuous systems, including feedback, provides a capability that these accounts lack. Some of these accounts assume that a cause must precede its effects, which is not the case in QP theory, where within-state causal changes are simultaneous. Some of these formalisms are Bayesian, which provide a way to test proposed causal models against data (Pearl & Mackenzie, 2018).

I view such research as complementary to the models of causality developed in the qualitative reasoning community. For example, work by Pearl (Pearl, 2009; Pearl & Mackenzie, 2018) and Halpern (2016) have created elegant formalisms for causal reasoning based on structural equation models. This line of research is agnostic with regard to how causal mechanisms are formalized. QP theory's notion of continuous process provides one such formalism. In other words, the graph of influences describing the QP causal structure of a domain could be treated as a set of structural equations, as per the discussion of Simon's causal ordering method. The structural equation causal models approach has focused heavily on evaluating potential interventions, with the stronger conclusions being licensed by stronger assumptions about the underlying form of the functional dependencies. Within a QP account, the occurrence (or nonoccurrence) of a process would be the natural place to intervene, rather than at a particular parameter, because there can be more than one directly influenced parameter, a source of coordination that would not show up in the structural equations alone. Thus, an integrated account could be even more comprehensive and powerful.

9.5 Summary

Research in qualitative reasoning has produced and refined models of causality concerning continuous changes. There is evidence that these models are psychologically plausible, in that systems created using them generate explanations that are perceived by human experts as reasonable, across a broad range of domains and tasks. Further empirical scrutiny is warranted: for example, the QP notion of causality predicts that continuous measurement causality, not incremental causality, will be widely found in human explanations of continuous causal systems, outside of analog electronics. In any case, the ability of these models to enable systems to produce explanations that seem reasonably natural provides evidence that they could be good models of human causal reasoning.

10 Qualitative Simulation and Reasoning about Change

The concept of qualitative simulation is central to qualitative reasoning, so it is worth considering its properties in more detail. Some of these properties have been extensively worked out and proven formally to apply to any theory of qualitative dynamics. Others are based more on experience with qualitative simulation techniques applied to a variety of problems and domains. We begin by summarizing the basic ideas, bringing together concepts introduced in chapters 5, 6, and 7. Then we discuss how existence and continuity interact with reasoning about change. Finally, we discuss some formal properties of qualitative simulation, ending with an open question.

10.1 Qualitative Simulation

We can think of a qualitative state as a situation describable by a set of propositions that hold over some instant or interval of time. The set of propositions includes qualitative values for all of the quantities in a situation, plus other statements needed to determine what model fragments exist and are active or inactive. For example, in thinking about our kettle on the stove, what we know about the temperature of the water (i.e., initially that it is lower than the boiling point and lower than the temperature of the stove) will be part of the constituents of any qualitative state for this situation. Similarly, statements in the conditions of instantiated model fragments are true, such as the stove being switched on, and will have to be known to be either true or false.

Being a bit more formal about qualitative states will help us gain insight into the nature of qualitative simulation. Let us define the *basis set* for the qualitative states of a scenario model to be the set of statements whose truth conditions must be known to define a qualitative state. For compactness, we compress possible ordinal relationships into a single statement with four possible values: >, <, =, or ⊥, the last indicating that one or both of the quantities in the pair being compared doesn't actually exist during that state. (The subtleties of how existence and change interact are discussed below.) For example, initially there is no steam in the kettle, so the comparison of its temperature with that of the stove is moot. Given a scenario model SM, we can define the basis set as the union of two other sets:

- Booleans(SM) = the set of nonordinal statements whose validity can vary over the course of the analysis.
- Ordinals(SM) = the set of pairs of quantities whose comparisons are needed in the analysis.

An example of a scenario model statement that is in Booleans(SM) is whether or not the stove is switched on. An example of a scenario model statement that isn't in Booleans(SM) is whether or not the stove is a stove: any analysis is assumed to have a fixed background or frame of reference that is held constant. Ordinals(SM) always includes every pair of quantities that appear in the conditions of instantiated model fragments of SM. To define Ds values, the comparison between the derivative of every quantity and zero is included in Ordinals(SM) as well. Moreover, as described below, qualitative simulation algorithms themselves can add new elements to Ordinals(SM) during the course of their operation. Part of the job of any implementation of qualitative process (QP) theory operating via first principles is to automatically compute BasisSet(SM) by gathering Booleans(SM) and Ordinals(SM) from the set of instantiated model fragments that constitute SM.

How many qualitative states can there be for a scenario model? Because there are two possibilities for every Boolean and four for every ordinal, an upper bound on the number of states is

$$|\text{States(SM)}| \leq |\text{Booleans(SM)}|^2 \times |\text{Ordinals(SM)}|^4$$

The inequality is used because equality would be the very worst case: every statement is completely independent of each other. But they invariably are not. For example, if there is no steam (i.e., the ordinal for its amount is equal to zero), then every other ordinal concerning it is moot (i.e., ⊥). Similarly, the heat of the water cannot be increasing while at the same time (in this situation at least) its temperature is falling. These mutual constraints implied by the causal influences and conditions of the model fragments greatly reduce the number of possible states.

Recall that envisioning is generating all of the possible states and transitions between them for some scenario model. There are two forms of envisionments:

- *Attainable envisionments* consist of all states that can be reached from some initial starting state (or set of starting states).
- *Total envisionments* consist of all states that can be reached by the system from any possible initial state.

Given our definition of BasisSet(SM), one can see, roughly, how to compute a total envisionment: find all consistent solutions to BasisSet(SM), which will constitute the qualitative states of the system. Then perform limit analysis on each state to find all of the transitions between them. Computing an attainable envisionment is less work: if one is given a complete initial qualitative state (i.e., all elements of BasisSet(SM) are known), then perform limit analysis repeatedly on it and the states that result until no more can be generated. In both cases, the results will be finite, because BasisSet(SM) is finite, but there could be an exponentially large number of states.

What if, given a description of an initial state, not all elements of BasisSet(SM) are known? Two strategies have been used in such situations. The first is to perform a search for *completions* of the state (i.e., values for the unknown elements of BasisSet(SM) that result in consistent, completely specified states). The second is to carry on with partial information about state (deCoste & Collins, 1991). The advantage of working with partial information is that it is often less work, but the disadvantage is that partial states may or may not be consistent (i.e., they have no consistent completions). Envisioners aimed at engineering applications generally search for completions. Their job is to come up with consistent possibilities, and they have no other means of checking the reality of their conclusions, so they wring as much information out of the qualitative model as possible.

When are envisionments compact, and when does the number of states explode? Envisionments are at their most compact when there is one interconnected set of influences that are tightly coupled. The spring-block oscillator and kettle on the stove are examples of tight coupling. The primary factor that causes state explosions is when subsets of a system are completely independent. For example, if there are N states in the envisionment of a simple situation, and we stitch together M of these simple situations, without any means for them to interact, to form a new, larger situation, then the number of states in the combined envisionment will be N^M. This is a classic example of a combinatorial explosion. There are several remedies to such situations. One is to use the causal analysis of the scenario model to factor the simulation, so that independent components (called *p-components* in QP theory) are simulated independently and combined states produced only on demand. Another is to shift perspective, so that one or more of the components drop out (e.g., choosing an appropriate time scale for modeling, as discussed in chapter 11). A third is to simply carry on and generate the whole thing: for some applications, that is worth it, as discussed in chapter 19.

Note that the finiteness of the envisionment rests on the existence of bounds on BasisSet(SM). All of the qualitative representations for value in chapter 5, except one, have this property. The exception is allowing landmark introduction. Recall that a landmark is a specific value taken on by a parameter at a particular time. If we think about the maximum excursion of a spring-block oscillator with dynamic friction, at every cycle, that maximum gets a little lower. If we ignore static friction, then this oscillation will continue forever, always getting smaller and smaller, generating each time yet another landmark whose relationships with its neighbors must be added to BasisSet(SM), thereby growing it without bound. Thus, envisioning under landmark introduction can result in an infinite number of states.

10.2 Existence and Continuity

One of the subtleties of reasoning about change concerns changes in existence. Phase changes in fluids provide one source of examples: when water begins to boil, steam is produced—a new individual whose properties are quite different in many ways from the water that it came from, even though the substance is the same. Modeling populations is another: when the population of a species reaches zero, it becomes extinct. How should we think about the continuous properties of objects that don't exist? And how can we accurately predict changes in existence?

Any solution should both fit human intuitions and be mathematically consistent. We start by distinguishing between logical existence and physical existence. Logical existence is simply that it is not inconsistent for there to be some state of affairs in which something exists. Coffee in my mug can certainly logically exist; a two-dimensional square circle cannot. Physical existence means that a particular individual actually does exist during some period of time. The arsenic in my coffee is an individual that logically exists but hopefully never physically exists. People have little trouble distinguishing between the two forms of existence. It is a practical necessity: in troubleshooting, for example, one postulates entities like "the leak in my car's brake system"—an instance of liquid flow—which we know logically might be the cause of the car's brakes not working and whose physical existence we might set out to determine.

QP theory postulates that entities that do not physically exist cannot have quantities. This seems to fit well with human intuitions. Try describing the properties of arsenic in someone's coffee to that person, its concentration and warmth. Or, for a less discomforting example, the birth rate of dodos. To even provisionally consider such quantities, we must grant provisional existence to the entities to which they belong. This happens all the time when troubleshooting, with leaks and shorts being familiar hypotheses to anyone who has done troubleshooting on fluid or electrical systems.

Model fragment types can be divided into two kinds: *conceptual* and *physical*. An instance of a physical model fragment type logically exists whenever the constraints for instantiating it hold (i.e., there are individuals satisfying the participant constraints). It physically exists when it is active (i.e., when its conditions hold). When ordinal relationships are involved in determining when the conditions of a physical model fragment hold, we call this *quantity-conditioned existence*. Contained liquids are, of course, one category of individuals whose existence is quantity conditioned. When the amount of them becomes greater than zero, they exist. Before that, they were potentially, but not actually, there. Similarly, if careful inspection of our car's brake system shows no signs of leaks, we conclude that there is no leak and look elsewhere for the cause of the trouble. What about conceptual model fragments? These are epistemic, part of our conceptual structure, so being active means that the concept that they represent holds in that situation. Thinking of physical existence implies conceptual existence, of course.

Changes in quantity-conditioned existence can be predicated via limit analysis, because they ultimately boil down to changes in ordinal relations. However, to model the assumption of persistence of individuals, we need to tweak the algorithm described in chapter 7 slightly. Taking existence into account, it now looks like this (Forbus, 1985):

To find the new states $\{A_i\}$ for each limit hypothesis LH applicable in qualitative state B:

1. Let E be the set of statements that will constitute the seed of next states. Initialize E by assuming that individuals whose existence is not quantity conditioned remain in existence and that all static preconditions remain the same. (This preserves background assumptions of the analysis.)
2. Extend E by assuming the changes in ordinals for LH. Moreover, assume that all other proposed changes do not hold (i.e., for all $LH_i \neq LH$, the ordinals that they mention remain unchanged).
3. When consistent, assume that the quantity-conditioned individuals that exist in B still do so in A.

4. For all ordinals in Basis(SM) not already processed, generate consistent extensions of E to create $\{A_i\}$, the possible next states.
5. If E is inconsistent or E is consistent but no consistent extensions are possible, then LH is impossible from B.

Step 3 implements persistence of existence. If a limit hypothesis causes a change in existence, that change will be forced by the assumption of its ordinals in step 2. Persistence must be assumed before step 4; otherwise, ⊥ relationships can be assumed for ordinals, which (in any correct QP implementation) will rule out the existence of the individuals involved. There are two kinds of consistency tests applied to candidate states. The first kind are local to the state: is it logically inconsistent? This is ascertained by looking for contradictions. The second kind concern continuity: could a particular A_i follow from B, given the within-state changes happening in B? For example, if $(< Q_1\ Q_2)$ in B, and $(\mathrm{Ds}\ Q_1\ {-1})$ and $(\mathrm{Ds}\ Q_2\ 0)$ in B, then there is no way for Q_1 to become equal to Q_2 in A_i, because the two quantities are actually moving further apart. Similarly, a transition to $(> Q_1\ Q_2)$ would also violate continuity.

An interesting interaction between existence and continuity occurs when considering potential generalizations of the mean value theorem. For example, Williams (1984) proposed that transitions from

$(= Q_1\ 0)$, $(\mathrm{Ds}\ Q_1\ 0)$

to

$(> Q_1\ 0)$, $(\mathrm{Ds}\ Q_1\ 1)$

should be allowed. In the domain of analog circuits that he was focusing on, this does not seem to lead to problems. But if Q_1 is the population of a species, for example, this would amount to spontaneous generation.

It was mentioned in chapter 7 that qualitative simulation algorithms can themselves introduce new ordinal comparisons. For example, when considering conflicting direct influences, additional assumptions can be made to resolve them. If there is but a single positive and a single negative direct influence, then an ordinal relationship between their magnitudes can be introduced and varied to explore possible outcomes. Alternately, a pair of *net positive* and *net negative* direct influence parameters can be created when there are multiple conflicting direct influences, with the ordinal relationship between this pair determining the outcome. Of course, every new ordinal relationship that is introduced must also satisfy continuity. Because such new ordinals can be discovered during the course of processing, QP implementations sometimes go back and reexamine previous states to see if they can be consistently extended with the new ordinal relations. If they cannot, then those states previously thought to be consistent actually are not, which causes them to be marked as inconsistent. Moreover, as discussed in chapter 7, if all of the transitions from a state are ruled out, that state itself can be marked as inconsistent. Thus, sometimes multiple states can be discovered to be inconsistent once additional ordinal comparisons are considered.

The need to rigorously examine continuity across all of the ordinals, including those originally not thought relevant, is necessary because of the partial information provided by qualitative representations. Essentially, an envisioner must wring out everything it can from qualitative relationships, because that is all that it has available. Although logically correct, the explanations for pruning possible behaviors that such algorithms produce look little like any human-generated explanation of reasoning through behavior. Both informal observations of human explanations and examination of interview and protocol data from multiple experiments lead me to think that people probably aren't doing anything like this when they are reasoning through behavior qualitatively. Moreover, envisioners also tend to generate more states than people do when trying to understand the same situation. (This makes envisioners excellent for checking domain theories, however!) These extra states are logically possible, given the domain theory, and often make intuitive sense once one sees them. But because they are empirically rare, it seems that we don't think of them. This is an example of how experience guides human qualitative reasoning.

It could be that envisioners are idealized models of human reasoners and that working memory and other capacity limits degrade this capability in people, leading to us generating fewer states and transitions, as well as not reporting, for example, continuity calculations concerning net influences. Or it could be that people are doing qualitative simulation in quite a different way (i.e., via analogy). We explore this hypothesis further in chapter 12.

10.3 Correctness of Qualitative Reasoning

How should we judge if qualitative reasoning is correct? There are three commonly used criteria, gold standards for modeling, if you will:

1. Are the results accurate with respect to the real world? This is the modeling standard we use in everyday life, in science, in engineering, and in other professions.
2. Are the results psychologically plausible? This is the modeling standard we use in cognitive science.

3. Are the results consistent with mathematical models of differential equations? This is the modeling standard that a mathematician might use to evaluate the relationship between these two systems.

This book concerns the psychological plausibility of the qualitative reasoning, and hence the second criterion is our main concern. Real-world accuracy ignores the importance of being able to express models that represent misconceptions. That said, if the results of a reasoning scheme cannot be not sufficiently accurate with respect to the real world, then it would be useless, in either our everyday or our professional lives. (Chapter 19 will provide ample evidence that qualitative models that professionals find useful in their work can and have been built.) The third standard, comparing qualitative reasoning with differential equations, is important because differential equations have proven useful for modeling continuous phenomena for several centuries now. Because the coverage claims of qualitative models rest on the fact that they describe spaces of ordinary differential equations, it is useful to explore what is lost and what is gained by this abstraction. The rest of this chapter focuses on this issue.

10.3.1 Phase Space

A commonly used tool to visualize behaviors in studying dynamics is *phase space*. The axes of a phase space are the independent parameters of a system, because given those, the others are completely determined. For instance, in our spring-block oscillator, the independent parameters are the position and velocity, because these extensive parameters can change over time. (Unsurprisingly, they are also the directly influenced parameters of the system in most QP theory models of motion.) Every point in phase space corresponds to a state of the system. Figure 10.1 shows two phase spaces for a spring-block oscillator.

The left corresponds to a frictionless oscillator, and the right corresponds to an oscillator with dynamic friction. The difference in dynamics between the two is easily apprehended visually, which is why phase space is attractive. (Of course, for most real systems, there are more than two independent parameters, limiting its utility as a visualization tool. But it still has its uses, as chapter 19 explains.)

A behavior of a system is represented by a trajectory in phase space. Each point along the trajectory corresponds to a quantitative state of the system. Because qualitative representations are abstractions of quantitative representations, qualitative states typically correspond to regions, although sometimes they can be lines and points. For example, figure 10.2 illustrates how

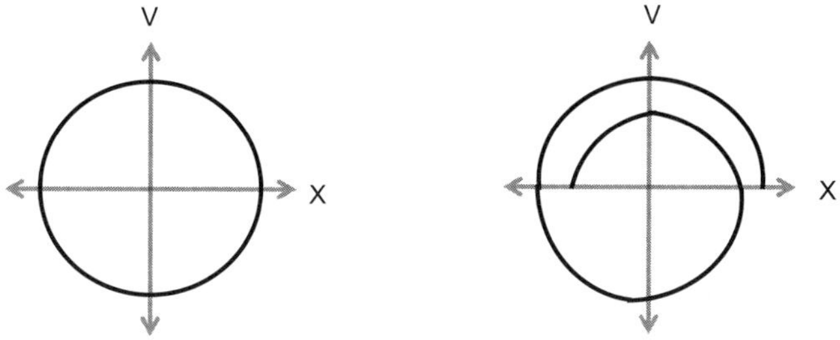

Figure 10.1
Phase-space representations of a spring-block oscillator. The axes are independent parameters of the dynamical system. On the left, the oscillator is assumed to be frictionless; on the right, dynamic friction is assumed.

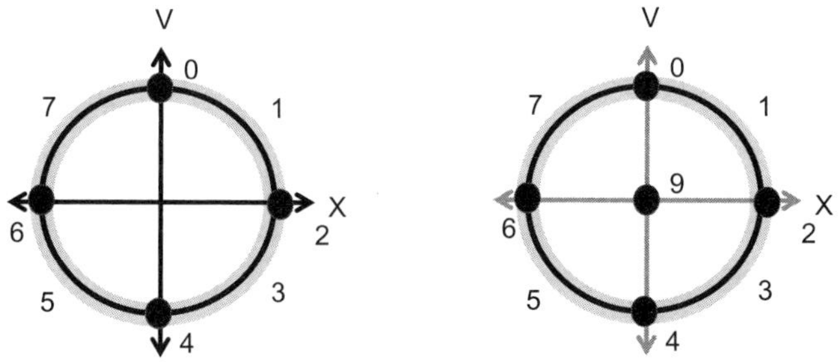

Figure 10.2
Qualitative representation of phase spaces from figure 10.1. Without landmark introduction, the difference between subsequent oscillations cannot be distinguished.

the phase spaces of the spring-block oscillator illustrated in figure 10.1 are decomposed into qualitative states, in terms of the model fragments we have seen earlier. The point in the middle of the phase space on the right, representing the oscillator with dynamic friction, is called an *attractor* because trajectories will ultimately end up there. In other words, we can think of qualitative states as regions of phase space, where transitions between qualitative states correspond to the boundaries between regions.

Now we have enough conceptual apparatus in place to see some deeper truths about qualitative representations. The arguments made here are intuitive and concern plausibility; readers who prefer formal proofs should read Kuipers (1994). The first property we would like from a qualitative representation is that for every qualitative state, there is at least one real behavior. This property certainly holds for our example of a spring-block oscillator without friction. When friction is added, the mathematical model approaches zero in the limit, so if we are willing to grant that this limit will ultimately be reached, this property is true for it as well. Suppose it were not true for some qualitative model. Then, arguably, there is something wrong with the domain theory that the model has been instantiated from. By adding conditions to the model fragments, one can always bound their applicability to appropriately shaped regions of phase space. (Perhaps this might require an impractical number of conditions, but that is a different argument—in principle, it seems it could be done.) So let us take this as a given.

The second property we would like from an envisionment is that every transition in the envisionment contains some path of states in the underlying phase space that actually can occur. In other words, each boundary between regions corresponding to qualitative states in phase space corresponds to some transition between those states in the envisionment, and there is some set of numerical parameters for the underlying system that leads to a trajectory that starts in one of the regions and then crosses that boundary into the other region. Can we construct a counterexample? Suppose we have a bifurcation (i.e., a region of phase space where trajectories split, leading to two basins of attraction). These attractors would be in distinct qualitative states, yet the boundaries between them would never be crossed. So we cannot even guarantee that individual transitions are all real.

The third property that would be useful in an envisionment is that every path through the envisionment corresponds to some possible behavior of the system. If we cannot guarantee that individual transitions are correct, then clearly we cannot guarantee this property. But the situation is even worse than that: even if all of the transitions are correct, we cannot guarantee that every path through the envisionment corresponds to a real behavior. Why? Consider again the humble oscillator, with dynamic friction. Suppose we use landmark introduction, so that each peak and trough of position gives rise to a new landmark value. For any particular cycle, locally, the system either reaches a smaller landmark or the previous one or goes past it to create a new, higher landmark. Energy considerations tell us that if it is decreasing on one cycle, it should be decreasing on the next—but that does not fall directly out of the qualitative representations. Therefore, if we locally select an arbitrary possible transition each time, we could generate a behavior that, in quantitative terms, looks something like figure 10.3.

What does this tell us? Recall that a reasoning method is called *sound* if, given correct inputs, it always produces correct outputs (chapter 3). Purely qualitative simulation is clearly unsound. A reasoning method is called *complete* if, given correct inputs, it always produces all of the correct

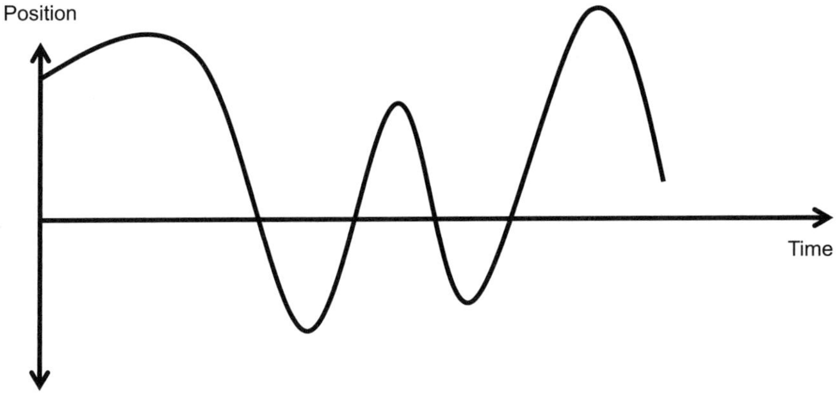

Figure 10.3
This behavior is physically impossible for a spring-block oscillator.

answers. Purely qualitative simulation is complete, in the sense that a well-constructed domain theory should always be able to generate appropriate qualitative states, and all of the transitions possible in the world can be found there. But it overgenerates, producing possibilities that simply cannot be, due to the abstract nature of the knowledge involved. It should be noted that Kuipers's papers on this topic describe qualitative simulation as sound but incomplete. Having been trained as a mathematician, he could not imagine working on an unsound algorithm, so with a deft application of deMorgan's law, the definitions were inverted. I believe that preserving the intuitive meaning of these terms is more important.

There is an interesting open question lurking here. Purely qualitative models produce local descriptions of state transitions that do not have or accumulate enough information to allow correct local choices every time. If we had a full quantitative model and numerical data, however, we know we could calculate what will happen, so there is a level of detail that, if we add it to the qualitative description, is sufficient to ensure soundness when generating behaviors from envisionments. What is the minimum level of detail necessary to ensure the soundness of qualitative simulation? Can anything less than fully detailed mathematical models do? This remains an open question for qualitative reasoning research.

Does the unsoundness of qualitative reasoning matter? It is a serious problem only if we were looking to reason just with purely qualitative information, without any access to either quantitative knowledge or experience. In applications where the original data are from sensors (either biological or industrial), the underlying consistency of the physical world ensures the consistency of state representations constructed from the data (assuming a reasonable domain theory, of course). In planning or design, completeness is more important: if real states or behaviors are missed, they correspond to missed opportunities or problems. For everyday commonsense reasoning, purely first-principles qualitative reasoning provides a set of expectations that includes an overabundance of ways to fail: with a rich enough domain theory, I cannot set my coffee cup down without considering the fact that it might break, for example. This is one of many reasons I have turned to an analogy-based account of qualitative reasoning: experience (either in the world or its stand-in, education) tells us the kinds of things that actually do happen in the world, and with analogical generalization, we can get estimates of probabilities for outcomes. Qualitative representations are essential for abstracting away the particular details of situations, to provide generalization, but purely first-principles representations need not bear the entire burden of human qualitative reasoning.

11 Modeling

Building models is at the heart of science and engineering. Modeling turns an unruly, messy situation or phenomenon into a clean, crisp description that can be analyzed to generate desired answers. Modeling involves selecting what is relevant and finding abstractions that are sufficiently expressive to capture what is being studied and yet simple enough to be analyzed tractably. A hallmark of a well-understood scientific or engineering field is a robust collection of models and consensus on when they are appropriate. However, modeling today is still something of an art.

Everybody models, not just scientists and engineers. But the problems involved are brought into sharp relief when considering the range of knowledge that scientists and engineers wield effectively. Consider the morning cup of coffee on the lab bench. In principle, quantum mechanics could be used to model how it is cooling and when it is ready to drink. But nobody does that. The amount of information required about the situation to initialize the model would be enormous and the calculations would be unwieldy, all out of proportion to the kind of question being asked. Indeed, given a common situation like this one, very likely analogical estimation (chapter 18) is used to harness experience directly to provide a rough but effective estimate. But what about less common situations? Consider, for example, an engineer designing a complex physical system. Such systems typically can be viewed from multiple perspectives. In designing a car, for example, an engineer working on the oil system might consider, for one analysis, only the volumetric properties of the lubricant: how much of it will be needed and how rapidly does it have to flow between different parts of the engine. Another analysis might center on thermal properties (e.g., will the oil ever get hot enough to break down?). Later analyses might combine both concerns, but only after initial efforts identify promising candidate designs, because such analyses require both more computation and for many more details to be specified.

If we turn from the thinking of an engineer or scientist to the role of a technician or instructor, we see multiple models are still a necessity. A technician monitoring or troubleshooting a complex system needs to formulate a manageable set of hypotheses, which often means starting with very abstract models (e.g., "there's a leak somewhere") and refining the hypothesis down to something quite specific through observations (e.g., "valve MS1A is leaking"). An instructor explaining how a complex system works needs to avoid overloading students, introducing them to various aspects of the system incrementally in ways that help them build effective models. Students, in turn, need to learn multiple models of systems and, more broadly, how to formulate new models for systems and phenomena that nobody has ever modeled before in order to become good scientists and engineers.

This chapter discusses ideas from qualitative reasoning that provide a methodology for the formal representation of, and reasoning about, models of continuous phenomena. They have been used in a variety of domains and, as chapter 19 illustrates, have been used in industrial applications, thereby demonstrating that they are capable of supporting human-level behavior in creating and using models. I start by illustrating some fundamental distinctions about models by introducing a rich example, a steam propulsion plant. The concepts used here are entirely qualitative and straightforward but, as you will see, are capable of generating interesting conclusions. Then the basic ideas of the *compositional modeling* methodology are introduced and illustrated by examples drawn from the steam plant model. Algorithms for *model formulation* are outlined, including a look at their computational complexity. Finally, we reexamine these algorithms to extract what they tell us about the nature of the problem of model formulation and how people are likely to do it.

11.1 Example: A Steam Propulsion Plant

Before the Industrial Revolution, ships were powered by wind or muscles. Harnessing steam to power ships brought about a revolution in nautical technology and practice. Steam-powered ships are still in use today, although their ranks are dwindling as newer technologies that are more compact and simpler percolate through fleets. On the other hand, the methods of extracting power from steam that they use are still prevalent in modern land-based power plants, most of which still use steam (heated by coal, oil, gas, nuclear reactions, or solar reflectors) to drive turbines that produce electricity to power our grids. Figure 11.1 shows a simplified schematic of a steam propulsion plant. Water is heated in the boiler to produce steam. The

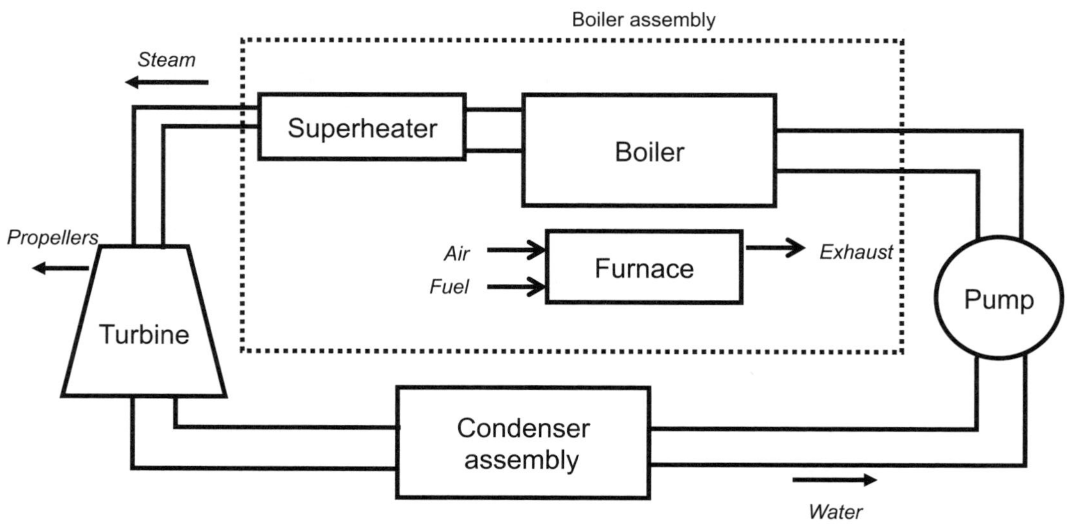

Figure 11.1
High-level view of a shipboard steam propulsion plant.

steam is expanded in the turbine, which spins, providing thrust for the propellers to drive the ship through the water (or alternately, drive the generators that produce electricity). The steam leaves the turbine and is cooled in the condenser, which returns it to the fully liquid state. The water from the condenser is pumped into the boiler, so that the whole process can begin again. This reuse of the same water is why it is often called a power *cycle*.

This schematic leaves out vast amounts of complexity. It summarizes what, in its physical instantiation, is a jungle of tanks, pipes, valves, and specialized components that, for a frigate, is the size of a small warehouse. The systems that are shown are greatly simplified. For example, there are valves that are only open when starting the system, to drain condensed water from pipes that soon will contain only steam. Many subsystems have been entirely left out. For example, water is inevitably lost from the system, so for ships that remain at sea for a long time, a distillation plant produces more fresh water.[1] Trainees learning to operate such systems do indeed spend time mastering the maze of pipes, so that, for example, they can diagnose and repair problems at sea. But abstract models such as this one provide a framework around which to organize their more detailed knowledge of a particular ship.

Even at this abstract level, we can still look at the system in several different ways. We can consider just the volumetric properties of the flows of liquid through the system to get an idea of how much liquid is involved and therefore how large the plant must be to produce the required power. This is an example of a *perspective* in modeling: focusing on just one type of phenomena in an analysis. Another example of multiple perspectives is in designing cell phones: the sensitivity of the radio, power consumption, and thermal properties are all interrelated, but designers initially focus on one or two aspects at a time to manage complexity.

Another aspect of modeling is *granularity*. Although the description in figure 11.1 is very abstract, for some purposes, even this is too detailed.

For example, to analyze the maximum possible thermal efficiency of a steam plant, we can abstract away the flows entirely and just look at the system as an abstract engine (figure 11.2).

To be sure, the plant's actual efficiency will be affected by the flows, but if the theoretical maximum is too low, then the actual efficiency is moot. On the other hand, we can look in more detail at subsystems of the plant. For example, an abstract model of what is inside the boiler is shown in figure 11.3.

The boiler itself is where the water being pumped in is turned into steam. Additional energy is added to the steam by the superheater, so that it starts at a temperature so high that there are absolutely no water droplets left in

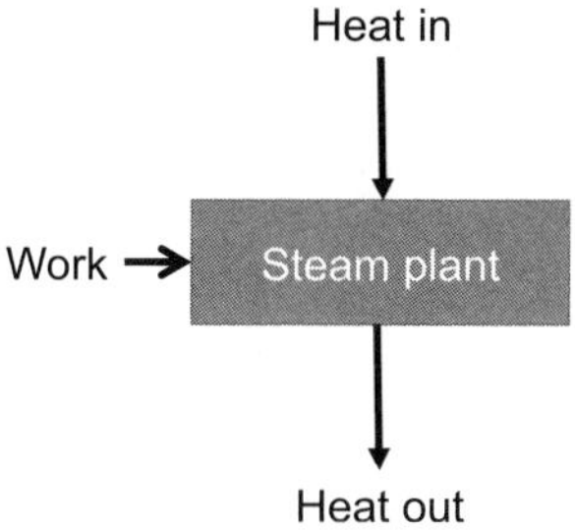

Figure 11.2
Viewing the plant as an abstract heat engine is useful when thinking about the overall efficiency of the plant.

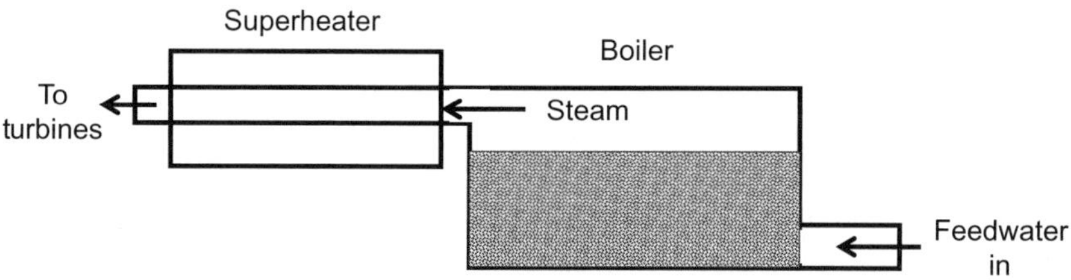

Figure 11.3
Boiler assembly.

the steam, even once energy is extracted by the turbine. (The steam leaving the turbine is moving so fast that, if there were any water droplets left, they would strike the turbine blades with the force of machine-gun bullets and shatter them.) If we want to analyze how changing the furnace's fuel/air ratio might affect the boiler's steam production, we need to think about the boiler and the furnace, but the other parts of the system outside this assembly (e.g., the condenser assembly, turbine, and pump) are irrelevant. We can even focus on just the furnace for some questions. For example, black smoke in the exhaust is a symptom of a problem (i.e., that the fuel/air ratio in the furnace is too rich). A model that captures this phenomenon does not even need to consider the boiler, let alone the rest of the steam plant's systems.

In addition to perspective and granularity, another aspect of modeling concerns *ontological assumptions*. In analyzing fluid systems, for example, scientists and engineers sometimes choose to individuate fluids based on where they are (e.g., the water in the boiler, as we discussed in chapter 8). In other cases, they think about a localized unit of fluid moving through the system (e.g., what happens to steam as it passes through the turbine).

In engineering, these are known as *Eulerian* versus *Lagrangian* methods. Within a particular ontology, additional *approximations* can cut complexity of analysis still further. For example, a common assumption for analyzing thermodynamic cycles is to ignore heat lost from the turbine. In real life, minimizing such heat losses is important for efficiency, but in many cases, the exact heat loss is not known and would be too expensive to measure. (The engineers who designed the turbines do not have this luxury, of course, for many of their analyses.)

Finally, another kind of assumption that is commonly made in modeling concerns default assumptions about behavior. For example, in reasoning about the steam plant in figure 11.1, we assume that there is water, steam, or both in all the parts that should have them and not, for example, that they are dry or filled with molasses. Although exactly what happens if the boiler runs dry might be of interest to model in considering disaster scenarios, it is not relevant to understanding its basic principles of operation. This greatly reduces the number of possible states of the system. Such assumptions are called *operating assumptions*. A very common operating assumption is assuming that the system is in steady state, that is, operating normally (as defined by all the expected physical processes happening where and when they should be), with parameters in their normal ranges. In systems like figure 11.1, there are often only one or two steady-state behaviors, which greatly reduces the number of possible states to consider.

What we are seeing in looking at this steam plant illustrates some general properties of modeling. There is almost never a single "correct" model for a physical system. To understand something, to design new artifacts, to safely operate artifacts, and to diagnose and repair them all require multiple models. Although each model can often be used for multiple purposes, our wide range of purposes means that multiple models are almost always needed. These models vary along five dimensions, which can be characterized by the types of assumptions that we have seen are used in making models:

1. *Perspective assumptions* concern what phenomena are included in the model.
2. *Granularity assumptions* concern what parts of the system are included in the model.
3. *Ontological assumptions* concern what organizational scheme is used to describe the phenomena.
4. *Approximation assumptions* concern simplifications that are made to ease analysis, given choices for the other assumptions.

5. *Operating assumptions* concern what behaviors need to be addressed in an analysis.

Next we discuss how the ideas used in chapters 6, 7, and 8 can be extended to represent and reason about such assumptions.

11.2 Compositional Modeling

Recall that in chapter 7, we introduced the following ideas:

- A *model fragment* is a logically quantified piece of knowledge that describes some aspect of an entity or phenomena. Examples of model fragments include representations of the thermal properties of an entity, the idea of a temperature source, and the process of heat flow.
- A *domain theory* is a set of model fragments that collectively describe an interrelated set of phenomena.
- A *scenario description* S is the description of a particular situation or system of interest, expressed in terms of entities, their properties, and relationships between them.
- A *scenario model for* S is the instantiation of model fragments from a domain theory implied by the entities and relationships in S.
- The process of creating a scenario model for a scenario description, given a domain theory, is called *model formulation*.

In that chapter, we always instantiated every model fragment possible on a structural description. That strategy obviously does not scale when multiple granularities can be used and fails when the domain theory contains mutually exclusive perspectives or ontological assumptions. The solution is to add more knowledge to enable the orchestration of the instantiation and assembly of model fragments into a model appropriate for the task at hand. The methodology of *compositional modeling* (Falkenhainer & Forbus, 1991) provides a means of doing this. In compositional modeling, the five kinds of assumptions discussed in the previous section are formalized and used to reason about what model fragments are relevant for a given analysis. Logical constraints between modeling assumptions become an important constituent of domain theories. We start by considering what criteria should be used to judge scenario models and how model formulation fits into the larger process of creating and using models. We then discuss how modeling assumptions can be formalized and the important kinds of constraints that need to be imposed on them to produce useful models. The issue of structural abstraction is described after that. Finally, we discuss model formulation algorithms, their trade-offs, and what this suggests about how people do modeling.

11.2.1 Modeling Criteria

Modeling, especially in science and engineering, is generally an iterative endeavor. Models are formulated and then used in one or more analyses. The results are evaluated as to suitability, and if they are good enough, the iteration ends. Otherwise, the problems with the model are used in making refinements or to generate an entirely different model, and the process continues until the questions driving it are answered. Model formulation, because it is separate from the use of the model, relies on the analysis of one or more *queries* to be made about the model to guide it. Examples of steam plant queries, in everyday terms, include the following:

"What is the maximum possible efficiency of the plant?"

"How much water is pumped through the system daily?"

"How does a change in water temperature in the boiler affect the temperature of the steam at the superheater outlet?"

"What causes black smoke in the exhaust?"

Notice that the first two require numerical information, whereas the second two are qualitative. All require focusing on different levels of granularity concerning the plant and its components. There are three key criteria for scenario models:

1. A scenario model must be *useful* for its intended purpose. That is, it must provide representations rich enough to enable whatever analysis techniques that are being used to derive an answer for the query (or queries) accurately.
2. A scenario model must be *coherent.* By coherent, we mean internally consistent and that it does not leave out relevant aspects of the phenomenon being modeled.
3. A scenario model should be *simple.* By simple, we mean that it does not contain aspects that are not necessary for answering the query. Simpler models are preferred because they generally require fewer resources to use. Resources here include both computational effort and amount and level of detail of input data and parameters.

The utility of coherence is obvious, but as we will see, this requirement has far-reaching implications for how domain theories are structured. As the need for revision and refinement indicates, judgments of usefulness are typically approximate. Simplicity and usefulness can trade off against each other: a qualitative model is simpler than a quantitative model, because it has fewer parameters and requires less detailed information to use, but if a numerical answer is required, typically a quantitative model is required. (But not always—for example, how fast is a stationary object moving?) In other cases, simplicity and usefulness go hand in hand: when explaining to a student how a steam plant works, a qualitative, causal explanation of steady-state behavior is more useful and far simpler than a numerical dynamical simulation.

11.2.2 Representing Modeling Assumptions and Constraints

To reason about modeling, we add a layer of control knowledge into domain theories. We further extend the logic of model fragment instantiation to take this new control knowledge into account. Given a potential instantiation of model fragment `MF` over entities $e_1, \ldots, e_n$ from the scenario description `S`, we allow an instantiation in the scenario model `SM` only if

1. `(considerMF MF)` holds in `SM` and
2. For each $e_1,\ldots,e_n$, `(considerEntity` e_i`)` holds in `SM`
3. `(ignoreMFI MF` $e_1,\ldots,e_n$`)` does not hold in `SM`

Intuitively, `considerMF` means that the phenomenon or aspect represented by the model fragment that is its argument is relevant for the scenario model. Similarly, `considerEntity` indicates that the entity that is its argument is relevant for `SM`. On the other hand, `ignoreMFI` indicates that a particular potential instantiation of model fragment MF is inappropriate in current circumstances. For example, in an initial thermal analysis of a basic steam cycle, heat flows are relevant but not heat lost by components like turbines. (Such inefficiencies are typically tackled in subsequent analyses.) We further constrain these predicates nonmonotonically as follows:

```
(<== (considerEntity ?e)
     (includeEntity ?e)
     (uninferredSentence (ignoreEntity ?e)))
```

That is, the original scenario mentions the entity (`includeEntity`), and we cannot infer that we should ignore it, based on the current modeling assumptions in force (i.e., cannot infer `ignoreEntity`). Similarly,

```
(<== (considerMF ?e)
     (includeMF ?mf)
     (uninferredSentence (ignoreMF ?mf)))
```

For instances, we rely on reification of binding lists:

```
(<== (considerMFInstance ?mft ?bindings)
     (considerMF ?mft)
     (uninferredSentence
       (ignoreMFInstance ?mft ?bindings)))

(<== (ignoreMFInstance ?mft ?given-bindings)
     (ignoreMFInstance ?mft ?other-bindings)
     (subsetOfBindings ?other-bindings ?given-bindings))
```

This enables systems to rule out, for example, all heat flows originating in a component, like a turbine.

Representing ontological assumptions Any specific model represents a particular point of view and hence must be based on a single set of consistent ontological assumptions. Some tasks involve multiple models that require different ontologies, but within each model, we assume that the ontological assumptions are compatible. An example of an ontological assumptions is that the contained fluids ontology should be used:

```
(considerOntology ContainedFluidsOntology)
```

Consistency in ontological assumptions is enforced by logical constraints between them:

```
(not (and (considerOntology ContainedFluidsOntology)
          (considerOntology PieceOfStuffOntology)))
```

Here we are assuming that logical environments (see chapter 2) are used in formulating scenario models, and hence the scenario model itself need not be an argument to the `considerOntology` relation.

Granularity assumptions Controlling granularity is one of the most important ways of controlling the complexity of a model. Crucial to the reasoning involved in understanding large systems is the fact that not all objects in the system need to be considered at once. There are two ways in which this happens. First, objects outside the current area of concern can be ignored. Second, abstractions allow collections of objects to be considered as a single, aggregate entity. To discuss granularity, we must introduce some conventions for part-whole relationships and for system boundaries. We will use a very simple model here, but the approach is compatible with more complex accounts. The relationship `(hasSystemPart ?system ?part)` holds exactly when system `?system` has, as part of it, `?part`. For example, the boiler and superheater stand in a `hasSystemPart` relationship to the boiler assembly. `?part` can be either a system or a primitive entity:

```
(iff (primitiveEntity ?thing)
     (not (exists (?p) (hasSystemPart ?thing ?p))))
```

The boiler, at least as we have described it, is simply a container and hence a primitive entity.[2] We stipulate that the `hasSystemPart` relation is not transitive, and hence a system contains as parts only its immediate parts.

The modeling assumption `considerSystem` means that the existence of the system's parts must be considered in any coherent scenario model:

```
(forAll (?s ?p)
  (implies (and (considerSystem ?s) (hasSystemPart ?s ?p))
           (considerEntity ?p))
```

This means that higher- and lower-level perspectives on that system must be suppressed:

```
(forAll (?super ?sub ?p)
  (implies (and (considerSystem ?super)
                (hasSystemPart ?super ?sub)
                (hasSystemPart ?sub ?p))
           (ignoreEntity ?p))
(forAll (?super ?sub)
  (implies (and (considerSystem ?sub)
                (hasSystemPart ?super ?sub))
           (ignoreEntity ?super))
```

So, for example, when thinking about the main steam cycle, the boiler assembly is treated as a black box. This requires that the domain theory contain one or more model fragments that contain laws that govern its overall behavior, in terms of the cycle properties. Suppose we decide to look next at the boiler assembly as the system under consideration. In that case, we need to model the inputs and outputs to the system that we are treating as exogenous for this analysis. The concepts of sources and sinks in scientific and engineering modeling serve exactly this role. These ideas can be formalized in terms of qualitative process theory, as we saw in chapters 7 and 8.

Perspective assumptions Perspective assumptions cover a wide range of modeling ideas. Two important kinds of perspective assumptions are *approximations* and *abstractions*. Approximations are used to construct simpler models, which are typically easier to use, although their results may be less accurate. One source of approximations is ignoring influences that presumably will be insignificant for the analyses to be performed. Ignoring evaporation in a bowl of ingredients being mixed for baking is an everyday example. An example from engineering is the *inviscid flow* approximation, which assumes that the viscosity of a fluid is zero, and hence dissipative effects can be ignored. Frictionless motion, inelastic objects, and incompressible fluids are other common examples of approximations. Abstractions, on the other hand, reduce the complexity of the model but without reducing accuracy. Valves, for example, can be modeled as either discrete (i.e., fully open or fully closed) or as a continuous, variable fluid resistance. Leaving out thermal properties in a volumetric flow analysis is another example of an abstraction.

Unlike ontological assumptions and grain assumptions, perspective assumptions are very domain specific. Moreover, they need to be localizable: some containers in a fluid system might need to be viewed as finite, whereas others might be viewed as sources or sinks. This means in formalizing perspective assumptions, we need to take their scope into account. We define `(considerPerspective` *<perspective><entity>*`)` to mean that, when true, the perspective *<perspective>* should be applied to entity *<entity>* in any scenario model being constructed. A domain theory needs to include laws that ensure that the consequences of these assumptions are appropriately specified. For example, if we are considering volumetric properties of a system, then we need to consider volumetric properties for all of its parts. On the other hand, we might choose to only worry about dissipative effects in one part of a fluid system while ignoring them elsewhere:

```
(considerPerspective DissipativeFlow FluidPath1)
(considerPerspective InviscidFlow FluidPath2)
(considerPerspective InviscidFlow FluidPath3)
```

`considerPerspective` statements can be used in the constraints on participants, so that they help control the instantiation of model fragments. For example, the simplest model of a fluid path would not even include a parameter for viscosity. Another model fragment, including a condition on participants that matches statements like the first one above, would introduce viscosity and fluid resistance parameters for the working fluid and constrain them via parameters describing the geometry of the fluid path. These constraints in turn can be decomposed into two model fragments, one consisting only of qualitative constraints (i.e., indirect influences), whereas the other would supply quantitative equations to be used if the additional perspective constraint of a quantitative analysis is added (see Falkenhainer & Forbus, 1991, for details).

Operating assumptions Engineers and scientists constantly use default assumptions about behavior to manage complexity. For example, a heat exchanger consists of a *hot leg* and a *cold leg*, and its purpose is to transfer heat from the fluid flowing in the hot leg to the fluid flowing in the cold leg.

When conducting most analyses, they assume that the fluid in the hot leg is hotter than the fluid in the cold leg, because that is how a heat exchanger is supposed to work, and thus the other two possible cases can be ignored. If the results of the analysis turn out to be wrong, this assumption (along with others made during the analysis) must be reexamined. With finite information and finite resources, mistakes are always possible: by understanding what assumptions have been made, recovery from mistakes is possible.

Operating assumptions have two roles. First, they focus analyses. In qualitative simulation, operating assumptions can rule out large subsets of an envisionment, thereby shrinking it radically. In analyzing the main steam cycle above, for example, we assumed that all of the parts had water in them and in the appropriate phases. That precludes, for example, any qualitative state where any component or path is empty. Second, operating assumptions provide reality checks for approximations. For example, in introductory physics textbooks, the analysis of a pendulum is typically simplified by assuming that $\sin(x)$ equals x, which is reasonable as long as x is small. Similarly, when quantitatively modeling liquid flow, it is useful to ignore turbulence when possible, because modeling it is quite complicated. The simplifying assumption of laminar flow is valid when the Reynolds number is less than 2,300. By including this constraint in the model and calculating the Reynolds number, a validity check can be built into the model itself (Falkenhainer & Forbus, 1991).

Three types of operating assumptions are common. The first concerns ordinal relationships (e.g., the constraint on Reynolds number above). The second concerns the normal mode of a system (e.g., the assumptions about relative temperatures in the legs of a heat exchanger). Modes can typically be described by conjunctions of operating assumptions on ordinal relations. The third are *steady-state assumptions*, which assume that all derivatives for some class of parameters are zero. Steady-state assumptions are very common in engineering analyses for two reasons. First, they radically constrain the space of possible behaviors, focusing on those that are a major subset of the behaviors of interest. Second, in many cases, there are not good transient models for phenomena of interest, or such models are radically more expensive to formulate or use than steady-state models.[3] Again, these can be localized: one might assume that the volumetric aspects of a system are in steady state while assuming that the thermal aspects are not. Domain theories must include constraints that propagate these assumptions appropriately across system boundaries (i.e., if a system is assumed to be in steady state for thermal properties, then that assumption should be made for the thermal properties of all of the fluids within it).

We will use `(operatingAssumption` *<statement>*`)` to indicate that *<statement>* is an operating assumption in the current model being formulated. This relationship is needed because, for some operating assumptions, the quantities involved may not always exist in the states generated by the analysis, whereas making the assumption of *<statement>* directly would forbid states where the quantities do not exist from even being considered.

Assumption classes Some collections of assumptions represent natural groupings that should be considered together, such as the alternative ways to model the same aspect of an object or phenomenon. For example, for any fluid path in an analysis that includes volumetric properties, we need to make some assumption about how to model flows in it. Otherwise, we would not have a coherent model. Such groupings are organized into *assumption classes*. An assumption class represents alternatives along a dimension for which a modeling choice must be made. A flowing fluid, for example, can be modeled as having zero viscosity (i.e., inviscid), a nonzero Newtonian viscosity, or a non-Newtonian viscosity (e.g., toothpaste). Not all dimensions are relevant in all contexts. Consequently, assumption classes are scoped by some condition that states when they are relevant. We assume the relation `(assumptionClass` *<condition>* *<list of assumptions>*`)` to describe an assumption class that is relevant when *<condition>* holds, where exactly one of *<list of assumptions>* must be true for a model to be coherent. (In other words, *<list of assumptions>* is considered mutually exclusive and collectively exhaustive.) For example, if `?pi` is an instance of liquid flow, then

```
(assumptionClass (FluidViscosity ?pi)
        (TheList (considerPerspective InviscidFlow ?pi)
                 (considerPerspective DissipativeFlow ?pi)
                 (considerPerspective NonNewtonianFlow ?pi))
```

Note that by making the assumption about the process instance, we can potentially take into account both the working fluid and the path when making this selection. For example, if the query driving the analysis is pressure loss, then `InviscidFlow` cannot be chosen, because it assumes no dissipation occurs. Such constraints must be expressed as part of the axioms in formulating the domain.

Assumption classes provide connective tissue for domain theories. Recall that model fragments are local: they may mention another type of model fragment as one of their participants if they serve to modify or combine its influences or constrain its quantities in some way. But otherwise, there is no global organization of model fragments in QP theory. Assumption classes impose just enough organization to guide the modeling process. The assumption classes instantiated for a model provide a kind of dynamic checklist of factors that must be considered, based on what has been done in building the model already.

11.2.3 Structural Abstractions

Scenario descriptions can be expressed at multiple levels of abstraction. Two levels are of particular interest:

- *Everyday ontology*: A description expressed in terms of the kinds of things one sees every day, easily recognizable, although perhaps in a novel configuration. A pan on the stove, an ice cube tray in a freezer or in an oven, and the lubrication system of your car's engine are all examples.
- *Structural abstractions*: A description expressed in terms of the conceptual entities found in a domain theory (e.g., contained liquids, fluid paths).

There need not, in principle, be a difference between these two levels. That is, the model fragments in a domain theory could indeed be expressed in terms of everyday entities and relationships. That may actually be a reasonable way to model aspects of people's everyday theories, as discussed in chapters 12, 17, and 18. But such knowledge would not be as general and hence not as broadly applicable as domain knowledge expressed in terms of more abstract concepts. Hence, in articulated, professional knowledge, abstract structural concepts are commonly identified and used to formulate knowledge. The cost is then recognizing occurrences of these structural abstractions when needed. This is called the *structural description to structural abstraction problem*. In the professional knowledge of scientists and engineers, learning these abstract concepts is an essential part of learning the domain: every physics student learns what a point mass is, for example. But every physics student also initially struggles with when it is appropriate to apply this concept. Thinking of a penny as a point mass works well when modeling what happens when you drop it off a building. But the same assumption fails completely when modeling what happens when you spin that same penny on a table. It seems likely that the knowledge of when to use particular structural abstractions is accumulated via experience, using analogical retrieval and generalization to incrementally acquire reasonable mappings between structural descriptions expressed in everyday terms and the structural abstractions of one's domain theory, as outlined below.

11.3 Model Formulation Algorithms

Methods for automatically creating models for a specific task are one of the hallmark contributions of qualitative reasoning. These methods formalize knowledge and skills typically left implicit by most of traditional mathematics and engineering.

The simplest model formulation algorithm is to instantiate every possible model fragment from a domain theory, given a propositional representation of the particular scenario to be reasoned about. This algorithm is adequate when the domain theory is narrow and tightly focused and thus does not contain much irrelevant information. It is inadequate for broad domain theories and fails completely for domain theories that include alternative and mutually incompatible perspectives (e.g., viewing a contained liquid as a finite object vs. an infinite source of liquid). It also fails to take task constraints into account. Just how simple a model can be and remain adequate depends on the task. If I want to know if the cup of coffee will still be drinkable after an hour, a qualitative model suffices to infer that its final temperature will be that of its surroundings. If I want to know its temperature within 5 percent accuracy after twelve minutes have passed, a macroscopic quantitative model is a better choice. In other words, the goal of model formulation is to create the simplest adequate model of a system for a given task.

More sophisticated model formulation algorithms search the space of modeling assumptions to control which aspects of the domain theory will be instantiated. The model formulation algorithm of Falkenhainer and Forbus (1991) instantiated all potentially relevant model fragments and used an assumption-based truth maintenance system to find all legal combinations of modeling assumptions that sufficed to form a model that could answer a given query. The simplicity criterion used was to minimize the number of modeling assumptions. This algorithm is very simple and general but has two major drawbacks: (1) full instantiation can be very expensive, especially if only a small subset of the model fragments is eventually used, and (2) the number of consistent combinations of model fragments tends to be exponential for most problems. The rest of this section describes algorithms that overcome these problems.

Efficiency in model formulation can be gained by imposing additional structure on domain theories. Under one set of constraints, Nayak (1994) showed that model formulation can be carried out in polynomial time. The constraints are that (1) the domain theory can be divided into independent assumption classes, and (2) within each assumption class, the models can be organized by a (perhaps partial) simplicity ordering of a specific nature, forming a lattice of causal approximations. Nayak's algorithm computes a simplest model, in the sense of simplest within each local assumption class, but does not necessarily produce the globally simplest model.

Conditions that ensure the creation of coherent models, that is, consistent models including sufficient information to produce an answer of the desired form, provide powerful constraints on model formulation. For example, in generating "what-if" explanations of how a change in one parameter might affect particular other properties of the system, a model must include a complete causal chain connecting the changed parameter to the other parameters of interest. This insight can be used to treat model formulation as a best-first search for a set of model fragments providing the simplest complete causal chain (Rickel & Porter, 1994). A novel feature of the Rickel and Porter (1994) algorithm is that it also selects models at an appropriate time scale. It does this by choosing the slowest time-scale phenomenon that provides a complete causal model, because this provides accurate answers that minimize extraneous detail. (Faster phenomena are modeled via indirect influences, and slower phenomena are represented by parameters that are constant.)

As with other artificial intelligence problems, knowledge can reduce search. Two kinds of knowledge that experienced modelers accumulate concern (1) the range of applicability of particular modeling assumptions and (2) strategies for how to reformulate when a given model proves inappropriate. Task requirements can sometimes guide the construction of qualitative representations from more detailed models dynamically (Sachenbacher & Struss, 2005). Model formulation often requires iteration. An initial qualitative model often is generated to identify the relevant phenomena, followed by the creation of a narrowly focused quantitative model to answer the questions at hand. Similarly, domain-specific error criteria can be used to determine that a particular model's results are internally inconsistent, causing the reasoner to restart the search for a good model. One approach is to formalize model formulation as a dynamic preference constraint satisfaction problem, where more fine-grained criteria for model preference than "simplest" can be formalized and exploited (Keppens & Shen, 2004).

11.4 How Might People Do Model Formulation?

The algorithms described so far work entirely on first principles. They do not take prior experience into account at all, except insofar as the contents of the domain theory represents a distillation of knowledge from both education and experience. On the other hand, anyone who does modeling in science or engineering will tell you about the importance of experience. In addition to individual experience, well-articulated standards for modeling are part of engineering practice in different disciplines, as well as conventions within particular scientific communities. Modeling knowledge thus also has a cultural component, ranging from the informal conventions used in a particular lab to the codification of engineering standards.

The experiential and cultural components of modeling knowledge can be combined with first-principles algorithms like those above to substantially increase their efficiency: if we know, for example, that we can safely ignore viscosity in a particular type of system, then that modeling assumption can be treated as fixed, and we thereby shave off one dimension of what might otherwise be an exponential search process. For example, Falkenhainer (1992) describes how accuracy measures from prior analyses involving specific model fragments can be accumulated in domain theories and reused to help make future modeling decisions.

Analogy seems like a plausible method to rapidly construct reasonable models. Let us examine how this might work, thinking about how a student might learn modeling in a domain. Examples accumulated by reading, ideally, can serve as a form of experience. Let us suppose that the student is reading a physics textbook, specifically an example involving a block on an incline plane. More deeply processed examples, the literature on self-explanation (Chi, Bassok, Lewis, Reimann, & Glaser, 1989) tells us, tend to lead to better learning. The involvement of analogy suggests two particular ways that this can happen. First, students may be making connections between their internal ontology and the types of entities mentioned in the problem. ("Okay, a block can be viewed as a point mass.") Second, students may be filling in the gaps in the explanation, adding relational structure that serves to constrain future matches. ("The sin(θ) in that equation is to calculate the vertical component of the force.") From their everyday experience, students know about sliding friction and know that when they slide real blocklike things on real surfaces, friction matters. But because the book is ignoring friction, the alert student reasons, it must not matter here. Thus, evidence about mappings from structural descriptions to structural abstractions (i.e., block to point mass), about how equations tie to geometry, and about modeling assumptions can all potentially be extracted from the analysis of examples.

As noted in chapter 4, when faced with a new problem, analogical retrieval automatically occurs in people to find similar examples. If the sliding block example is retrieved when relevant (a big if, as the literature on transfer indicates), then an analogy with the new situation gives us advice about how to model it. Moreover, there is evidence that analogical generalization can be used to learn mappings from structural descriptions to structural abstractions. Klenk, Friedman, and Forbus (2008) started with worked solutions to physics problems, expressed in predicate calculus. The worked solutions were at the level of detail that might be found in a textbook rather than being fully elaborated proof trees.[4] Each step is reified, with its antecedents formally represented. For example, here are two statements from a worked solution that indicate how specific entities in a problem are mapped to the more abstract concepts of the domain theory:

```
(stepUses Gia-2-10-WS-Step3
  (abstractionForObject Ball-2-10 PointMass))
(stepUses Gia-2-10-WS-Step3
  (abstractionForObject Drop-2-10
            ConstantTranslationAccelerationEvent)
```

In this representation, `stepUses` indicates that its first argument, the step (here, `Gia-2-10-WS-Step3`), uses the second argument as an antecedent. The relationship `abstractionForObject` indicates that, for the object that is its first argument, the concept that is its second argument is the abstraction used in this problem. Other statements in the problem, not shown, indicate that `Ball-2-10` is an instance of `Ball` and `Drop-2-10` is an instance of `DroppingAnObject`. Given a sequence of worked solutions, the model uses SAGE (chapter 4) to learn how to perform structural abstractions. In particular, there is a generalization pool for each specific structural abstraction, that is, there will be a generalization pool for

```
(abstractionForObject ?x PointMass)
```

along with every other concept used as a structural abstraction, as determined by its appearance in the second argument of an `abstractionForObject` statement. When processing a worked solution, a case is constructed for each `abstractionForObject` statement and then added to the appropriate generalization pool. The case consists of every fact that mentions the specific entity (e.g., `Ball-2-10`). Thus, it will include consequences about the ball, spatial relationships it participates in, equations that mention it, and so on. Thus, over time, each generalization pool will consist of some combination of generalizations and unassimilated examples. These generalization pools are used to make decisions about structural abstractions as follows:

1. Given an entity *e* from a new problem, create a case for it in the same way that entity cases are created when learning from a worked example. Being a new problem, this case will typically have less information about it

(e.g., no equations). However, it will have whatever events occur involving it (e.g., `DroppingAnObject`), because these are part of what it means to specify a problem.

2. For each generalization pool, examine the best match retrieved for the entity case. If it includes a candidate inference for an `abstractionForObject` statement that related *e* to the generalization pool's concept, then keep that concept as a candidate.

   a. Calculate a confidence score for this hypothesis by taking the normalized structural evaluation score for the match between the entity case and the best retrieved item.

3. Select the hypothesis with the largest confidence score as the concept to use in approximating *e*. If there are no hypotheses, do not select any abstraction for *e*.

With only two examples per structural abstraction, the analogical model built by SAGE was sufficient to enable the correct structural abstraction to be assigned 89 percent of the time. With eight examples per structural abstraction, the accuracy rises to 99.5 percent. This provides evidence that analogical generalization can provide a good model for quickly learning to make decisions about structural abstractions in modeling.

A similar process might be used by people in making modeling decisions more generally. Suppose our student, moving on in the curriculum, reads about a sliding block again, but this time in a mechanics textbook, where sliding friction is taken into account as well as gravity. Generalization pools for assumption classes could provide data-efficient learning for rapid decision making about individual modeling choices for new problems. This is consistent with both informal observations of the rapidity with which experts make such decisions and that they can't always clearly articulate why one decision was made versus another. The partially generalized relational structures in SAGE generalization pools may be too large to easily articulate, particularly if they are not recognizable as a named pattern that is part of the expert's technical vocabulary. Having such named patterns (e.g., laminar flow, turbulent flow) provides ways of simplifying and regularizing experiential knowledge and connecting it to book learning. This is compatible with the use of analogical learning to learn new relational terms rapidly.

12 Analogy in Dynamics

There is a growing body of evidence that people use analogy throughout human cognition (Gentner, 2010). Thus, it would be surprising if people did not use it for qualitative reasoning. This chapter explores the idea of runnable mental models, trade-offs between first principles and analogical reasoning, and a similarity-based model of qualitative simulation. Analogy, I argue, provides a psychologically plausible processing account of mental simulation and qualitative reasoning.

12.1 Mental Models and Runnability

Mental models (Gentner & Stevens, 1983) are the conceptual models that people use in reasoning about the world. They are similar in kind to folk theories (Clark, 1987; Rozenblit & Keil, 2002): they are stored in long-term memory and are retrieved and used to reason about particular situations. There is a sense about mental models that they are easier to use than doing deduction: the intuition is that one "runs" the mental model, easily inferring behaviors and their consequences. This sense of being easier than step-by-step deduction is shared by the other notion of mental models in cognitive science (i.e., that of Johnson-Laird, 1983). However, the Johnson-Laird sense of mental models is about mocking up specific situations in short-term memory that are then used to answer questions about simple discrete situations (the Wason task described in chapter 3 being a favorite) via counting. This is quite different from our focus here on reasoning about continuous systems and behaviors over time and quite different from the way that the term is used in research on conceptual change and cognitive development. Examples of mental models reasoning in the genre that I mean concern how people reason about basic phenomena, like electricity and evaporation, and about complex systems, like steam plants and photocopiers.

An appealing intuition is that mental model reasoning is like watching a movie of a physical system with your "mind's eye." This intuition has been explored computationally many times (Battaglia et al., 2013; Funt, 1980; Gardin & Meltzer, 1989; Kosslyn & Schwartz, 1977). There are certainly reasons to believe that visual perception is used in human spatial reasoning, as discussed in part III. However, there are three fundamental problems with quantitative simulation accounts of mental models:

1. *There isn't enough information.* Consider a badly placed spray can of paint that gets knocked over and rolls onto the stove, where you are cooking pasta. You have never seen this situation before (I hope), but nevertheless you know it is dangerous, and quick action is needed to get the can off the stove before something really bad happens. Almost none of us have enough knowledge about the physics of this situation to write a mathematical model of it that would allow the prediction of an explosion. And even if we did, that model would require a host of numerical parameters, all of which have to be determined by looking at that situation and estimated with sufficient accuracy that the simulation produces reasonable results. This is clearly not psychologically plausible.
2. *There isn't enough computational power available.* Brains are powerful computers, to be sure. But they are slow, and computational fluid dynamics makes staggering computational demands. Could there be neural structures that acted as a parallel medium to do such computations? The problem with dynamic simulation is that what happens at later times depends on accurate models of what happened earlier, so temporal behavior has to be derived sequentially. To achieve even moderate accuracy requires elements that compute each step in 10^{-9} seconds.[1] This is many orders of magnitude faster than neurons operate. Doing Monte Carlo simulation, to try to generate alternate answers and overcome lack of knowledge of parameters, as proposed by Battaglia et al. (2013), makes the computational burden even more unrealistic.
3. *The answers it produces are not as useful.* Knowing a specific velocity for the pieces of shrapnel that will decimate your kitchen is not terribly useful. (Such quantitative results are useful in physics simulations for computer games, of course.) What one wants to know are the types of events that are likely to happen, quickly enough, with enough causal information that one can figure out how to prevent undesirable events. Sifting through high-resolution streams of temporal data to detect events adds an additional computational burden to using quantitative simulation.

Qualitative representations of behavior are a better fit for mental simulation. They do not require massive amounts of accurate data to produce predictions. The kinds of outputs produced by qualitative simulation are more relevant for mental models tasks. Specifically, qualitative simulation explicitly represents events, includes multiple possible outcomes, and includes causal information that can be useful in ascertaining how to change situations for the better.

Could qualitative simulation be a computational model for runnability? It depends on the notion of qualitative simulation. As chapter 10 described, first-principles qualitative simulation is worst-case exponential. What does this mean for everyday reasoning? It means that, to use first-principles reasoning, it has to remain extremely focused, so that the number of quantities in a system remains very small. Otherwise, the number of possible states explodes, and hence the load on memory becomes implausible.

An example makes this clearer. If we think about cooking a meal in a kitchen, there are a lot of things going on. We may be heating water for pasta on the stove, roasting garlic in the oven, and grinding ingredients for pesto in a food processor. If we take the entire kitchen and everything happening in it as a single system, it is large and there will be many states. On the other hand, if we use our knowledge about causal mechanisms (i.e., continuous processes) to carve it up into a larger number of subsystems, each of which is relatively small, first-principles qualitative simulation becomes more plausible. We can think about what is happening on the stove, in the oven, and in the food processor more or less in isolation. (In theory, there could be interactions [e.g., the extra humidity caused by boiling water for pasta could affect the roasting of the garlic]. But in practice, it's just not something that people seem to consider.) Each of these operations is relatively simple, with only a handful of outcomes. Thus, first-principles qualitative simulation could be done on them quite quickly. In planning these operations, to be sure, we want to be sure that the pesto-making is finished before the pasta-making, so they do interact, but only at a more abstract level. So it could be that runnability comes from first-principles qualitative simulation but restricted to very small systems so that it is still reasonably rapid.

Another explanation is that runnability comes from qualitative simulation but not first-principles qualitative simulation. What makes a simulation qualitative is the use of qualitative representations, including events, states, and causal relationships within and between them. Analogical reasoning seems like a perfect fit for this kind of reasoning. Consequently, we compare first-principles reasoning and analogical reasoning from the standpoint of modeling human mental models reasoning next.

12.2 Human Qualitative Reasoning: First Principles or Analogical?

Qualitative reasoning captures several important properties of mental model reasoning:

- *Handling incomplete and inexact data.* Qualitative information is easily extracted via perception, and such rough distinctions are more likely to be easily remembered than precise details. For example, signs of derivatives are easier to perceive than absolute magnitudes of quantities, and part III provides examples of how qualitative visual and spatial representations provide a bridge between visual perception and cognition.
- *Support for simple inferences.* Everyday "obvious" inferences can be carried out easily. For example, if nothing is happening, nothing is changing.
- *Representation of causal knowledge.* Qualitative representations make causal knowledge explicit. They provide vocabularies for expressing partial knowledge about causal theories and mathematical relationships, as well as methods to assemble this partial knowledge on demand for reasoning.
- *Representation of ambiguity.* We easily imagine multiple alternate behaviors in everyday reasoning. Such a capability is essential for imagining what might happen in a situation, both to see if a desired outcome can be obtained and to plan for what could go wrong.

First-principles qualitative reasoning has some important properties that make it interesting as a psychological model. First, it is generative. People can often reason well about novel situations and systems. Many researchers in qualitative reasoning have the goal of building a kind of idealized physical reasoner, a system that can reason with sophistication about the world at the level that the best human scientists and engineers do but without their frailties. This goal has led to focusing on conceptual models that maximize generality. The laws of qualitative physics, as it were, are expressed in domain-independent terms, and knowledge of particular domains is expressed in situation-independent forms. Second, people can and do articulate generic, general-purpose causal laws. To be sure, this happens more often with people who have more education and training, but nevertheless, it is part of the constellation of human capabilities and therefore part of the range of phenomena to be explained.

Unfortunately, purely first-principles qualitative reasoning has several serious drawbacks as a model for human reasoning. The first is that, as discussed in chapter 10, it is worst-case exponential in both time and memory used. This means that it does not scale well. Adding one object to a situation can sometimes double or triple the size of the envisionment, for example, if it requires splitting each existing state via a new distinction. By contrast, human mental models reasoning is rapid and appears to scale well as the size of problems increases. There are several standard ways to handle exponential processes in situations where performance bounds are in place. One is to simply use a time-out and stop computing when a resource bound is exceeded. Unless the underlying calculation is cleverly arranged, there is no guarantee with such schemes that the behaviors of interest will actually be among those that it manages to generate. Whether or not first-principles qualitative simulation could be organized in something like a "salience first" order of computation is an open question. It does not seem very likely to me. A more sophisticated approach is to use metareasoning to estimate the value of particular computations (Horvitz, 2001; Russell & Wefald, 1991). Such estimations would rely on a prediction that exploring the consequences of a particular state would be interesting, which is the sort of prediction that one is doing qualitative reasoning to generate in the first place. A third approach is to simply limit the size of the inputs, so that the exponential resource usage is tolerable, as suggested earlier.

The second problem with first-principles qualitative simulation is excessive detail. To correctly predict behavior, a first-principles qualitative reasoner must rigorously and relentlessly apply continuity and the mean value theorem. This can lead to states with more distinctions than tend to be reported in verbal protocols (e.g., Kuipers & Kassirer, 1984), as well as many more states than a person would generate. Comparisons between rates, for example, are needed for continuity calculations that rule out inappropriate state transitions. But we have never seen such comparisons mentioned in verbal protocols, except when someone is reasoning through a dynamic equilibrium (e.g., inflows balancing outflows). This does not by itself rule out their use internally. It could be that such information is simply underreported verbally. However, these discrepancies are grounds for asking whether such calculations are psychologically frequent. We all know that what goes up must come down. But when we think about that, are we always forced to consider that, for an instant, the ball's velocity must be exactly zero at the peak of its travel? A first-principles qualitative simulator must always generate that state to produce locally correct results. But even with what may seem like an obsessive level of detail, first-principles qualitative simulation can still result in spurious behaviors (chapter 10). To be sure, human qualitative reasoning is full of impossibilities, as the history of perpetual motion schemes demonstrates (Ord-Hume, 2006). Although some of the coarse properties of unsoundness in envisioning can be removed by energy considerations—the same factor that reveals fallacies in perpetual motion schemes—not all of them can be, and it is far from clear that the underlying sources of the errors involved are the same.

The third problem with the first-principles account of qualitative simulation is that it relies exclusively on abstract, generic knowledge. Human knowledge includes concrete, specific information as well as general-purpose knowledge. Experience plays a major role in human model formulation, for example (chapter 11). But qualitative reasoning research has tended to eschew such knowledge. This may seem odd, but there is an understandable reason. The exclusive focus on situation-independent domain knowledge in qualitative reasoning research arose out of a desire to avoid ad hoc models. For example, the *no function in structure* principle (de Kleer & Brown, 1984) was motivated by earlier systems whose models incorporated knowledge about how a system as a whole was intended to function into models of its component parts (Rieger & Grinberg, 1977). Such intermingling of structure and function violates compositionality, leading to models that are narrow and brittle. If the correct functioning of the whole system is implicitly built into the models for its components, then how that system behaves when its operating environment is different or a component is broken cannot be derived from that model.

On the other hand, the fact that people store and remember behaviors of specific physical systems is uncontroversial. We often build quite elaborate yet concrete mental models of balky appliances, for example. It also seems likely that people's mental models include laws and principles that are situation specific (Brown, Collins, & Duguid, 1989). Such experiences are fuel for analogical reasoning. How can generativity arise from analogical reasoning and learning? There are four ways:

1. *Qualitative representations promote transfer.* Assuming people store and use qualitative representations of situations and behaviors, then two situations that vary only in quantitative details will look identical and hence match well.

2. *Analogical reasoning can produce inferences even when matches are partial.* For commonsense reasoning, within-domain analogies (e.g., reasoning about what happens when pouring coffee into a new cup based on experiences pouring coffee into another cup) typically provide reliable guides to action.

3. *Multiple analogies can be used to piece together models for complex systems*. In understanding the heart, for example, multiple analogies can contribute different inferences (Spiro, Feltovich, Coulson, & Anderson, 1989). Moreover, analogical reasoning can be chained, with commonsense conclusions building on each other (Blass & Forbus, 2017).
4. *Analogical generalization produces incremental abstractions*. Analogical generalization provides a method for learning more general models from accumulated experience, thereby improving transfer and hence generality.

This last point is particularly important for the structure of domain knowledge. If, as we suspect, analogical processing is central to human reasoning and learning, then it will be used in reasoning about continuous systems as well. And this has profound implications for the kinds of mental models people have. Specifically, it suggests the following:

1. Most people's knowledge about continuous systems is highly context specific. We have a great deal of experience with such systems and situations.
2. People's knowledge includes a range of models, varying almost continuously from highly context specific to very abstract.

From the perspective of a traditional qualitative reasoning researcher, these ideas are anathema: if one is hand-building domain theories, then first-principles models have the advantage that every statement added potentially can be used in an indefinite number of ways. On the other hand, as we learn how to build systems that learn by reading, sketching with people, and interacting directly with the physical world, we can create systems that accumulate and learn from massive bodies of experience. This may be the best way to create humanlike domain theories.

Let us look at the continuum of models implied by this view, using a simple everyday experience: filling a cup with coffee, a bit too enthusiastically, leading to a spill. Each type of model supports prediction, albeit in different ways.

12.2.1 Remembered Experience Model

Remembered experience is a memory of a behavior involving a specific cup at a specific time (e.g., more coffee pouring into your favorite cup, leading to it flowing over the top and spilling on your desk). The description of the behavior includes many concrete details, such as visual properties of the objects and their behaviors.

How much of our experience we store and in what forms is still something of an open question. Nevertheless, it seems likely that the vast knowledge we have about the world at least starts out with remembered concrete experiences. Even concrete experiences can be used for prediction: pouring wine into one glass is very much like pouring wine into another.

12.2.2 Partial Generalization Model

Although most visual properties may be gone, some aspects of the situation can still be very concrete (e.g., coffee cups instead of containers). No logical variables have been introduced. Instead, the entities involved are almost all generalized entities of the form that SAGE (chapter 4) constructs.

We call these models *protohistories*, because they are prototypical descriptions of behaviors (Forbus & Gentner, 1986b; Friedman & Forbus, 2008). Protohistories are constructed via analogical generalization. The example behaviors can either be observed directly or transmitted culturally. All of the times you have poured something with at least a modicum of intent and attention constitute examples of direct observation. The formation of solar systems, on the other hand, is something that many people, including some professionals, have models of but is not something that any of us have directly observed. Protohistories created via observation are likely to far outnumber those created via cultural transmission, and their "heft," in terms of the number of examples that went into them, is likely to be much higher. On the other hand, assuming culturally transmitted explanations are understood correctly, their abstract nature provides a form of high-octane fuel for matching and generalization. Providing a new term for a behavior (i.e., "pouring") is an especially strong incentive for people to learn more about the concept (Christie & Gentner, 2014).

12.2.3 Causally Annotated Experience Model

We can add relationships to remembered experiences that specify causal attributions among aspects of the behavior. For example, you might have noticed that the overflow was caused by you continuing to pour coffee once the cup was full. More broadly, these causal attributions might come about by experimentation or someone explaining the situation to you, by analogy with an explanation for another situation, or by the application of a more general abstraction. Additional qualitative relationships might be included (e.g., the rate of overflow depending on the rate of pouring).

Although still very concrete, these annotated experiences provide a simple form of abductive explanation. If a sufficiently similar behavior is observed, the explanatory relationships in the previous situation will be conjectured by analogy to hold in the new situation. If the match is based on the first state in a remembered qualitative sequence of states, then the visualized potential undesirable outcome could be avoided by changing causal inputs (i.e., stopping before overflow occurs and pouring more slowly, to make this easier to do).

Descriptions like these can be thought of as rulelike, in that they describe commonalities and causal hypotheses that are applicable to a broad variety of experience. (Gentner and I have sometimes called them *situated* rules, because they may explain some of the signature phenomena of situated cognition.) Nevertheless, they are still retrieved and applied via analogical processing. Their generality means that they will match in a broader set of circumstances, thereby providing more predictive and explanatory power.

When and why causal annotations are introduced for protohistories are interesting questions. Bolstering predictive abilities related to planning (i.e., how to bring about desired outcomes and avoid undesirable outcomes) is one obvious motivation. Having multiple predictions about a situation or system is another.

Note that treating causal models as annotations on either concrete experience or protohistories is a *distributed* model of causality. A causal model might not be retrieved and applied when appropriate if it was formulated for a protohistory that only differs in surface properties from the current situation. This distributed nature provides an explanation for *pastiche models* (Collins & Gentner, 1987), where a single person has been observed to have multiple, somewhat incompatible models of the same phenomenon. A computational model of conceptual change that captures such phenomena (i.e., Friedman's [2012] *assembled coherence theory*) is discussed in chapter 17.

12.2.4 Generic Domain Theory

A generic domain theory is a set of model fragments, logically quantified, including processes for flow and a model fragment for overflow, so that limit analysis can be used to predict that overflow is one possible outcome of a filling situation.

Let us step back and consider these four states of knowledge. The first state of knowledge represents pure memory. The last state of knowledge represents the sort of knowledge used by first-principles qualitative simulators. The examples in between illustrate that these extremes do not exhaust the possibilities. That is, we should think of domain knowledge as being spread over a continuum of generality. The intermediate levels of generalization and explanation, where partial explanations have been constructed in a conservative fashion, are both very powerful and probably very common in human knowledge.

How can these intermediate levels of knowledge be used for prediction and explanation? In the same way as specific examples: by within-domain analogies. Let us examine a simple computational model of this process and then return to its implications for trajectories of learning and for expertise.

12.3 Similarity-Based Qualitative Simulation

Similarity-based qualitative simulation relies on a library of remembered experiences and generalizations drawn from them, applied via analogical processing to understand new situations. The tasks it supports are as follows:

- *Prediction*: Given a new situation, similarity-based retrieval and analogical comparison is used to map a remembered behavior onto the situation.
- *Abduction*: Given a new behavior to be explained, an explanation is constructed by mapping explanations for remembered behaviors onto the new behavior.

The predictions produced by analogy, we conjecture, correspond to the content of mental simulations. Because multiple behaviors can be retrieved and applied, branching predictions are possible, just as they are with first-principles qualitative simulation. (This abduction model is one of two analogical abduction models explored here; chapter 17 describes the other.) We start by summarizing a simple implementation of this model (Yan & Forbus, 2005) and then discuss how it could be generalized to encompass more of the phenomena.

12.3.1 A Prototype Similarity-Based Qualitative Simulator

Figure 12.1 shows the architecture of a simple implementation (the Similarity-Based Qualitative Simulator [SQS]) of the general model.

The input is a situation, and the desired output is a prediction of the state (or states) that might happen next. Processing begins by using analogical retrieval via MAC/FAC (chapter 4) on a library of experiences. Recall that MAC/FAC returns between zero and three remindings; if there is no reminding, then no prediction is possible. If there are multiple remindings, the reminding with the highest structural evaluation score (i.e., the closest match to the situation) is selected for processing.[2] Second, the match between the retrieved situation and the current situation is scrutinized by the re-representation system and tweaked if necessary to ensure that there are candidate inferences concerning state transitions, because these are what will provide predictions. Here, all re-representation methods that might improve the match are be carried out, exhaustively—another simplification that, in the absence

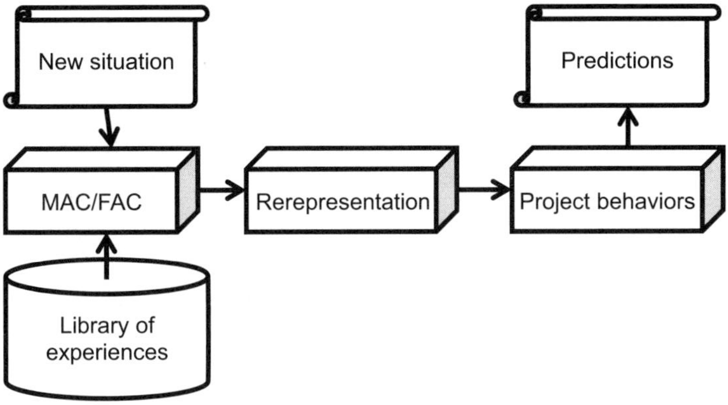

Figure 12.1
Information flow in the SQS.

of more specific task constraints, is not unreasonable. If re-representation fails, the system returns to the original match.

The third step is to use the correspondences and candidate inferences of the mapping to project possible next states. Let S be the initial situation and R_s be the retrieved state mapped to it. Transitions from R_s to another state (say R_n) are part of the description of R_s, and because S has no transitions (by assumption), information about transitions will appear as candidate inferences. These inferences will contain an analogy skolem, a placeholder representing "something like" R_n. SQS creates a new entity, say S_n, to represent the analog of R_n with respect to S, then retrieves facts from the experience library about R_n and projects them onto S_n by extending the mapping with a correspondence of $R_n \leftrightarrow S_n$. Figure 12.2 illustrates.

The substitution process for generating new predictions is likely to lead to other analogy skolems (i.e., additional unknown objects conjectured for the target domain), which need to be resolved if possible. This means either identifying (or conjecturing) suitable entities in the target to be aligned with that base item. The conditions that the skolem must satisfy are extracted from the candidate inferences and solved for by reasoning. If no existing entity is found, then a new entity is created and the candidate inference constraints are applied to it.

Finally, each expression proposed about the target is checked for consistency and adapted if necessary. Two tests are used to determine consistency: (1) argument constraints associated each predicate are enforced, and (2) each proposition should not be provably false (Falkenhainer, 1987). An alternate

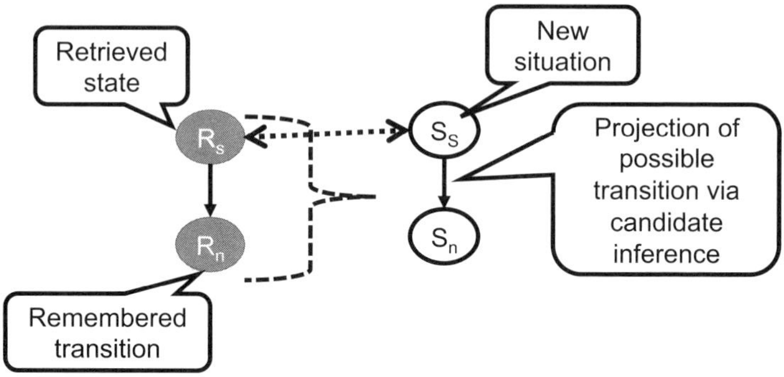

Figure 12.2
State transition computed via candidate inference projection. The dotted arrow indicates the corresponding states, and the dashed line indicates the projected information.

target correspondent will be sought when an inconsistency occurs. If the inconsistency cannot be resolved, the system returns to the next best reminding to restart the behavior projection process, until a consistent predicted behavior is formed for the current situation.

Although the prototype implemented only a subset of what we think is happening in human similarity-based simulation, the system still showed some intriguing behaviors and possibilities (Yan & Forbus, 2005). Given descriptions of simple qualitative states and transitions generated by a first-principles qualitative simulator, it was able to make accurate behavior predictions about new situations. For example, to test its ability to work with partial descriptions, it was given several variations of everyday feedback systems (i.e., home heating systems, flush toilet) as memories and as probes. There are two different ways of controlling home heating systems. In *discrete control* systems, any noticeable difference between the desired temperature (called the *set point*) and the actual temperature leads to the furnace being turned on full blast. In *proportional control* systems, the level of heat supplied by the furnace depends on how different the set point temperature is from the actual temperature—the larger the difference, the higher the rate of corrective heat flow. As it happens, although most home heating systems use discrete control, many people believe that they use proportional control, in some cases citing an analogy with the gas pedal of a car (Kempton, 1986). When given the description of a home heating system involving proportional control, SQS incorrectly suggested that the furnace would be on at its maximum rate, based on an incorrect analogy with a discrete-control home heating system. Although in the reverse direction of people's misconceptions, this example illustrates that analogy, unexamined, can easily lead to misconceptions. In a cognitive system that was able to gather data (from the physical world via experimentation, from reading articles, or from talking to people), this incorrect prediction could be detected, and the actual new behavior stored as a new case. If SAGE were used to build generalizations, these two situations might be near-misses, and in that case, it would construct discrimination hypotheses to distinguish them (McLure et al., 2015).

As an example of re-representation, consider comparing a general schema for a home heating system to a specific home heating system using an old-fashioned analog thermostat. Here is a small fragment of the representations involved in both situations:

```
B: (senses SensorX (Temperature RoomAirX))
   (compares ComparatorX (Temperature RoomAirX)
                         TemperatureSetpointX)
T: (senses ThermostatY (Temperature RoomAirY))
   (compares ThermostatY (Temperature RoomAirY)
                         TemperatureSetpointY)
```

The statements from the base indicate that, in the abstract functional model serving as the base, the functions of sensing and comparison are handled by separate components (`SensorX` and `ComparatorX`). The statements from the target indicate that, for this specific system, the thermostat handles both functions. This is not uncommon in older houses, where a bimetallic strip, curled into a springlike coil, is used to provide both functions. A mercury switch, that is, a sealed glass container with a bead of mercury in it, turns the furnace on when the switch is tipped. The switch is connected to the center of the coil. Because the different metals expand at different rates with temperature changes, the angle of anything connected to the center of the coil changes. The strip is coiled to amplify the effect of the temperature change on angular position. To set the desired temperature, the orientation of the coil is changed, typically by turning an analog dial. Thus, the thermostat handles both the function of sensing and comparison of the current value to the set point.

These fragments are an example of a problem that arises when applying schema via analogy to a new situation. If the base were a traditional schema, each functional role would be a variable (e.g., `?SensorX` and `?ComparatorX`) instead of constants. The schema would be applied via instantiating these variables, and there is no problem with two different variables having the same value. But in structure mapping, the 1:1 constraint means that we cannot simply map the two constants `SensorX` and `ComparatorY` both to `ThermostatY`. This is where re-representation comes in. Re-representation is about improving a match by changing the representation of the base, target, or both. Here we have a re-representation opportunity involving *rival matches* (Yan, Forbus, & Gentner, 2003), which are violations of the 1:1 constraint that lead to the structural inconsistency of at least one match hypothesis with the best mapping. It is often caused by the same entity playing multiple roles in the same representation, the situation we have here.

In the SQS prototype, the match produced by the best reminding was scrutinized to look for opportunities to improve it via re-representation. Its memory included a decomposition of the thermostat's physical properties into functional properties (i.e., the curvature of its bimetallic strip measures the temperature, and the angular distance between the bimetallic strip and the dial's angle provides the comparison). This is an example of an *entity-splitting* re-representation strategy. In general, entity splitting requires identifying ways to divide an entity into distinct parts or aspects and rewrite its roles in the description to use one or the other of these parts or aspects. In the case of the bimetallic strip, different properties of it are used for different functional purposes (a common design tactic). The curvature of the strip changes with temperature, and hence that property is what is being used to sense temperature. The angle that the mercury switch makes, which determines if the furnace will turn on, is what is used to compare the temperature to the desired set point. After re-representation, each of the aspects of the thermostat can match to distinct functional descriptions from the retrieved schema, leading to a much better match:

```
T′: (senses (CurvatureFn BimetallicStrip)
            (Temperature RoomAirY))
    (compares (AngleFn BimetallicStrip)
              (Temperature RoomAirY)
                                TemperatureSetpointY)
```

This redescription of the target provides a better match and predictions.

Implications of the SQS prototype This simple prototype had a number of limitations compared to human mental models reasoning:

1. The use of a single retrieval means that it didn't capture branching behaviors unless the prior explanation involved them.
2. The library of experiences was quite small: significant expansion would be needed to stress-test retrieval and re-representation.

3. More first-principles reasoning needs to be integrated to filter candidate behaviors that would otherwise be physically impossible and to enable small behaviors to be patched together to explain larger ones.
4. Learning strategies, in the form of storing back the results of re-representation and using SAGE to construct generalizations, also need to be explored.

Nevertheless, it illustrated some important properties of similarity-based qualitative simulation:

- It produces qualitative predictions of behavior rapidly.
- It operates in multiple domains.
- It operates across situations that are not identical (i.e., *near transfer*).
- Logically possible behaviors that are rarely observed are not predicted.

One gap between this model's capabilities and people is that people are capable of composing what they've learned about simple examples to reason about more complex systems. One explanation for this could be that they immediately learn generic, first-principles domain theories. But another explanation is that they combine multiple retrievals, in essence gluing together behaviors learned about simpler systems to explain more complex ones (Bylander, 1991). Consider trying to explain the three-container liquid flow situation in figure 12.3 with the fully analyzed description of flow involving two containers, as shown in that same figure.

This behavior can map onto the three-container scenario in two different ways, each corresponding to a different mapping constructed by SME (figure 12.4).

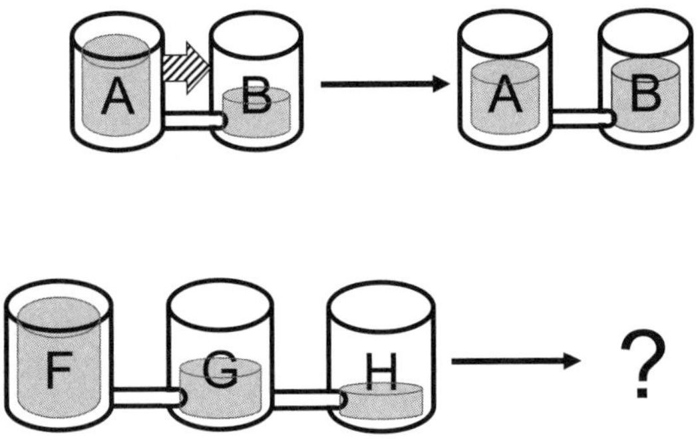

Figure 12.3
How might people compose behaviors to explain what happens in a more complex system?

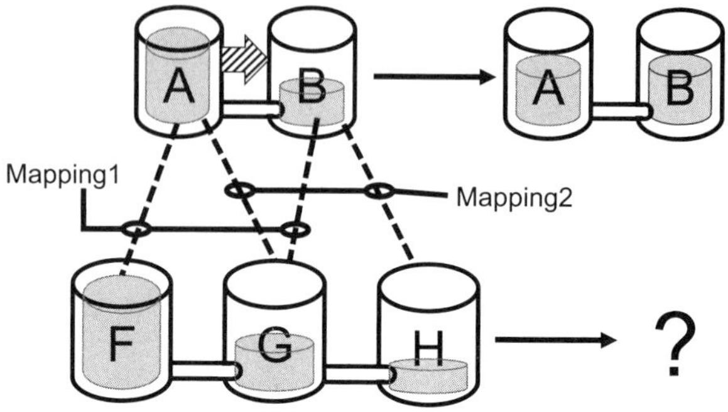

Figure 12.4
Multiple mappings can project processes, with results integrated via influence resolution.

The two mappings cannot be directly combined into one, because they violate the 1:1 constraint of structure-mapping theory (i.e., G would map onto two different containers). But what if we selectively combine inferences from these mappings? Specifically, what if we project the process structures and influences from both mappings but not how the influences resolve? (We know that these will be among the candidate inferences because there are initially no causal explanations in the target.) Influence resolution could then be used to ascertain how the effects combine and whether or not the projected individual transitions could still occur, given the potential differences in signs of derivatives. There are several variations of this scheme worth exploring. For example, projecting all inferences and then removing those that are contradictory would maximize the amount of reused experience. But in any case, it seems that similarity-based qualitative simulation could be extended to capture our ability to piece together smaller experiences to explain larger systems, thanks to our reliance on compositional, causal models in our explanations.

The other big gap concerns how well SAGE would work as a model of learning to predict behaviors. As experience accumulates, SAGE should construct more abstract descriptions of behaviors. The ontology in which these behaviors are expressed will start out being the ontology that is used in encoding the entities involved. The destination of a pouring event, for example, will start out being classified as cups, glasses, sinks, floors, hands, barrels, and so on. The destination will also have material properties. For example, the destinations are mostly rigid but not always (e.g., pouring water into a plastic bag). Over time, those relationships that are shared in common will persist (i.e., have high probability), and those that are accidental will fade into the background (i.e., have low probability).

But nothing in this process as described so far will get us to the concept of a container per se. There are probably two routes to this more general concept. The royal road, far easier and more commonly used, is language. The very word *container* is an example of a relational category (Gentner & Kurtz, 2005), a category of object defined relative to some more complex structure (in this case, holding fluids and hence a reasonable destination for a motion involving fluids). The tougher trail is to introduce a new concept based on criteria within the process of assimilating new examples itself. There are several ways that this might happen, although all are speculation at this point. One way would be driven by communication needs. For example, the very lack of a good word that is applicable to a useful generalization (or set of generalizations) suggests adding a concept for it and using the generalization to construct encoding criteria for it. Another way would be driven by noticing that similar generalized entities were showing up in multiple generalization pools—for example, the destinations of pouring and flowing start looking a lot alike (as measured via SME). Sufficient overlap among the parts of generalizations from multiple generalization pools could be a signal that this concept is worthy of introducing on its own. This would allow the ontology used in making predictions to coevolve with the evolution of the descriptions of the behaviors themselves.

Another aspect of this process that bears closer investigation is how more abstract models are created. If QP theory is correct as a psychological account, it provides a strong inductive bias on the nature of theories about continuous phenomena. That is, people look for processes and influences, guided by both experience and language. (How QP theory interacts with natural-language semantics is discussed at length in chapter 13.) Note that influences are flexible enough to be used even in the earliest stages of making causal hypotheses about behaviors. (This is elaborated in chapter 17.) The point is that analogical mapping and retrieval provide powerful ways to reuse experience, and analogical generalization provides a mechanism for refining that experience into forms that are more directly transferable, because they have fewer concrete specifics, and that are easier to form causal hypotheses about, because irrelevant information has been deemphasized.

Experiences of surprise and novelty provide clues as to what sort of models are being used by someone. In the early stages of understanding a phenomenon, even partial first-principle models are unavailable. Nevertheless, one can still be surprised: when a situation matches a set of well-worn generalizations that provide a set of possible outcomes, and something else happens instead, that is a surprise. At earlier stages of understanding, a surprise suggests novelty (e.g., one is sampling a new portion of the space of possible behaviors). More accumulation is the appropriate response, and the circular reaction of children to novelty (Piaget, 1952) is certainly a productive strategy for this. At later stages, there are additional hypotheses to be considered: a causal relationship proposed earlier might not actually be correct (or applicable in the current situation), or an inappropriate modeling assumption might have been chosen (e.g., ignoring gas pressure when choosing a vessel for fermentation). Later stages should be more articulable.

One of the valuable aspects of the earlier stages of models is that they still incorporate some amount of concrete information (e.g., visual and haptic information). The examples and generalizations that SAGE retrieves in response to a new situation will also project such information onto the new situation, albeit modulated by the match to the existing visual, haptic, and other sensorimotor information available about the situation. This, I believe, explains the seeming concreteness of mental simulation. That a doll will fall, and what general direction it will fall in, when pushed off a perch is something we are likely to have strong opinions about. We might have a concrete fleeting impression of exactly how long it will take to fall or what pose it will land in. However, if events come out otherwise, we are not unduly disturbed. (Unless, of course, there is a cognitive developmentalist working in the wings to produce an impossible behavior to test us, as in the experiments outlined in chapter 18.)

12.4 Psychological Implications

Qualitative reasoning is not an island; it should use the same mental processes used in other aspects of cognition. Consequently, properties of analogical processing that have been found in other areas of cognition should appear in reasoning about mental models as well. Here are three further predictions that Gentner and I have made (Forbus & Gentner, 1997).

12.4.1 Distribution of Reliance on Memory with Expertise

We conjecture that the use of memories in predictions with experience may vary as a U-shaped curve. That is, when little is known, memory use dominates, because comparison with previously observed behaviors encoded in perceptual terms is all that is available. As more is known, memory use may drop in favor of more abstract representations, such as situated rules. This may be especially likely in domains where learners need to articulate their models (e.g., situations where they are collaborating with others). It may be the case that as the domain becomes very familiar, memory use increases again, because the learner has experienced a large number of samples from the distribution of situations that occur. The theory-laden vocabulary learned by this stage may also greatly increase the frequency of relevant remindings (see below).

12.4.2 Differences in Novice/Expert Retrieval Patterns

The usual pattern in similarity-based retrieval (Gentner, Rattermann, & Forbus, 1993) is that retrieval is heavily based on surface properties (i.e., information about appearance and attributes of participating objects) rather than relational properties (i.e., causal arguments or abstract principles). In experts, however, the frequency of relational remindings increases (Novick, 1988). A possible explanation for this phenomenon is that an expert's ability to encode phenomena in theory-laden terms provides additional overlapping vocabulary that helps the MAC stage find appropriate matches. For example, in solving physics problems, it has been observed that experts sort problems based on similarity in underlying principle, whereas novices sort problems based on similarity in the kinds of objects involved (Chi, Feltovich, & Glaser, 1981). Additional support for this explanation is provided by results suggesting that inducing subjects to encode materials more deeply increases the proportion of relational remindings (Faries & Reiser, 1988). The same phenomena should be observable in teaching people to make predictions in novel domains.

12.4.3 Factors That Should Promote Expertise

Research on the role of comparison in development suggests two ways to speed up learning:

1. *Progressive alignment*: By exposing people to multiple very similar examples, their conservative learning mechanisms are more easily able to create the abstractions needed for transferable knowledge than if the same examples are interspersed with very dissimilar examples (Gentner, Rattermann, Markman, & Kotovsky, 1995). Kotovsky and Gentner (1996) showed that experience with concrete similarity comparisons can improve children's ability to detect cross-dimensional similarity. Specifically, four-year-olds' ability to perceive cross-dimensional matches (e.g., matching size symmetry with color symmetry) was markedly improved after experience with blocked trials of concrete similarity (blocks of size symmetry and blocks of color symmetry) compared to control groups who received no training.
2. *Inviting comparison with relational language*: Giving learners language for expressing a shared relational system can dramatically improve their ability to learn it via comparisons (Gentner & Namy, 2006; Gentner &

Rattermann, 1991). For example, Kotovsky and Gentner (1996) taught four-year-olds labels for the relations of monotonic change ("more-and-more") and symmetry ("even"). During the training task, children learned (with feedback) to classify the stimuli as to whether they were "more-and-more" or "even." After this training, the children who were successful in the labeling task scored far better on a cross-dimensional version of the task than children without such training.

Applying these results to qualitative mental models yields three suggestions for how they might be learned more easily:

1. Show learners many situations varying in quantitative details but with identical qualitative behaviors before showing them behaviors with a different qualitative structure. For example, those learning about heat and temperature might first be exposed to a number of situations involving only heat flow before showing them a situation where heat flow is involved in phase changes, because in the latter, the temperature of the object changing phase generally remains constant instead of increasing.
2. Name patterns of behavior (heating, cooling) first, and then move on to naming the physical mechanisms underlying them (heat flow, boiling).
3. Teach the compositional primitives of qualitative reasoning explicitly to give learners a richer vocabulary for expressing their partial but growing knowledge.

12.5 Discussion

The intuition that we just "run" our mental models is extremely strong. This seemingly effortless visualization of possibilities cannot be due to some kind of internal quantitative simulation, because we do not have the data, the models, or the computational capacity to carry out such simulations even for one behavior, let alone multiple possible behaviors (Davis & Marcus, 2016). Qualitative representations, because they abstract away details, provide more robust matching across a wide range of situations. Moreover, they can be constructed with little initial data and provide explicit representations of both causal mechanisms and ambiguity. Qualitative representations provide a lens through which we see the continuous world. By accumulating experience in terms of qualitative encodings of observed behavior (both from senses and from cultural transmission), analogical generalization constructs intermediate representations that can be applied to a wider range of new situations and systems and, in doing so, provide a combination of both concrete and abstract information, giving us both a "feel" for what will happen but also, when we have them, explanations that can be used for planning and diagnosis. The ability of the same analogical reasoning mechanisms to handle both within-domain and cross-domain analogies should provide a flexibility and smoothness to prediction and abduction that is more in accord with human behavior.

Analogy has also been explored by Ashok Goel as central in reasoning about engineering systems using functional representations in a variety of domains (Goel, 2013). His group's *structure-function-behavior* models focus on symbolic descriptions of specific behaviors of components, which can then be composed into descriptions of how systems work. Although human inspired, their focus to date has been on creating useful automated reasoning systems. But their emphasis on the use of design patterns in adapting designs, for instance, seems like a plausible way that analogy is used by engineers, and so investigating that connection could prove useful.

12.6 Summary

Qualitative reasoning should be delimited by the kinds of representations used, not by the particular reasoning techniques. This processing account of qualitative reasoning is very different from that adopted by most of the qualitative reasoning research community. As noted earlier, the hypothesis of the sufficiency of first-principles qualitative reasoning was a valuable one to make, because it forced the development of representations that are capable of supporting the reasoning involved in a wide range of human expertise. But as we move into trying to understand how people actually do such reasoning, the processing account needs to be compatible with what we know about human cognition more broadly. I believe that in the long run, this will lead to even more robust accounts of the expert reasoning of scientists and engineers because they, too, continue to have and use these multiple levels of models in their mental warehouses. Moreover, it will enable us to create accounts of qualitative representations and reasoning that make stronger ties to research in vision, robotics, and other areas where interacting with the physical world directly becomes important.

13 Dynamics in Language

If qualitative representations are our default level of representation for continuous aspects of the world, then we should expect that language—which, after all, is a symbolic system for communicating about the world—would have important connections with qualitative representations. In this chapter, I argue that qualitative representations do in fact play an important role in the semantics of natural language by examining how the qualitative representations of dynamics introduced in previous chapters manifest in language. This account comes out of Sven Kuehne's (2004) PhD thesis and subsequent work with C. J. McFate (e.g., McFate, Forbus, & Hinrichs, 2014). Although it is not yet complete, it illustrates the explanatory power of qualitative representations when understanding what is happening when one is writing, talking, or reading about change and its causes in continuous systems.

I start by describing the motivation in more detail. Then I discuss how we bridge from the representation conventions you have seen so far to the even more incremental representations commonly used in natural-language semantics research. Then I examine, construct by construct, an account of how the ideas in QP theory manifest themselves in English. Two forms of evidence are provided: a corpus analysis and descriptions produced by an implemented learning by reading system. Finally, I examine how this account compares with other accounts in the cognitive science literature.

13.1 Motivation

There are three reasons to view QP theory as a candidate conceptual component in natural-language semantics. First, the notions of continuous process and qualitative causal relationships it defines can be used to draw psychologically plausible conclusions (chapters 7, 8, and 12). Because descriptions of continuous processes are abundant in language concerning physical phenomena (more on this below) and are routinely used in metaphors (e.g., Gentner, Bowdle, Wolff, & Boronat, 2001; Lakoff & Johnson, 1980), this means that qualitative reasoning can be used to help derive expectations and entailments of information gleaned from language. Second, the causal account QP theory provides is consistent with human causal explanations in most continuous domains (Forbus & Gentner, 1986a, 1986b, 1997; also chapter 9, this volume). Thus, it is not just the conclusions that seem reasonable but also the structure of the explanations for them that seem compatible with human reasoning. Third, the abstract level of information that qualitative representations support seems a natural fit for the level of specificity commonly found in natural-language descriptions of continuous principles and situations. One does not need to understand differential equations or carry out detailed simulations to understand physical metaphors (e.g., "her anger increased until she exploded").

13.2 Recasting Qualitative Representations as Linguistic Frames

Language builds information incrementally, which means that the representations often posited in natural-language research are somewhat different from what we have been using up until now. The representations we have used so far are based on positional notation. That is, statements consist of predicates with arguments, and their arguments are indicated by position within the structure of the assertion. This is very convenient for reasoning systems, but it is not so convenient for representing linguistic phenomena. The reason is that any particular piece of language may provide only a partial representation of a complex event or relationship. The pieces must be assembled from information decoded piecemeal from multiple phrases and sentences. This is one of the underlying motivations for the use of frame systems in conceptual and linguistic representations (Fillmore, 1976; Fillmore & Atkins, 1994; Minsky, 1974). Thinking in terms of assertions, the strategy is to reify events and relationships, using role relations to connect their pieces to the thing itself (what is also known as a *neo-Davidsonian* representation [Parsons, 1990]). The idea of frames (or schemas, as per chapter 2) also incorporates a bundling of facts to be considered together, making them available for rapid reasoning.

The largest purely linguistically motivated compendium of frame semantics is FrameNet (Fillmore et al., 2001), so we use conventions from it below. In frame semantics, meaning is expressed in terms of systems of structured representations, *frames*, whose parts (called *frame elements*, abbreviated FE, which can be thought of as roles in a schema) are bound to parts of a text and have associated with them inferences that provide meaning. The packaging of physical knowledge and principles in QP theory (inspired in part by Minsky's [1974] notion of frames) suggests a natural alignment with frame semantics. There is a basic *continuous process* frame, whose structure provides the fundamental aspects of continuous processes. Subframes describe particular categories of physical processes, with differences in their participants and consequences being the differentia that set them apart. Instances of these frames are combined with frames from other aspects of the semantics to create the frame system describing the meaning of a text. The qualitative causal mathematics of QP theory is expressed through another collection of frames. In addition to their role in continuous process descriptions, these qualitative causal frames can be used for other domains with continuous parameters, such as economics or metaphorical extensions of physical concepts. To provide a broad vocabulary of everyday concepts and relationships, including role relations, in our systems, we use a mapping of FrameNet to the OpenCyc ontology. However, the discussion below does not require any knowledge of that ontology.

13.3 How QP Theory Manifests in English

This section outlines a set of frames and frame elements that suffice to encode much of the structure of QP theory. We start with quantities and build from there. In addition to defining the frame elements, we discuss how they are manifested in language. This is a work in progress. The set of frames is, at this writing, incomplete, because state transitions are still being incorporated and the mapping is being extended to type-level qualitative representations (chapter 21). Moreover, the catalog of constructions that map frame elements to language is even less complete. Nevertheless, the formalism has already reached the point where it has been productively used in computational experiments.

13.3.1 Quantities

The Quantity frame represents continuous properties. Here is an example sentence to introduce its elements: "The temperature of the brick is 35 degrees Celsius."

- Entity (required): Represents what this quantity belongs to. In the example above, this is "brick."
- Quantity Type (required) Specifies the type of parameter that this quantity is. In the example, this is "temperature."

- Value (optional): Specifies the numerical value of the parameter. In this example, it is "35."
- Units (optional): Specifies the physical units of the numerical value. In the example, this is "degrees Celsius."
- Ds (optional): Specifies the sign of the derivative of the parameter, as one of {`-1`,`0`,`1`}. In "The temperature of the brick is increasing," "increasing" indicates this element, with a value of `1`.

The first two constituents identify a quantity. We will refer to a QP frame that consists of just these as a *quantity reference*. This frame representation is underspecified, in the sense that the same structure is used to denote a statement about particular values (as in the example sentence) and as a fluent used in more indirect statements (e.g., ordinal relationships that hold across whatever values may hold for a pair of quantities over the interval for which the fluent is true). There is also ambiguity as to whether or not the entity is a specific object or a generic. The same frame would be used for "The temperature of boiling water is 100 degrees Celsius." Making these distinctions is deliberately left to later stages of semantic interpretation, which can use context to make more informed choices.

The choice of a unique filler for the entity means that new conceptual entities have to be introduced to discuss quantities that are compared with some standard. Examples include pressure and temperature differences, which in traditional mathematical notation are typically denoted as binary functions, with the arguments being the individuals involved. For example, pressures in pneumatic systems are often measured as "gauge pressure" (i.e., with respect to the background atmospheric pressure).

Quantity types can be specialized via compound noun phrases; for example, "radiation heat" refers to the flow rate of heat transferred by radiation (i.e., not via conduction or via convection). Handling the range of compound noun phrases that are used in English is still a difficult and open problem, and this affects the extraction of qualitative knowledge as well (e.g., "the water pressure in the reactor's inner chamber").

Quantities manifest in English in a variety of ways. When both are referred to explicitly, the entity is typically connected to the quantity type via a prepositional phrase (e.g., "temperature of the brick," "speed of the ball"). But either can be implicit, with the appropriate context:

"The water entered the system. Its temperature rose."

"The amount of water in the holding tank rose. In the flash chamber, it dropped."

Verbs can introduce quantities indirectly. For example, the sentence

"The increased temperature lengthens the bimetallic strip"

introduces the length of the bimetallic strip. Adjectives can also do this. For example, the sentence

"Iron is denser than wood"

introduces density for both iron and wood.

The examples so far all explicitly mention a quantity, or at least words that can be directly connected to particular types of quantities. However, quantities implied by statements also need to be introduced to better understand them. For example,

"As the temperature rises, the liquid expands."

Expansion implies an increase in physical extent. Depending on context, we may be able to further infer particular changes in quantities. If the liquid is in a cup, then the appropriate quantity to use is probably volume. If the liquid is in a thermometer, then height may be what the writer is attempting to communicate. If the liquid is pressed between two panes of glass, then area is more likely to be appropriate. Thus, the correct choice of implicit quantities can rely on both context and background knowledge.

Some adjectives encode quantity type as well as information about value. For example, the sentence

"The heavy hot brick"

tells us something about the mass and temperature of the brick. Adverbs provide quantity type information, too, but they can be more polysemous:

"The gas molecules are moving faster than the liquid molecules."

"The mercury is expanding faster than the water."

In the first sentence, "faster" suggests that the introduction of an ordinal comparison between the velocities is warranted. In the second sentence, it is the rate of expansion (i.e., the rate of the expansion process instances associated with each piece of liquid) that is larger for one than the other.

Possession provides another means of introducing quantities. For example,

"The brick has mass."

"The city's food production…."

In these cases, the entity is the possessor and a quantity of the given type is what is possessed. Rate phrases that modify verbs can also introduce quantities. For example, in a strategy game manual, one might see

"A citizen consumes food at a rate of 2 units per turn."

Numerical values and units can be expressed in straightforward ways. For example,

"<quantity reference>" is "<numerical value><units>."

They are also commonly mentioned in passing in subordinate clauses. For example,

"The steam, now at 850°F, ..."

Quantity-specific verbs (e.g., "costs," "weighs") are also commonly used to connect the quantity reference to its value. Symbolic values are also common. For example,

"The brick is hot."

As per our discussion of symbolic values in chapter 5 (and revisited in chapter 18), such terms are associated with a scale (often implicit) that can be used to estimate ranges of plausible values. For example, something described as "hot" in a discussion of cryogenics typically means a temperature that is much lower than something described as "cold" in a discussion of stellar dynamics.

Limit points are also expressed as quantity references (e.g., "the boiling point of water," "the maximum excursion of the lever"). One common set of patterns uses range indicators as modifiers of a quantity (e.g., "maximum," "minimum," "average," "resting"). Another common pattern is to combine "point" or "threshold" with the phenomenon or condition that the limit point is about, e.g. "boiling point," "melting point," "detection threshold." As with other quantity references, limit points can be used as fluents in ordinals or in the expression of values. Limit hypotheses are expressed by statements involving temporal conditions over limit points (e.g., "When the temperature reaches the boiling point ...").

Signs of derivatives are signaled in several distinct ways. The simplest is to use generic terms such as "increasing," "constant," or "decreasing" in one of the ways used to express values. For example,

"The temperature of the water is increasing."

For some quantities, "rising"/"steady"/"constant" also works, but not for all (e.g., "rising length" is inappropriate).

Noun forms tend to be used for differences, especially when additional information needs to be stated about the change. For example,

"The increase in price is significant."

"The drop in pressure caused him to become unconscious."

Some verbs (and adjectives) provide information about quantity types along with signs of derivatives. For example,

"The oven is cooling."

"The cooling oven is still hot enough to bake the bread."

13.3.2 Ordinal Relationships

The Ordinal frame represents ordinal relationships.[1] The frame elements are as follows:

- Q1 (required): One term of the comparison, a quantity reference.
- Q2 (required): The other term of the comparison, a quantity reference.
- Reln: The relationship between Q1 and Q2. This is one of $<$, $>$, $=$, $\geq$, $\leq$, $\neq$, `sameOrder`, or `negligible`.

The intended meaning of most of the values for Reln is obvious. `sameOrder` and `negligible` involve qualitative order-of-magnitude comparisons (chapter 5). `sameOrder` indicates that Q1 and Q2 are within the same qualitative order of magnitude of each other and implies that when one is considered, the other cannot be ignored if they are part of the same situation or system under analysis. On the other hand, `negligible` indicates that Q1 is at a different and smaller qualitative order of magnitude than Q2 and hence can be ignored. For example,

"The heat loss from the handle is negligible compared to the heat supplied by the stove."

Just as QP theory proposes ordinal relationships as the mainstay of qualitative representations for values, English provides a variety of comparative terms that encode quantity type and therefore provide more compact statements. For example,

"The oven is warmer/hotter/cooler than the stove."

"The brick is heavier/taller/wider than the egg."

Some require context (e.g., "bigger than" could be length, area, or volume, depending on context). One can also say, for some quantity types,

"The gold costs/weighs more than the concrete."

but such constructions are not available for all quantities (e.g., area or volume), and this construction seems especially inappropriate for intensive quantities. Finally, on the more general-purpose end, there are the general comparatives. For example,

"<quantity reference1> is greater than/less than/equal to <quantity reference2>,"

where the quantity types of the two references are the same.

Adjectives are often used to introduce implicit comparisons. For example,

"The cool dough is placed inside the hot oven."

tells us that, whatever the temperature of the dough is, it is lower than the temperature of the oven.

13.3.3 Influences

Recall that there are two types of influences, corresponding to the direct effects of a process and the other representing the indirect effects they have on the rest of the world. We can capture their commonalities by introducing a generic Influence frame, which will then be specialized to handle the two cases. The frame elements for the Influence frame are as follows:

- *Constrained* (required): Specifies the dependent quantity (i.e., the effect in this relationship).
- *Constrainer* (required): Specifies the quantity that acts as a cause in this relationship.
- *Sign* (optional): Specifies the direction of effect, which can be + or –.

The sign is optional because not knowing the direction of effect is a common intermediate state of knowledge when learning a domain or system. To be sure, predictions become extremely weak, but even change/no change predictions can be useful, and predicting change and then observing its sign is progress toward improving one's model.

As noted above, there are two specializations of the Influence frame. DirectInfluence frames are used to represent `I+`/`I–` statements and are often introduced via verbs (i.e., as part of processes). `Qprop` frames are used to represent `qprop+`/`qprop–` statements.

Influences manifest in language in a number of ways. Indirect influences are especially prolific in terms of the number of surface forms that entail them. Kuehne (2004) found seven distinct patterns, summarized in table 13.1. The first pattern, "The/The," corresponds to what Culicover and Jackendoff (1999) call the comparative-correlative construction.

Changes are either signs of derivatives (e.g., "increases") or comparatives (e.g., "larger," "more"), as described above. Signs are expressed by words such as "up," "down," "positively," "negatively," which map to either + or –. In patterns where there are two changes, if the changes are in the same

Table 13.1
Surface forms that suggest qualitative proportionalities.

The/The	"The larger the surface area is, the more convection heat is lost from the surface." THE <*Comp*$_1$><*Quantity*$_1$> [<*Change*$_1$>], THE <*Comp*$_2$><*Quantity*$_2$> [<*Change*$_2$>]
As	"As the volume increases, the density decreases." AS <*Quantity*$_1$><*Change*$_1$>, <*Quantity*$_2$><*Change*$_2$>
When	"The liquid in a thermometer expands when it is heated." <*Quantity*$_1$><*Change*$_1$> WHEN <*Quantity*$_2$><*Change*$_2$>
Depends	"The amount of heat depends on the amount of motion." <*Quantity*$_1$> DEPENDS ON <*Quantity*$_2$>
Affects	"The area of the flow path affects volume flow." <*Quantity*$_1$> [<*Sign*>] AFFECTS <*Quantity*$_2$> <*Quantity*$_1$> AFFECTS <*Quantity*$_2$> [<*Sign*>]
Influences	"The supply of money positively influences inflation rate." <*Quantity*$_1$> [<*Sign*>] INFLUENCES <*Quantity*$_2$> <*Quantity*$_1$> INFLUENCES <*Quantity*$_2$> [<*Sign*>]
Causes	"Heat gain causes air temperature to rise." <*Change*$_1$><*Quantity*$_1$> CAUSES <*Change*$_2$><*Quantity*$_2$>

direction, then the sign is +. If the changes are in the opposite direction, then the sign is –.

Note that four of the patterns explicitly mention directions of change for both quantities. This forces an interpretation of their relationship as being functional, and hence they are indirect influences. On the other hand, three of the patterns (Depends, Affects, and Influences) do not. These patterns are also used in text to indicate the existence of some underspecified causal relationship between the quantities they connect. It might indeed be a single indirect influence, but it might be an entire chain of both types of influences. Thus, the process of interpretation must be done with some care and seems to involve considerable domain knowledge.

In QP theory, direct influences are always parts of processes. In English, there is a similar close connection in how they are manifested. They are always connected to some change (typically indicated by a verb), which can be viewed as a partial specification of a process. For example, in

"Heat flows from the hot brick to the cool ground."

there are two direct influences, a positive one on the heat of the ground and a negative one on the heat of the brick, as indicated by the prepositional phrases indicating the source and destination for the flow. The general pattern that covers this sentence is

"<*Qtype*><*Change*> [<*from*><*Entity*$_1$>] [<*to*><*Entity*$_2$>] [<*via*><*Path*>],"

where <*Change*> is a verb indicating a process like "flows," "moves," "leaks." Elements in [] are optional. <*from*> can be "from," but other phrases are commonly used as well (e.g., "out of," "away from," with similar variations for <*to*>). <*via*> can be "via" but also "along" or even "in," as in "in the trough." The constraints that hold between these patterns for English still need further explication. Kuehne (2004) has identified additional patterns for direct influences that can be viewed as variants of this pattern, involving optional agent arguments and multiple changes. For example, in "Within a closed container, the heat lost by one substance must equal the heat gained by another substance," the first substance is treated as an agent, with two direct influences to represent the gain and loss of heat.

13.3.4 Model Fragments and Processes

The frame ModelFragment has the following frame elements:

- *Type* (required): Indicates the type of model fragment (e.g., `Contained-Liquid`).
- *Participants* (optional): Each value is an individual that participates in the model fragment.
- *Status* (optional): Either active or inactive.
- *Conditions* (optional): One or more frames or statements whose conjunction determines whether or not the model fragment is active.
- *Consequences* (optional): Each value is either a QP frame whose entities are drawn from the participants of this frame or a statement involving the participants of this frame.

Most of these are optional because any particular statement will contain only one or two frame elements. The frame ContinuousProcess is a specialization of ModelFragment. As per QP theory, only ContinuousProcess frames are allowed to have direct influence frames among their consequences.

Processes manifest in English as particular classes of verbs. For example, the verbs of fluidic motion identified in FrameNet[2] include the following: bubble, cascade, course, dribble, drip, flow, gush, hiss, jet, leak, ooze, percolate, purl, run, rush, seep, soak, spew, spill, splash, spout, spurt, squirt, stream, and trickle. Although all of these make sense for liquids, only a subset makes sense for gases: gases do not dribble, for instance. Heat, despite the rejection of the caloric theory, continues to be treated metaphorically in English as something that flows like a fluid. "Flow" is commonly used, as is "leak," but "bubble," "hiss," and "splash" clearly don't work for heat. FrameNet treats flow as a specialization of motion, which makes sense, given the overlap in source, destination (called Goal in FrameNet), and path. Similarly, FrameNet includes the following verbs in its expansion frame: contract.v, contraction.n, dilate.v, enlarge.v, enlargement.n, expand.v, expansion.n, explosive.a, grow.v, inflate.v, lengthen.v, shrink.v, stretch.v, swell.v

Participants are connected to model fragments via prepositional phrases. For example,

- "from" *<p>* indicates the participant that is the source.
- "to" *<p>* indicates the participant that is the destination.
- "of"/"in" *<p>* indicates the entity that the process is acting upon.
- "along"/"in"/"through"/"around" *<p>* indicates the entity that is a path for some variety of motion.

Notice that although QP theory puts no limits in principle on the number of participants, the small number of available prepositions puts limits on the number that can easily be communicated via language. This may be a constraint on the types of model fragments that are easily learnable, given how important cultural transmission is for constructing this kind of knowledge.

13.4 Evidence

The set of patterns presented here should not be regarded as complete. They were developed through the analysis of a corpus of chapters from science books, both textbooks and those intended for the general public. Nevertheless, they are encouraging, in that they show that QP theory does seem to capture some aspects of natural-language semantics. Four questions naturally arise at this point:

1. How common are such patterns in texts where one would expect to find them (e.g., science books)?
2. Does this account of semantics for continuous aspects of the world fit smoothly with accounts of semantics for other aspects of language?
3. Can qualitative models in fact be extracted from such texts as part of processes assumed for natural-language understanding?
4. Are the inferences produced by qualitative reasoning useful in deriving the entailments of text?

The fourth question is in some ways the easiest: many of the problems described in the literature, and in this book so far, are initially posed in natural language. Descriptions of everyday situations have been used to show that qualitative reasoning can produce answers compatible with what people produce for such questions. Therefore, although more detailed comparisons between human reasoning and particular qualitative reasoning methods will still lead to new insights (e.g., chapters 12, 17, and 18), we will not consider the fourth question further here.

Next let's look at the evidence so far concerning the first three questions.

13.4.1 Corpus Analysis

The natural way to explore how common such patterns are is to conduct corpus analyses. Sven Kuehne and I (Kuehne & Forbus, 2002) examined four chapters of a book on solar energy, *Sun Up to Sun Down* (Buckley, 1979). We chose this book because it is very clearly written, and it uses both diagrams and analogies heavily. We chose chapters 2 through 5 because they provide a basic exposition of heat, temperature, and types of heat flow. Two evaluators familiar with the theory independently scored each sentence. Then they compared their results, discussing divergences until they came to agreement.

We looked at the linguistic realizations of physical processes in the text. Based on the QP frame semantics, we defined nine types of information about processes: the process name (P), information about subclasses of a process (i.e., a specialization) (SC), and participants (PA); about conditions: antecedent activations (AA), antecedent ordinal relations (AO), and antecedent relations (AR); and finally about consequences: indirect influences (CII), direct influences (CDI), and consequence relations other than influences (CR). Multiple pieces of information can appear within a single sentence, so we scored the number of phrases of particular types in addition to the number of sentences in which they occurred. Sentences can contain multiple types of information, so the same sentence can appear in multiple categories. We also distinguished between information from examples (identified through a preliminary analysis) and general information, because we have hypothesized (chapter 12) that commonsense physics arises from within-domain analogies involving concrete descriptions. Tables 13.2 (general information) and 13.3 (example-specific information) show our results.

Notice that the example-specific data contain more than twice the number of processes, about five times the number of participants, and a lot more information about the consequences of the mentioned processes. However, the amount of information about the conditions of a process (categories

Table 13.2
General statements using QP theory concepts.

Type	P	SC	PA	AA	AO	AR	CR	CDI	CII
#Sentences	10	1	8	15	5	1	9	8	15
#Phrases	11	4	14	16	5	1	18	16	18

Table 13.3
Use of QP theory concepts in examples.

Type	P	SC	PA	AA	AO	AR	CR	CDI	CII
#Sentences	26	0	28	15	6	5	26	19	14
#Phrases	26	0	74	15	7	5	53	38	17

AA, AO, and AR) is nearly the same. As expected, any information about specialization of processes (SC) is found only in the general information.

What kind of coverage does QP theory provide? Of the 216 sentences, 94 mention at least one element from the QP frame system proposed here. That means that QP theory can account for roughly 43 percent of these chapters. This is reassuring, given that these chapters contain precisely the sort of material that QP theory was designed to handle. The material that is not concerned with QP theory grounds these concepts in everyday experience (e.g., ovens, houses, insulation). One would expect a smaller fraction of QP-relevant material in other types of text.

13.4.2 Compatibility with Other Aspects of Semantics

Although many aspects of the world are continuous, many aspects are not. Thus, QP theory at best covers a subset of the conceptual structures needed in a full account of natural-language semantics. Is it broadly accountable with linguistic accounts of other areas of semantics? For evidence on this question, we look at how the analysis of physical processes in QP theory compares with those in FrameNet. Although there is not universal agreement about FrameNet's accuracy as a model of lexical semantics, and it does not by default tie into a deeper ontology intended to support reasoning, it is nevertheless by far the largest and most carefully motivated database of lexical semantics currently in existence. This is why we chose to use its conventions in the first place. Nevertheless, because a number of the QP models for basic physical processes were developed before FrameNet existed, it is interesting to see if linguists ended up making compatible distinctions when considering overlapping phenomena from their perspective.

Let us begin with FrameNet's analysis of motion. Their motion frame has frame elements of Theme, Source, Goal, and Path, all of which can be viewed as specializations of the participant frame element of the PhysicalProcess frame type. In QP theory models of motion, there is a quantity Position that is defined with respect to the Path from Source to Goal. A DirectInfluence frame, with its Constrained FE being Position and its Constrainer FE being Velocity, is one of the values for the Consequence FE of the Motion frame. An Ordinal frame with Q1 = Velocity, Q2 = zero, and OrdReln = ≠ is the Condition for the Motion frame. Thus, the two analyses dovetail nicely. The FrameNet analysis provides additional configural information that is implicit in the QP Frame description, whereas the QP Frame description provides information about the mechanism that is implicit in the FrameNet analysis. FrameNet treats flows of all sorts as specializations of motion, so a similar mapping can be created between the participants for liquid flow, gas flow, and heat flow and the FrameNet Fluidic_Motion frame. For a more detailed analysis, including how FrameNet's valence patterns can be used, see McFate and Forbus (2016).

13.4.3 Natural-Language Understanding Examples

Can QP frames be extracted automatically from text via methods normally employed in natural-language understanding? Again, there is currently no universal agreement on how language understanding works cognitively, so the best we can do is show that at least one of the existing accounts can do the job. The Explanation Agent Natural Language Understanding system (EA NLU) was originally developed to answer this question (Kuehne & Forbus, 2004) but has since been extended to handle stimuli found in computational social science (Tomai & Forbus, 2009), as discussed in chapter 18. EA NLU uses a traditional pipeline model, with syntactic analysis being performed by Allen's (1994) parser and a general-purpose semantic analysis system based on discourse representation theory (Kamp & Reyle, 1993). Knowledge resources used included the COMLEX lexicon (Macleod, Grishman, & Meyers, 1998) and ResearchCyc knowledge base contents, augmented with representations for qualitative reasoning, including QP frames. That version of the system extracted model fragment instances from simplified English texts. For example,

"Heat flows from the hot brick to the cool ground"

yielded the analysis shown in figure 13.1.

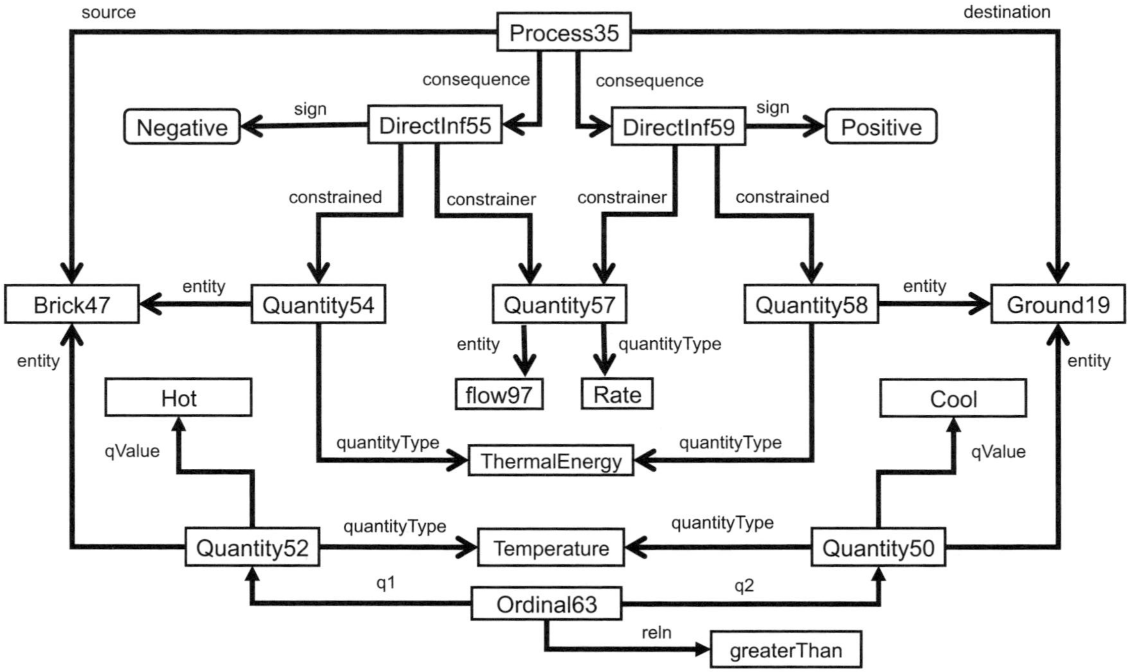

Figure 13.1
QP frames constructed for the sentence "Heat flows from the hot brick to the cool ground" (simplified to only show the most important relationships).

The reason for using simplified English was to factor out complex syntax, in order to better focus on semantic analysis. For example, for every QP construct, at least one of the patterns by which it manifests in language was implemented, but not every pattern available in English.

This language system has been used to enable Companions to learn via reading. For example, McFate et al. (2014) show how qualitative knowledge can be learned from science books and from strategy game manuals.

Extracting QP information at scale from open texts remains an interesting research problem. In addition to extending the range of syntactic coverage, interesting issues of semantic interpretation need to be addressed, many of which are shared with other areas of linguistics. For example, identifying generics (e.g., "Heat flows from something hot to something cold.") is needed to extract intended general knowledge from text, rather than just example-specific explanations. How should the inferences that QP theory sanctions be used in semantic interpretation, both locally and over broader scales of understanding texts? By combining forces, there is an opportunity to build a rich inferential semantics for the continuous world that could be an important contribution to models of human conceptual structure.

13.5 Other Accounts

Talmy (2000) proposed that an abstract, metaphorical notion of force forms the basis for the semantics of causal reasoning. Building on Talmy's ideas, Wolff (2007) has further argued that these ideas can be extended to social causation as well as physical changes. Although the metaphor of force and direction is certainly interesting, these accounts have never been formalized in the way that qualitative representations have. Moreover, force dynamics accounts do not capture the subtleties concerning continuous causation that qualitative representations handle (e.g., feedback systems, immediate vs. delayed effects, and the mechanisms that underlie change). Nevertheless, it would be interesting to explore how these two accounts could be integrated.

III Space

Space permeates our thinking. Every organism must navigate the physical environment to find what it needs. Many organisms reorganize their environment, as illustrated by ant colonies, beaver dams, and cities. Understanding how to make and use tools, as crows and primates do, requires understanding interrelationships between space, shape, and forces. Understanding science, engineering, and mathematics deeply involves spatial thinking. Thus, understanding the representations that underlie human thinking about space is one of the most important problems in understanding human cognition more broadly.

In this section, I argue that qualitative representations are crucial in human spatial cognition. As with other domains, people often use multiple representations to reason about space. From vision, hearing, and touch, we get quantitative information about particular entities and events in space. These quantitative inputs are, I argue, used to construct qualitative representations of space and shape dynamically. Although there are some default choices for qualitative representations, often the specific representations constructed depend on the kind of task being performed. Understanding what the appropriate qualitative distinctions are for a task constitutes part of the expertise needed for that task.

Chapter 14 provides a theoretical framework by combining two lines of research from different areas of cognitive science. In artificial intelligence, work on spatial reasoning has explored the interaction of qualitative and quantitative representations in a variety of tasks, ranging from navigation to expert reasoning in science and engineering. Specifically, this chapter describes the *metric diagram/place vocabulary* model of qualitative spatial reasoning. Metric diagrams are a functional model of the quantitative aspects of human visual processing and are responsible for two things: (1) handling processing of quantitative, coordinate representations and (2) constructing qualitative representations to support reasoning. Place vocabularies are the qualitative decompositions of space for particular problems. This model has been used to create systems that can do humanlike reasoning about simple motion problems, expert-level reasoning about mechanical systems, and reasoning about spatial dynamic systems (e.g., weather patterns). In cognitive psychology, a complementary line of research has explored categorical and coordinate models of spatial cognition, focusing more on particular spatial skills and memory effects. I will argue that these lines of investigation have come up with essentially the same idea. That is, qualitative representations of space and shape are essentially spatial categories. The combined evidence from these lines of research makes a very strong case for this framework.

Decomposing space into regions is only one aspect of qualitative spatial reasoning. Researchers in qualitative spatial reasoning have also identified a rich set of vocabularies of relationships that capture relevant aspects of topology, orientation, and relative position. These *qualitative spatial calculi* provide additional candidate hypotheses for the vocabulary of human qualitative representations. Chapter 15 summarizes some of the more promising calculi, outlining how they have been used in tasks and supporting psychological evidence, where available.

I return to the integration of qualitative and quantitative representations and reasoning to look at sketch understanding in chapter 16. Drawings have long been used to explore aspects of human spatial cognition. I describe CogSketch, a sketch understanding system that embodies a model of human visual and spatial representations. The explanatory power of this model is illustrated through multiple computational experiments. Because qualitative representations are a form of symbol system, one might expect, as with dynamics, that they interact heavily with language, another powerful human symbol system. I examine an experiment suggesting how spatial prepositions might be learned via analogical generalization. To show that these representations can model human visual problem solving, I outline simulations of three visual problem-solving tasks (i.e., geometric analogies, Raven's Progressive Matrices, and an oddity task). These simulations all automatically construct their representations from the kinds of PowerPoint figures used for psychological experiments and function at human levels of performance. They use the same representations and reasoning engine, with only high-level *spatial routines* differing between tasks. They predict ordinal differences in human reaction times, and ablation studies on them yield new insights into aspects of spatial cognition.

Taken together, I believe the evidence in this section provides a strong case for qualitative representations being a bridge between visual perception and cognition.

14 Qualitative Spatial Reasoning: A Theoretical Framework

Our knowledge of space starts in the raw signals of sensation, is organized through perceptual processing, and ends up being used in conceptual reasoning, both overtly and covertly. Sensation, especially vision, involves massive amounts of signal processing and attentional processes that constantly shift and select small subsets of this wealth of information for further processing. Conceptual reasoning, on the other hand, seems to operate with a much smaller number of elements. The summarization of the results of perception into elements that are amenable to conceptual reasoning requires a bridge, a means of discretizing continuous space into meaningful units. In other words, my hypothesis is that qualitative representations of space and shape provide a bridge between perceptual and conceptual processing.

We start with a simple but concrete example to bring the issues into focus. Then we argue on functional grounds that the *metric diagram/place vocabulary* model (Forbus, 1983) is essential for complex spatial reasoning, via the *poverty conjecture* (Forbus, Nielsen, & Faltings, 1991). Additional work based on this model is outlined to provide evidence that it is capable of explaining human-level reasoning in a variety of tasks. We then shift to looking at evidence from work in cognitive psychology and neuroscience, which has converged on similar ideas (specifically, the *categorical/coordinate* distinction, which is, I argue below, simply another way of saying qualitative/quantitative), but while exploring a different set of issues and tasks. Finally, we summarize by pulling together these disparate lines of research into a unified account.

14.1 Reasoning about Motion through Space

What are qualitative spatial representations like? Let us consider a simple example, shown in figure 14.1.

We have two balls that are constrained by surfaces below, moving in directions indicated by the arrows. Let us say that if the balls leave the dashed lines

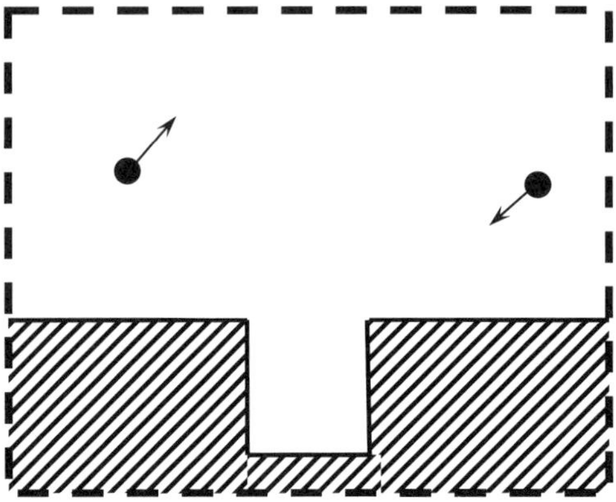

Figure 14.1
Can these two balls collide?

indicating our area of interest, they never return. Can these two balls ever collide? Our intuitions tell us that they might, but of course we cannot be sure unless we watch (or simulate) for a while. But some variations on this problem suggest that we are capable of doing more subtle reasoning than simple simulation. Consider the four scenarios illustrated in figure 14.2.

In figure 14.2a, we can see that the two balls will not collide, because they both leave our region of interest and hence, by assumption, never return. In figure 14.2b, if the balls are bouncing perfectly straight up and down, we know that they cannot collide, because they can never be in the same place and hence never in the same place at the same time, which is what it means to collide. Figure 14.2c carries this spatial intersection idea further: if we assume that A is trapped inside the well, and B never enters it, then we can again infer that they cannot collide. Finally, in figure 14.2d, a collision is certainly possible unless the two balls don't bounce (e.g., they are eggs instead of balls).[1] In that case, given their headings, they will stop moving once they first collide with a surface.

What representations and reasoning suffice to come to these conclusions? In the FROB model (Forbus, 1980, 1983), space and surfaces were described qualitatively by automatically carving them into regions and edges with identical functional properties with respect to this task. Such a description is called a *place vocabulary*. Figure 14.3 illustrates the place vocabulary for this particular problem.

There are four kinds of elements:

1. *Free-space regions* are places where a ball is free to fly or fall.
2. *Surfaces* are the edges of an object where a ball might collide.

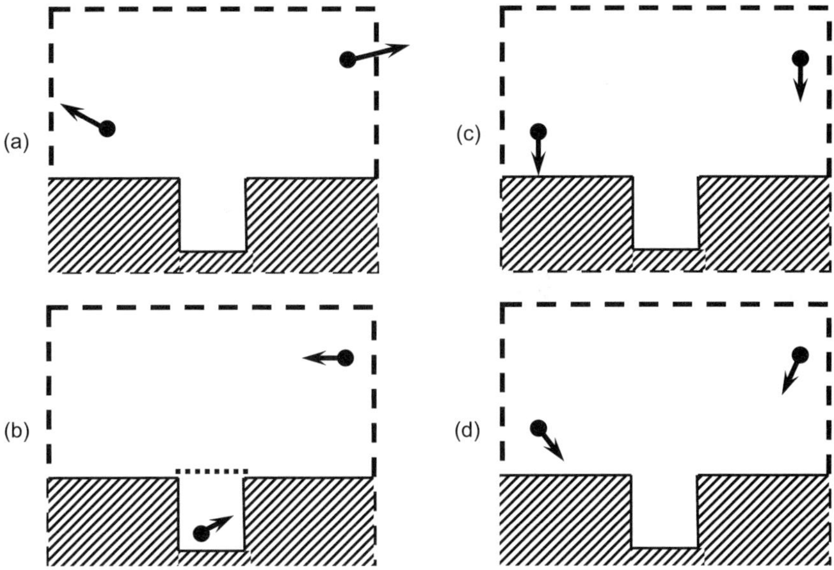

Figure 14.2
Four scenarios in the FROB's bouncing ball world.

3. *Transition edges* are the boundaries between free space regions.
4. *Exit edges* are the boundaries between our region of interest and the broader world.

Why this particular set of regions and edges? Why not just use one region for free space and single multisegment edges for the surfaces and exits? Because that representation is too ambiguous to support the kind of reasoning that we did concerning the scenarios in figure 14.2. For instance, in figure 14.2, the well is implicit in the diagram. This qualitative representation makes its existence explicit, thereby enabling reasoning about the possibility that a ball might get stuck there. Furthermore, by splitting surfaces at corners, we respect both a perceptual distinction and the fact that a ball hitting on either side of the corner could end up going in quite different directions. Because gravity acts vertically, dividing free space into regions above surfaces makes sense (consider again figure 14.2c). Because balls can lose energy, identifying wells by splitting free space with horizontal edges projected from corners provides another useful set of distinctions (e.g., figure 14.2b).

There are some important properties to note about this representation. First, it is grounded in the diagram. In many spatial reasoning situations, some quantitative representation, which we take to be perceptual, can support

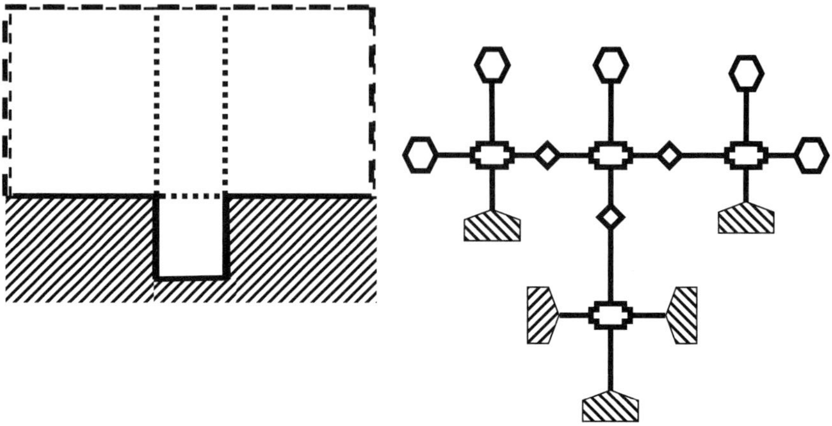

Figure 14.3
Place vocabulary (right) computed from the metric diagram (left), assuming gravity and ignoring shapes and sizes of moving objects. Shaded places are surfaces, notched rectangles are free space regions, diamonds are borders between free space regions, and hexagons are exits outside the region of interest. Arcs represent connectivity and are annotated with qualitative directions (e.g., up, down, left, right), not shown in this illustration.

quantitative calculations necessary to compute place vocabularies. We call this general class of quantitative representations *metric diagrams*. We consider an important part of the purpose of human perception to be providing such representations for grounding qualitative spatial reasoning. Second, the elements are nonoverlapping, so that each position of a ball maps to a unique element of the place vocabulary. Third, it encodes direction, in that relationships can be defined about which place will be encountered next if a ball is moving in a particular direction. These three properties enable the model to map from a given scenario into a qualitative state for each ball. The spatial aspect of state is simply which element of the place vocabulary the ball is in.[2] Direction of motion can be quantized by signs of direction: (0,0) is not moving, (1,0) is moving horizontally to the left, (1,–1) is moving horizontally to the left and down, and so on. The kind of activity a ball is doing can be captured by the interaction of its direction of motion and its place, as shown in table 14.1. The aftermath of a COLLIDE state is REST if the ball is perfectly inelastic (e.g., it's really an egg) or FLY otherwise, in a direction determined by a simple table that converts the incoming qualitative description of velocity to a description of the outgoing velocity. (This simple world does not include the possibility that

Table 14.1
Mapping from type of place to kind of activity possible for a ball in that place.

Free space region	FLY
Exit	LEAVE
Transition	TRANSIT
Surface	COLLIDE, REST, or FLY, depending on the relative directions of the surface normal and the direction of motion

balls spin, which would add additional entries to the table for determining post-collision directions.)

A set of qualitative simulation rules can be defined that, given a qualitative state, derive all of the possible next states for a ball. As with qualitative dynamics, such rules predict multiple possible outcomes. For instance, a ball flying upward to the left in the rightmost free space region could leave the diagram still going upward, transition to the region on the left, or stop rising, depending on where it actually is within that region and what its velocity actually is.

These qualitative simulation rules can be run exhaustively on a scenario for each ball. Notice that, because there are only a few possible states for each place and a fixed set of places per diagram, the size of each simulation is polynomial in the number of places. (Contrast this with envisioning for qualitative dynamics, where the number of states is worst-case exponential in the number of quantities in the situation!) Given additional assumptions about a scenario (e.g., that a ball cannot enter a well), we can prune states that violate that assumption, plus all of the states that are no longer reachable given what has been directly ruled out. In addition to reachability, FROB also used two other constraints on motion to prune possibilities. One is that, unless a ball is perfectly elastic, it will either stop or leave the diagram. The other is that a ball bouncing in transverse on a horizontally oriented surface (or between two vertically oriented surfaces) can only stop within that place, leave going in that direction, or change transverse direction. Otherwise, FROB falls prey to the original form of Zeno's paradox.

This, combined with the observation that ruling out being in the same place permits ruling out the possibility of collision, provides a mechanism for the FROB model to automatically derive the conclusions about the scenarios in figure 14.2. Figure 14.4 shows the place vocabularies for each scenario after such pruning to illustrate.

If quantitative information is available, it can often be projected into the qualitative representation to provide further constraints. Suppose, for

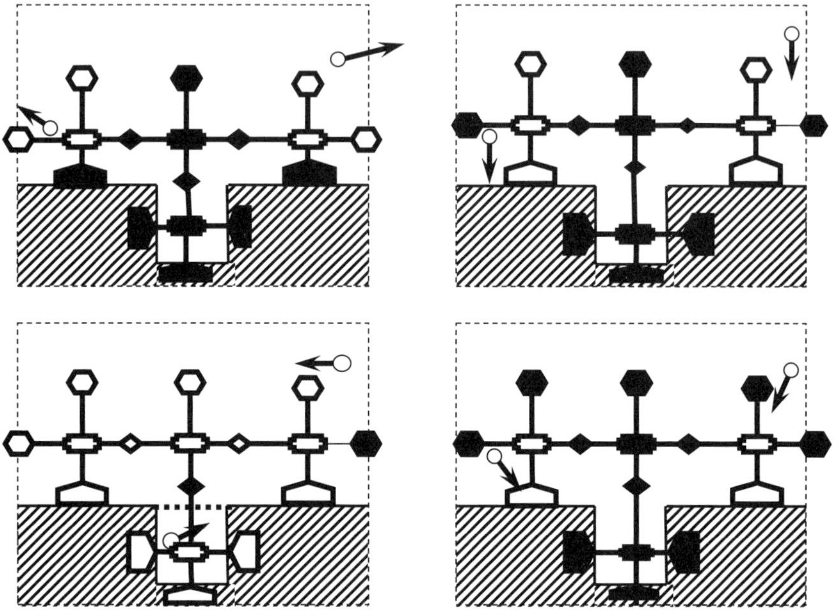

Figure 14.4
Qualitative constraints suffice to rule out possible collisions in each of the scenarios from figure 14.2. Blacked-out places are excluded by the implications of the constraints of that particular scenario. The lack of overlapping places between the two balls is sufficient to rule out collisions.

example, that we knew the height and velocity of the ball, so that we could calculate its total energy. If all of that energy were converted to potential energy, that would reflect the maximum height that the ball could reach. Such a line can be encoded in the metric diagram and its implications propagated through the place vocabulary—in figure 14.5, for example, the dashed line represents the maximum height, and hence we can conclude that the ball will never leave out the top.

14.2 The Metric Diagram/Place Vocabulary Model

FROB provides a simple example of the metric diagram/place vocabulary (MD/PV) model of human spatial reasoning. Quantitative representations ground spatial reasoning, and we assume that processes rooted in perception can be used as oracles for many questions involving visual relationships. The metric diagram component constitutes a functional perspective on these

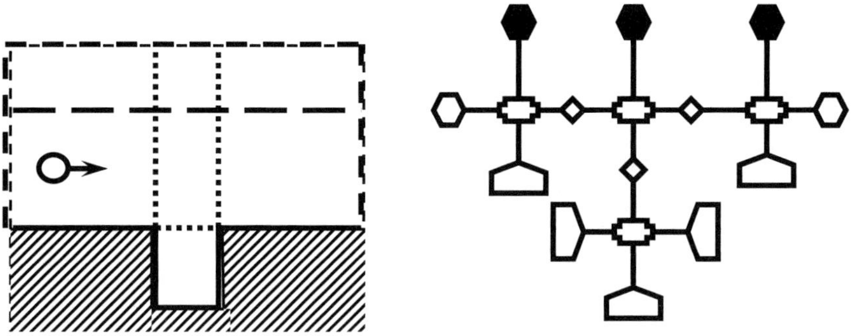

Figure 14.5
Applying quantitative knowledge to prune qualitative possibilities. Here, numerical information about the velocity and height of the ball is available, allowing its maximum possible height to be calculated. Places above that height cannot be reached, as shown by the pruned place vocabulary on the right.

capabilities, not a claim as to the specifics of implementation. A key role of the metric diagram is to compute qualitative representations of shape and space (e.g., place vocabularies for representing space). These symbolic representations provide compact explicit summaries of key aspects of the situation. What is important in a situation varies with the reasoning one is doing, of course. If we were reasoning about moving objects of varying sizes, for instance, the "one size fits all" place vocabulary that FROB used would be inappropriate. (Later examples will involve more complex schemes, e.g., for reasoning about kinematic mechanisms such as clocks and reasoning about movement of vehicles over terrain.) Thus, qualitative spatial relationships need to be computed based in part on what sorts of reasoning are to be done with them.

14.2.1 The Poverty Conjecture

The need to ground qualitative spatial representations in quantitative representations may seem surprising, in contrast with the account of qualitative dynamics presented in part II. Although the representations of quantity presented there are compatible with numerical information, none of the reasoning presented there relied on more a handful of numbers, and most required none at all. Why can't we do the same thing with spatial reasoning?

I believe that there is a fundamental reason why this is so. The *poverty conjecture* is the following: there is no purely qualitative, general-purpose representation of shape or spatial properties. What does this mean? Suppose there were such a representation. Being general purpose, it should support reasoning about how to assemble a complex object, like a bicycle. We should be able to take each part, compute its representation in isolation, and then throw the quantitative information away, working solely with the qualitative representation. Anyone who has assembled something should be feeling skeptical at this point. Given a pair of parts, one can easily see whether or not they will fit together. But to create an a priori scheme for representing each part to support such reasoning would require, in essence, very detailed quantitative information (e.g., the kind of representations found in computer-aided design models).

Consider a general class of spatial problem, which is important in designing mechanical systems: given two two-dimensional shapes, can one roll smoothly across the other? Some qualitative representations work for simple cases (e.g., two squares cannot, whereas two cylinders can). But suppose one cylinder has a protrusion and the other cylinder has a notch? Then the outcome depends on the relationships between the size and shape of the protrusion and notch (e.g., figure 14.6).

And that requires a quantitative comparison of the two parts, thereby robbing us of generality. Although qualitative representations are useful—indeed, critical—for spatial reasoning, there does not seem to be a qualitative representation of shape or space that can be computed independent of information about the particulars of a task and typically even information about the specific problem within that task.

It is one thing to say that such a representation has not yet been found versus that one does not exist. Human ingenuity is impressive, and people

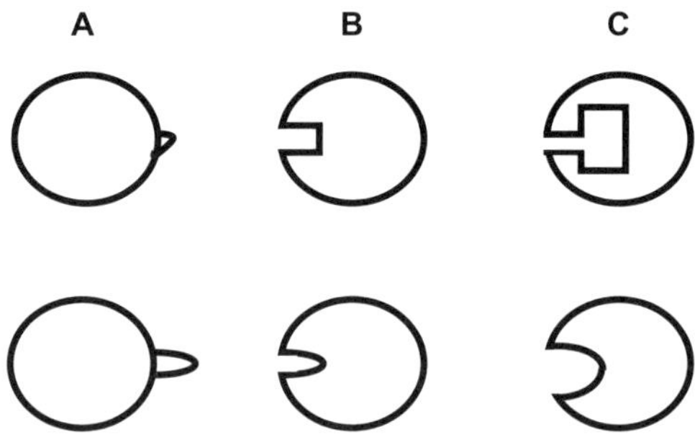

Figure 14.6
Examples of the Rolling Problem. Can the shapes in column A roll smoothly against the shapes in columns B and C? Quantitative information is needed to answer this; qualitative descriptions of the parts computed in isolation cannot solve this.

have tried for decades to create such representations and failed, but that does not by itself rule out someone discovering one tomorrow. Similarly, the fact that people rely heavily on sketching, diagrams, and models when working on spatial matters provides only weak evidence for the conjecture. After all, people may find it simply more convenient to harness their powerful perceptual systems for spatial reasoning. The strongest argument is mathematical. The power of qualitative representations of number comes from the existence of a total order on numerical values, a property of their being one-dimensional.[3] In higher dimensions, a unique total order does not exist. Therefore, we do not get the same kind of power out of qualitative spatial representations—they are inherently more ambiguous than qualitative representations of number.

This has been a controversial claim. Much of the work on qualitative spatial calculi (outlined in chapter 15) has been motivated by the desire to prove it wrong (Cohn & Renz, 2008). However, I think the weight of accumulated evidence continues to support it. For example, the sheer variety of qualitative calculi developed to date suggests that no single calculus can cover the range of spatial reasoning that people perform. And, as the other examples below indicate, systems based on the MD/PV model are capable of human-level reasoning in multiple tasks. As always, we learn more from triangulation with evidence from multiple areas of cognitive science. AI research provides evidence as to what kinds of information suffice for particular tasks. Cognitive psychology provides evidence as to what kinds of information people actually use in tasks.

14.3 Other Examples of the MD/PV Model

Although a number of systems have provided evidence for the utility of this model, I focus on just two areas of research, because they illustrate just how powerful the interplay of qualitative and quantitative can be. The first is reasoning about mechanical mechanisms, and the second is reasoning about fields. I discuss each in turn.

Understanding how moving objects interact is a crucial aspect of spatial reasoning. Consider a mechanical clock. Understanding how such mechanical systems work requires combining an understanding of *kinematics* (i.e., the geometry of motion) and dynamics. Dynamics can be represented using ideas discussed in part II; here, we focus on kinematics. What should a representation of a kinematic state include? Research on AI programs that understand mechanical systems indicates that the two constituents of state are (1) connectivity, because contact is required for one object to affect another's behavior, and (2) shape, because the shapes of objects determine their connectivity. The reasoning involved in understanding how mechanical systems work can be broken down into a set of basic inferences, paralleling those of qualitative dynamics:

1. *Finding potential connectivity relationships*. The place vocabulary for kinematics must make explicit the pairwise contacts between objects in the system.
2. *Finding kinematic states*. The pairwise contact relationships must be assembled to form a complete kinematic state. Some quantitative information is still required at this point to calculate relative positions and sizes, for example, but the symbolic representation computed in this step suffices for the remaining inferences.
3. *Finding mechanical states*. This involves identifying the forces on objects, what directions they are free to move in, and how they are actually moving for a given kinematic state.
4. *Finding state transitions*. Motion can eventually lead to changes in connectivity, thereby providing kinematic state transitions. Dynamical state transitions must also be found (e.g., pendulums exhausting their kinetic energy).

The CLOCK system (Forbus et al., 1991), for example, started with scanned images of the parts of mechanical systems and automatically produced envisionments of systems like mechanical clocks, describing how they work (figure 14.7). The original physical-space descriptions of the parts were automatically translated into *configuration space*, where the axes are the angles of each of the components of the systems. Qualitative representations of configuration space are extremely powerful for mechanical reasoning, enabling a wide range of designs to be analyzed (Faltings, 1992; Sacks & Joscowicz, 2010).

The second line of research concerns a very different type of spatial reasoning. Many spatial phenomena are distributed, such as interpreting measurements of air pressure over a region to construct a description of weather fronts. Yip and Zhao (1996) developed an elegant model, the *spatial aggregation* model, to handle such problems. It uses techniques from computer vision to extract qualitative structure from spatially distributed data points. Because these qualitative structures are grounded in quantitative representations, processing is often recursive, using the analysis at one level to provide a new set of higher-level entities that are analyzed at subsequent levels. Thus, it is a recursive form of the MD/PV model. Ideas from their Spatial

Figure 14.7
A mechanical clock. By scanning photographs of the parts, complex mechanical systems such as this can be analyzed qualitatively.

Aggregation Language (SAL) have been used to analyze the operation of the heart, understand cell phone propagation patterns, and construct descriptions of weather fronts (Zhao, Bailey-Kellogg, Huang, & Ordomez, 2007).

These results provide evidence that the MD/PV model suffices for building systems that can perform at human levels of competence for several important spatial reasoning tasks. This, in turn, provides evidence that it is a reasonable model of human spatial reasoning, at least in terms of overall performance. The precise details of how spatial relationships are computed are almost certainly not the way human visual systems work, because these software systems are designed to be efficient on today's computers. But the broad representational capacities they provide are evidence about what it is useful for visual systems to compute. We return to this issue in chapter 16.

14.4 Categorical/Coordinate Models in Psychology

In parallel with qualitative reasoning research in AI, cognitive psychologists also began exploring an analogous distinction between different forms of spatial representations. Kosslyn (1994) proposed that spatial relations might be split into two kinds. *Coordinate relations* make explicit reference to quantitative information (e.g., that a cup of coffee is six inches from the edge of a table), whereas *categorical relations* are more abstract, indicating a broad equivalence class (e.g., that a cup of coffee is on a table). Functionally, it makes sense for an organism to compute both types of representations: knowing that one can place a cup on the table only requires knowing that regions of open space are larger than the cup's bottom, whereas actually grasping it requires knowing more precisely where it is. Interestingly, this is essentially the qualitative/quantitative distinction previously identified by AI research. A spatial category is simply the region that satisfies a given relationship (i.e., points in that space are functionally equivalent for some purpose, which is equivalent to the definition of a qualitative representation of space).

This convergence is encouraging, and the complementary perspectives and methods applied by both fields yield deeper insights than either alone. Two issues that have been explored extensively in cognitive psychology and neuroscience are how these representations interact with memory and where such processing might occur in our brains. We summarize each in turn.

Suppose we look at a configuration of objects, which we must later reconstruct (perhaps to fill in a missing piece or to recognize whether or not we are seeing another instance of it). For AI systems, the representations stored in long-term memory (when models include it, they often do not) are often perfect reflections of what was originally encoded. AI systems, for example, can store both the original floating-point coordinates and spatial relationships computed for a stimulus, although particular systems might store only one or the other. People appear to work differently. The category adjustment model of spatial memory (Huttenlocher, Hedges, & Duncan, 1991), for example, proposes that both qualitative and quantitative information are stored, but quantitative information is stored with reduced accuracy. To estimate a location from memory, both the qualitative and quantitative information are combined via a Bayesian method that uses distributions of errors associated with each type of information. For example, the most common stimulus used in such experiments is a circle, and participants are asked to identify the location of a point previously seen within. The spatial categories are hypothesized to be the quadrants of the circle, with the quantitative estimate for that representation corresponding to the center of that quadrant. Participants do indeed distort their estimates toward the category. This effect is quite robust, and there is also evidence that it occurs in location estimates for natural scenes (Holden, Curby, Newcombe, & Shipley, 2010).

Neuroscience can provide evidence about what kinds of processing are connected or dissociated. Experiments by Kosslyn and others (e.g., Kosslyn et al., 1989) suggest that the right hemisphere of our brains is more heavily used in coordinate processing, whereas the left hemisphere is used more heavily in categorical processing. The tight integration between qualitative and quantitative reasoning found in both AI models and the category adjustment model suggests that such processing must be coordinated, and indeed Chatterjee and his colleagues (Amorapanth et al., 2012) find activation in both hemispheres, although they may be differentially activated depending on the nature of the relationships being queried. Moreover, they suggest that there are distinct neural systems for handling what they call spatial schemas—which, in essence, are qualitative spatial representations. This suggests that the MD/PV model may indeed be neurally plausible.

14.5 A Unified Account

Ideas and evidence from AI, cognitive psychology, and neuroscience paint a surprisingly unified picture. We can summarize this picture as follows:

1. Symbolic vocabularies of shape and space are central to human visual and spatial thinking.
    - These vocabularies include both relationships between entities (e.g., above) and new entities introduced to symbolically represent functionally equivalent regions of space (e.g., the space above a surface).
    - Their organization reflects task-specific conceptual distinctions as well as visual distinctions.
    - *Qualitative spatial representations* and *spatial categories* are two terms commonly used for these vocabularies.
2. Qualitative spatial representations provide a bridge between conceptual and visual representations.
    - They are computed by our visual systems.
    - They provide the spatial representations for natural-language semantics.
    - They enable rapid reasoning with little information.
    - They support expert-level reasoning in highly spatial domains, such as understanding mechanical systems and interpreting spatially distributed data.
3. Qualitative spatial reasoning is tightly integrated with quantitative representations and reasoning.

- Unlike qualitative dynamics, qualitative spatial representations are typically extracted from visual or other perceptual information.
- Qualitative representations frame questions that may need quantitative processing to answer.
- Memory for spatial situations often includes both qualitative and quantitative information, which are used in reconstructing quantitative properties on retrieval.

15 Qualitative Spatial Calculi

Representing a situation qualitatively involves creating a web of relationships to make explicit the relevant aspects that are implicit in the quantitative, coordinate representations. But what sorts of relationships? The subtlety of space suggests that a variety of systems of relations will be needed to capture people's abilities to reason about it. Research in AI on *qualitative spatial calculi* provides candidate vocabularies for these systems of relations, and there is already evidence for the psychological plausibility of some of them. I begin this chapter by examining the basic idea of a qualitative calculus, using the region connection calculus (RCC) as an example. Then I examine a variety of calculi for different aspects of space, including topology, distance, and orientation.

15.1 Example: Region Connection Calculus

One of the most basic sets of qualitative distinctions that can be drawn is topological. Topology, in mathematics, is about the connectivity of entities. Imagine two blobs on a plane. They may or may not be touching. Perhaps one is inside the other, or they are just touching. These possibilities, and closely related others, are the distinctions that a calculus for topology must make explicit.

A qualitative spatial calculus consists of a set of relations. This set must satisfy two properties. First, the set of relations must be mutually exclusive. In terms of our imagined blobs, this means that only one relation from the set of relations can hold between them at a time. Second, the set of relations must be collectively exhaustive. For our imagined blobs, that means some relation from the set holds between them. Taken together, these two properties enable reasoning by exclusion: if we have ruled out all but one relation, then that relation must hold, for instance.

The most well-known and widely used qualitative spatial calculus is the region connection calculus (RCC), so we start with it. This is actually a family of calculi, which describe topological relationships at varying levels of detail. RCC8 has eight relationships, listed in table 15.1. There are elegant mathematical formalizations of these relationships, but they would take us too far afield for our purposes.[1] From these intuitive definitions, you can see that these relationships are mutually exclusive. The two inverse relationships are necessary to make RCC8 collectively exhaustive. DC, EC, PO, and EQ are all symmetric; for example, (DC O_1 O_2) holds if and only if (DC O_2 O_1) holds, so they are their own inverses.

Suppose we have a network of RCC8 relationships that describe some but not all of the topological relationships involving a set of objects. Even partial information has implications that can significantly constrain the rest. For example, if object A is disjoint from object B and object C is inside object B, then we know that objects A and C must be disjoint. *Transitive reasoning* provides a formalization of such intuitions. Suppose R_1 and R_2 are relations from RCC8, such that (R_1 A B) and (R_2 B C) hold. What do these relations tell us about the relation that holds between A and C, let's call it R_3? (Because RCC8 is collectively exhaustive, we know that some R_3 must exist. Whether the information we have is sufficient to uniquely determine it is another question.) We can construct a table that provides, for all possible combinations of R_1 and R_2, what values of R_3 are consistent with it. Table 15.2 shows some illustrative entries.

Table 15.1
The RCC8 relations.

Relation	Intuition
DC	Disconnected. The two blobs are completely distinct.
EC	Edge connected. The two blobs just touch, either at a line or a point, but their insides are completely distinct.
PO	Partial overlap. The two blobs partially overlap, in that their edges intersect and some of the edge of one is inside the other and vice versa.
TPP	Tangential proper part. One blob is inside the other, but there is still a shared edge.
NTPP	Nontangential proper part. One blob is entirely inside the other, with no overlap in their edges.
EQ	Equal. The two blobs are entirely coextensive.
TTPi	Inverse for tangential proper part
NTTPi	Inverse for nontangential proper part

Table 15.2
Partial description of the transitivity table for RCC8.

(R_1 A B)	(R_2 B C)	(R_3 A C)
...	...	...
NTPP	TPP	NTPP
...	...	...
PO	PO	No information
...	...	...

The first concrete entry corresponds to the logic underlying our initial example. Some cells of this table are nicely constrained; for example, NTPP (nontangential proper part) with TPP (tangential proper part) yields NTTP, meaning completely inside. Alas, many of the entries are not so constrained. Given only PO and PO relationships between regions, we know absolutely nothing about the relationship between the first and third regions. That is, all relationships are possible. In fact, half of the entries in the table are ambiguous. Out of the sixty-four pairs of possible relationships, eight (13 percent) lead to two possible answers, nineteen (30 percent) lead to three possible answers, two (3 percent) lead to four possible answers, and three entries (5 percent) yield no information at all.

Nevertheless, transitivity reasoning can be useful. Think again of a qualitative description as a web of relationships. Multiple transitivity inferences can sometimes be combined to provide more precise answers by eliminating possible relationships that do not show up in all disjunctions. This kind of processing is known as *constraint propagation* in AI (Mackworth, 1977). It is an attractive method of inference because it can be very fast,[2] and even when it does not yield a unique answer, it whittles down the space of possibilities so that solutions can be found more easily via backtracking search. (A possible explanation of the Necker cube phenomena in vision is that a constraint solver in our visual system finds two solutions to the network of relationships that describe the three-dimensional [3D] orientations of the lines of the cube.) From a cognitive science perspective, constraint propagation is interesting because, when applicable, it tells us that a computation can be done cheaply. Moreover, constraint networks are amenable to local, parallel computations and hence of interest to those working on neural models.

RCC8 has been used in many ways. For example, RCC8 relationships can be derived automatically from images via computer vision systems. Moving to this qualitative level of representation helps ensure that scenes that are semantically close have similar descriptions and can help overcome

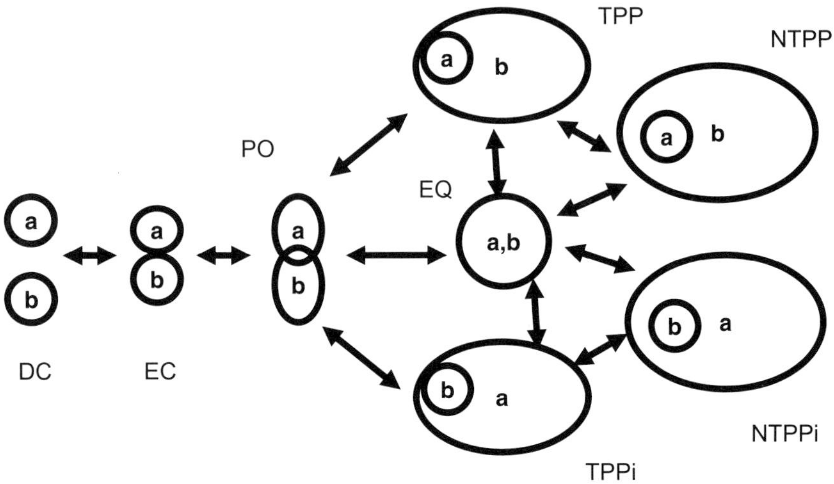

Figure 15.1
Continuous transitions between RCC8 relationships. This adjacency structure can also be thought of as the conceptual neighborhoods for these relations.

the effects of sensor noise (Cohn, Magee, Galata, Hogg, & Hazirika, 2008). RCC8 relations are also useful in guiding the computation of other spatial relationships, which we return to in chapter 16. They can also be used for a kind of qualitative spatial simulation. Consider figure 15.1, which depicts the sequence of transitions that two spatial entities can go through as they move and transform on the plane.

The RCC8 relationships between pairs of objects define (perhaps just one component of) a qualitative state of a system. Assuming the transformations of the entities themselves are continuous, then a pair of entities that starts out as `DC` must first go through `EC` before reaching `PO`. Similarly, before they reach `NTPP`, they must go through `TPP` or `EQ` from `PO`. Thus, the arrows define possible transitions in qualitative state, based on spatial transformations of the shapes. This has been used to reason about, for example, the spatial aspects of what happens when a virus invades a cell (Cui, Cohn, & Randell, 1992).

This idea of continuity as a higher-order constraint on sequences of relations has been formalized as the *conceptual neighborhood* of a relation (Freksa, 1991). In addition to providing continuity, conceptual neighborhoods have been proposed as a measure of similarity on relations. That is, a situation where two objects are `PO` is closer to one where they are `EC` than it is to a situation where they are `DC`.

Is RCC8 part of people's mental vocabularies of qualitative representations? Experiments by Knauff and his colleagues suggest yes (Knauff, Rauh, & Renz, 1997). They had participants group pictures that were made up of circles in various relationships, similar to the pairs in figure 69, but all circles to eliminate shape as a factor and varying metric and orientation information. The groupings that participants made were compatible with RCC8. On the other hand, Klippel, Li, Yang, Hardisty, and Xu (2013) argue that further experimentation suggests that a coarser set of relationships might be more robust as a psychological model. Do people really distinguish between tangential proper part and nontangential proper part, as well as between edge connected and partially overlapping, for example? And does it make sense to assume a single level of representation that is universally used, given the richness of human conceptual structure? Klippel et al. (2013) also argue that topology is often not the most salient spatial property that people extract from a situation: relative size, for example, is often salient, yet size is not a topological property. Thus, a wider range of qualitative calculi, covering a broader range of spatial phenomena, is worth exploring. We turn to this next.

15.2 A Collection of Calculi

RCC8 is just one of many qualitative spatial calculi that have been created. Most share the same basic features: defining a set of mutually exclusive and collectively exhaustive relationships, defining their implications via composition tables, defining conceptual neighborhoods, and applying them to one or more tasks. An exhaustive compendium of existing calculi would take us too far afield (but see Dylla et al., 2017), so this section focuses on calculi that seem to be of most interest to cognitive scientists as potential models of aspects of human spatial representations.

15.2.1 Intersection Models of Topology

An alternate approach to a qualitative formalization of topology has been developed by Egenhofer and his colleagues (Egenhofer & Franzosa, 1991). Consider again our two abstract blobs in the plane. Intuitively, we can think about their interior, exterior, and the boundary between them. Now think about whether or not each of these spaces intersects. This gives rise to a 3×3 set of possibilities, which leads to 512 possible combinations of relations. But most of these relationships are not possible, given reasonably constituted blobs. Having no intersections between these spaces at all, for example, would imply that they are on different planes. On the other hand, for most pairs of blobs, their exteriors will intersect. So the set of relationships that is actually possible is drastically smaller. The size of the set of relationships depends on additional assumptions that can be made about the blobs. If they are both two-dimensional, then the relationships produced include those of RCC8.[3] If one of them is a line segment, additional relationships become possible, especially taking into account that a line segment has two ends that may have varying relationships with the interior, exterior, or boundary. This is the insight behind Egenhofer's spatial calculi.

It is useful to have a mathematical way of characterizing a set of possible spatial relationships. But are these psychologically plausible? Do people use them? How does the set of distinctions that people make compare with those possible under this kind of generation scheme? To investigate this, Mark and Egenhofer (1994a, 1994b) conducted several behavioral experiments. A sorting task was used, where participants were given cards illustrating various configurations of a region and a line. The region was always the same; only the line varied. Two (and in some cases three) examples were given per relationship, one with a straight line and one with a curved line. They were told that the region was a state park and the line was a road. They were asked to sort the cards into groups, based on whether or not their verbal description of those configurations would be the same. Across twenty-eight participants, there was considerable variation. There were nineteen mathematically distinguishable relationships, given the assumptions of their analysis. A few participants did indeed differentiate between all nineteen relations. Most participants identified between nine and thirteen sets of relationships. Interestingly, in the examples shown in the paper, none of the participants violated the conceptual neighborhoods of these relations. In other words, using the mathematically distinguishable relationships as the most specific level of detail, abstractions made over this most specific set were always continuous. To pin this down further, the researchers also asked participants to rate how well the same forty stimuli fit two statements: "the road crosses the park" and "the road goes into the park." Consensus among participants was strong, which suggests that there are stable conventions for mapping from language to these spatial relations, at least when the context is held constant.

How general are these findings? The authors are appropriately cautious, noting that, in addition to individual differences, culture and contextual factors also can play a role. To examine the effect of language, they have used a combination of English and Chinese speakers, with participants from other language groups as available. Roughly, the variation they see between speakers in a language is higher than the variation that they see between languages. On the other hand, they argue that the domain of discourse might matter considerably. All of their examples were geographical. (Their motivation is to understand people's naive geography, in part to make future geographic information systems [GISs] more compatible with human conceptual structures [Egenhofer & Mark, 1995].) If participants were told that the illustrations show circuit layouts, for example, would they produce the same clusters? Examining how spatial relations are tied to language in several other spatial domains could provide valuable insights into how general our mappings between language and space are.

Metric information also plays an important role in spatial language terms. Metric information can be used to make additional distinctions and in some cases even override topological information. Returning to Mark and Egenhofer (1994a, 1994b), the situations best described by the phrases "the road that exits the park" and "the road that ends just outside the park" are identical topologically, only differing in metric properties. To explore these issues, in one experiment, participants were given the task of drawing spatial configurations involving a region and a line that exemplified fifty-nine natural-language English terms, such as along edge, bypasses, cuts across, cuts through, runs along, runs into, and within. The drawings were analyzed to identify both the topological relations that they contained but also metric refinements that made sense for particular topological configurations. Three types of metric refinements were considered. The first were *splitting ratios*—for example, what are the relative sizes of the subregions when a line entirely crosses a region, or what fraction of a line is along the boundary of a region versus outside the region? A case analysis of the topological relationships yielded seven distinct splitting ratios that could be defined, each of which is relevant to multiple topological relationships. The second type were *closeness* measures (e.g., how far a line's interior is from the region's boundary, when they don't overlap, versus how far into the region the line extends, when it does overlap). The third type was *approximate alongness*, which combines the first two. Thirty-four participants were given cards that included a shape describing a park, as well as an English sentence that described the relationship between a road and the park, and told to draw a road that fits that relationship. Their analysis suggests that the spatial terms could be divided into two broad categories. The first consists of terms that could be identified with a single topological relation. The term *avoids*, for example, always has the entire line in the exterior of the region. The other category consisted of terms that had several topological relations associated with them. A cluster analysis was used to identify ranges of metric parameters associated with each of the terms, thereby providing a kind of dictionary that represents a translation between language and space for this type of situation. To check it, human-subject ratings for drawings illustrating five spatial terms from a prior experiment (described above) were used to ascertain if the terms chosen by the dictionary agreed with those ratings. The parameters predicted to be significant (e.g., the metric refinements) were indeed significant in agreement ratings.

15.2.2 Distance Calculi

We readily distinguish between things that are near and far away, without knowing detailed, metric information about distances. However, what we mean by near and far varies with context: Northwestern is near downtown Evanston when one is walking, near Chicago if one is driving, and near Minneapolis if one is flying. San Francisco, even if flying, is far away, and Shanghai is really far away. We can know these relative measures of distance without having specific distances in mind. Clementini, Di Felice, and Hernandez (1997) proposed a formalization that uses finite symbolic algebras but defined in terms of numerical intervals with properties that help constrain their composition.

15.2.3 Orientation Calculi

Direction matters in spatial configurations. When dealing with large-scale space, we often describe directions with respect to cardinal geographical directions (e.g., north, south, northwest). When dealing with close visual space, we can similarly use global reference frames (e.g., above, left of). All of these simple systems of relations can be considered qualitative spatial calculi. There are edge cases, especially when moving into 3D, as Coventry and Garrod (2004) and others have observed. In the worst case, relationships might be collectively exhaustive but not mutually exclusive due to the effects of spatial extent. If we are standing on the twelfth story of a twenty-four-story building, the building itself is both above and below us. To calculate such relationships from metric information typically involves (sometimes implicitly) carving up space into regions based on a bounding box around the object representing the reference object and using the region(s) that the located objects are in within that grid to assign a relationship (figure 15.2).

The grid of directions imposes a simple qualitative representation on the space around it, albeit a crude one. Considering the landmarks visible from a location provides a basis for more fine-grained representations that take into account the complexity of the environment in a location. Suppose we look around ourselves, scanning from left to right, and noting the sequence of objects around us. This gives rise to a set of objects that

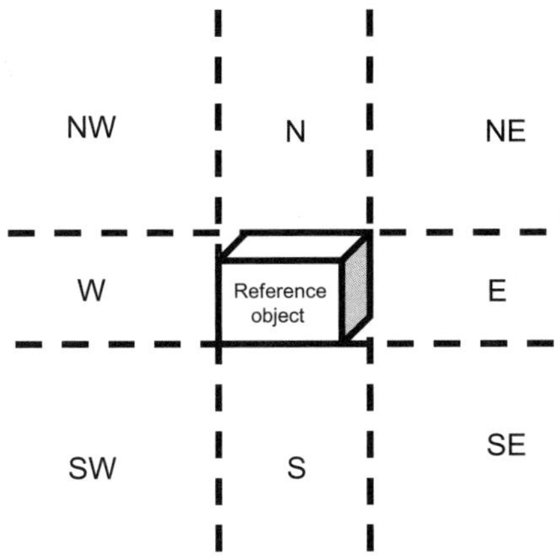

Figure 15.2
Global grid of directions around a reference object.

implicitly divides space into a set of qualitative regions, based on visibility (Röhrig, 1994). Figure 15.3 illustrates two such schemes.

Figure 15.3a shows how three different locations would be encoded in terms of a sequence of landmarks. Here {`Bank`, `Store`, `Home`, `Bar`} and {`Store`, `Home`, `Bar`, `Bank`} are equivalent, because they represent the same ordering of landmarks, but with different starting points. However, landmarks can be of varying sizes and at differing distances from our location, which means that they can overlap as we are scanning, and one can even contain another, with respect to how we see them. These limitations of the cyclic order algebra led to development of the cyclic interval algebra (Osmani, 2004), illustrated by example in figure 15.3b. A set of relationships analogous to those used for temporal intervals but modified to be applicable to locations around a reference point is used to describe a position. As one walks by the trajectory from `A` to `B` to `C` in figure 15.3b, the relationships describe the relative location of the two buildings for each position.

Any calculus that provides a qualitative representation for position can be used to decompose space into qualitatively distinct regions (i.e., a place vocabulary), based on constructing regions where all the points within them have identical qualitative representations. This provides a way to link quantitative spatial representations, such as maps, models, and diagrams, to qualitative models. In finding one's position on a map, sans GPS and street signs, for example, visual search of the map can be governed by the constraints imposed by these cyclic constraints. (It would be surprising to me if, given the history of exploring the psychology of navigation, there

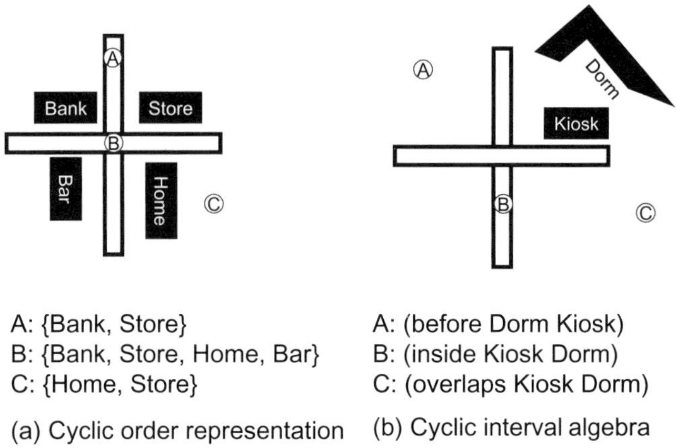

Figure 15.3
Cyclic relational algebras. These representations describe positions in terms of what can be seen as one scans around a position.

were not already existing data that bore on whether or not people actually use qualitative descriptions to guide visual search.)

Global frames of reference are often not sufficient for spatial reasoning. Consider recognizing occurrences of a particular spatial layout, such as the placement of equipment in a brewery. If the brewery were to be rotated (carefully), the description of the internal layout of its parts should remain the same. This requires using relative orientation relationships. A relative orientation relationship always has some reference object and the object being located with respect to that object. For example, if we take a body-centric (egocentric) perspective, we have a reference object with an associated direction (i.e., our front). In that perspective, we commonly use front/back and left/right as qualitative distinctions to describe the positions of things around us. More fine-grained approaches divide orientation into sectors, such as the Oriented Point Relational Algebra (OPRA) family of sector-based representations (Lucke, Mossakowski, & Moratz, 2011; Moratz, Dylla, & Frommberger, 2005). Another way of increasing granularity is to add an additional reference point, as in the single-cross calculus (Freksa, 1992), for example. Understanding the utility of different representations for different tasks is something that is still being explored. For example, in the Sketch-Mapia project (Schwering et al., 2014), a variety of qualitative spatial calculi are used to analyze hand-drawn sketch maps to determine how to integrate them into a GIS that has quantitative information. The researchers found, for example, that cyclic interval algebra can yield high accuracy in aligning street junctions and that fewer directional relations provided higher accuracy than a finer-grained representation with more directional representations in aligning street segments. Exploring the trade-offs between representations is important because it sheds light on the underlying nature of the information-processing problems involved in spatial reasoning tasks.

15.3 Reasoning Issues

As noted in chapter 2, any representation needs to be scrutinized from the following three perspectives:

1. What do its elements mean in terms of intended models? Carefully described theories are obviously better than representational vocabularies where new primitives are pulled out of the air or whose meaning is at best suggestive due to the use of natural-language sounding terms for predicates (McDermott, 1976).
2. How can descriptions in this representational vocabulary be computed from inputs? In some cases, it might be presumed that they are directly computed by some subsystem in an organism (e.g., a specification of the representations produced by a vision system or natural-language parser). But in most cases, an account of how these representations might be computed from more basic information needs to be provided.
3. How can the representation be used for reasoning? What kinds of reasoning does it support? What is the computational complexity of that reasoning?

The qualitative calculi described in this chapter provide exemplary models of the first point, in part because many of them are grounded in mathematical formulations of topology and mereology. There is also ample evidence that these representations can be automatically computed from reasonable inputs (e.g., the use of RCC8 to aid in video analysis). The qualification operation in the SparQ toolkit (Wallgrun, Frommberger, Wolter, Dylla, & Freska, 2007) provides algorithms for mapping from quantitative data to many qualitative calculi. Therefore, although such mappings do raise interesting issues concerning noise and setting appropriate thresholds, they can be derived from quantitative information. Moreover, some of these representations have been proposed as elements to be used in the semantics of natural language involving space (Mani & Pustejovsky, 2012). Thus, we know that these descriptions can indeed be generated, in multiple ways, and support incremental accumulation of information. We have already discussed the third point, reasoning, with regard to the use of mutually exclusive and collectively exhaustive sets of relationships to define an algebra and the use of constraint propagation techniques to perform transitive inferences across sets of relations. Here we return to the issue of computational complexity in more detail.

Unfortunately, the ambiguity we saw with RCC8 can get even worse as the number of qualitative distinctions grow. In fact, reasoning with several of these calculi is NP-hard, meaning that they cannot be performed in polynomial time.[4] That does not necessarily rule them out as a cognitive model: people can and do tackle NP-hard problems, such as packing dishes in a dishwasher and planning trips with many stops daily, with great success. But it does mean that we should be careful in assuming too much in terms of how much we can wring out of constraint-based reasoning about them.

Do these complexity results mean these representations are not useful in cognitive models? Not at all. First, they only concern squeezing new information out of multiple statements within the same representational scheme. But that is not the only way that they can be used, as the applications to visual processing and aligning sketch maps with a GIS indicate. More broadly, spatial relationships from these calculi can be used in axioms that combine them with other types of information to reason about other aspects of the world. (For example, if I have been painting the floor behind me as I move, I should plan my trajectory so that it ends at the door rather than a corner, unless I was willing to stand there for quite a long time.) Moreover, such relationships can be used in analogical reasoning, where they help establish the similarity (or differences) between two situations being viewed or between a current situation and a remembered situation. What matters the most is that they make distinctions that correspond with the distinctions that people tend to make. Tractable inference within a qualitative calculus itself, if it is available, is icing on the cake.

15.4 Summary

We have seen that there are a variety of qualitative spatial calculi, capturing various aspects of human spatial concepts. Topological representations have received the most attention because they are fundamental and extremely useful. Alas, there is an unfortunate tendency in the literature to treat topological and qualitative as identical. This is especially true in the literature on the connections between language and space. For example, Talmy (2000) describes relationships as topological if they are invariant with respect to size, distance, and shape. However, topological, in the usual mathematical meaning, also implies invariance with respect to position. Hence, swapping the position of two objects, changing which is above or below the other (or to the left or right of each other), leads to a situation that is identical topologically but not psychologically. What Talmy (and others exploring language) typically mean is qualitative, not topological. Topological relations are indeed qualitative, but not all qualitative relations are topological, as research into positional relations and calculi for distance, direction, and orientation above demonstrates. The self-described mantra of the Maine group, "Topology matters, metric refines" (Egenhofer & Mark, 1995), might be better expressed as "Qualitative frames, metric refines," because it is hard to believe that distance, direction, and orientation do not matter—their own evidence indicates otherwise. The idea of qualitative representations provides a bridge between different areas of cognitive science, as well as between perception and cognition.

16 Understanding Sketches and Diagrams

Sketches and diagrams are heavily used by people to communicate with each other and to think through ideas. The act of sketching engages our visual and motor capabilities, helping us think things through, especially when the matter under consideration includes spatial aspects. Providing directions via a map, explaining how a system works, and creating a new design are all facilitated by a pen (analog or digital) and media to draw upon. But drawings alone are not enough: diagrams in textbooks include captions, maps involve cultural conventions and legends, and sketches are typically accompanied by dialogue where participants declare and clarify the intended meaning of what is drawn. Thus, understanding what is needed to understand diagrams, as well as the processes needed to produce them, requires a combination of visual, spatial, linguistic, and conceptual representations and processes. This makes it a fascinating area for cognitive science research.

It will be useful to clarify what I mean by particular terms. Although every sketch can be viewed as a diagram, I will follow the common convention of reserving "diagram" to mean professionally prepared drawings, typically accompanied by a caption and/or other text, which must be interpreted without interaction with whomever produced it. Everything else will described as sketches. The distinction matters, as we will see below, because hand-drawn sketches provide an additional set of problems in their interpretation.

This chapter examines sketch understanding research to see how qualitative representations and reasoning have been used to create systems that capture aspects of human sketching and understanding. We start by briefly looking at how several fields of cognitive science have investigated sketching. We then describe the *nuSketch* model of sketch understanding, using the *CogSketch* system (Forbus, Usher, Lovett, Lockwood, & Wetzel, 2011) as an example. The basics of the representations and reasoning used in CogSketch are outlined next. Then we describe some experiments that use CogSketch to model cognitive phenomena—namely, learning spatial prepositions, reasoning about depiction conventions, and visual problem solving. The latter is especially interesting because it shows that CogSketch, as a model of high-level vision, provides human-level performance on several tests, including Raven's Progressive Matrices.

16.1 Investigations of Sketching and Diagrams

Drawing is a topic that has long fascinated cognitive scientists. Children's drawings have been used to provide insights into their understanding of the world and how it develops over time (Piaget & Inhelder, 1956; Van Sommers, 1984), including how cognitive capabilities change with age and disease (Lange-Kuttner & Vinter, 2008).

What do diagrams do? The metric diagram/place vocabulary model, presented in chapter 14, suggests that they serve two purposes. First, they enable our visual system to serve as an oracle for a class of spatial queries—instead of having to reason as to whether one object is above another, we can simply look if they are both depicted in the diagram. Second, they provide an external anchor for qualitative representations of space that our visual system computes, as we saw with FROB's world of bouncing balls. Larkin and Simon (1987) suggest a third purpose, extending our short-term memories by offloading references to external media. All three purposes are good reasons for people to use diagrams, especially given how powerful our visual systems are. And the poverty conjecture suggests that this adaptation is probably important for any intelligent system, given the value of the conciseness of qualitative representations and the need to compute them dynamically from quantitative spatial information.

In AI, there have been two almost nonoverlapping lines of research involving sketching. One focuses on diagrams, including modeling how people understand them, building systems that can understand them and use that understanding for some application, and developing new diagram-based representations and reasoning schemes (e.g., Cox, Plimmer, & Rodgers, 2012). The other focuses on sketch recognition, with the goal being to identify the entities that someone is drawing. Typically, this is done to provide input to some other software system (e.g., a geographic database or computer-aided design program). A vocabulary of entity types is created as part of an application, and statistical machine learning techniques are used to learn how to identify, for a new sketched entity, which type it is. Such systems vary in what they require as input. Some use images of drawings, gathered via cameras, such as PARC's ZombieBoard, which translates hand-drawn whiteboard images into PowerPoint diagrams, tables, and bullet points (Saund, Fleet, Larner, & Mahoney, 2003). Some use digital ink, gathered via mouse or digital pen, which includes temporal information for each point as well as its position, allowing systems to use velocity information as if they were intently watching the person draw (Hammond & Davis, 2005; Valentine et al., 2012). Some complement digital ink with text generated via automatic speech recognition, thereby providing another channel to express intent. Because both sketch recognition and speech recognition are noisy, having both can provide robustness. For example, in QuickSet (Cohen et al., 1997), the lists of the top *N* hypotheses of the two recognizers are compared, with the highest-ranking hypothesis that shows up on both lists being chosen as the output, which improves accuracy.

The recognition approach focuses on naturalness of input. If one's goal is to make existing software applications easier to use, this is a terrific approach. However, the state of the art has several significant limitations. The first is that it only works with a predefined vocabulary of entities. This requires engineering the domain and training users to stay within it.[1] The second is that today's machine-learning techniques tend to require large amounts of training data to achieve reasonable accuracy. For some applications (e.g., a specialist who works in a single domain much of the time, like engineers or military commanders), the recognition approach can provide considerable fluency and can be much better than traditional graphical user interfaces. However, when considered as a model for human sketching, it falls short. We can recognize when something sketched isn't anything close to what was expected, dynamically adjusting our expectations. We set up local conventions during sketching (e.g., one blob may represent a building, and a similar-looking blob may represent Godzilla approaching the building). In other words, we draw freely on our knowledge of the world, which is several orders of magnitude larger than any fixed-domain sketch recognition system can handle (typically a few dozen). Thus, I believe a different approach is needed to model how people understand sketches. This new model, described next, is, I believe, bringing us closer to creating systems that we can sketch with as if we are interacting with another person.

16.2 The nuSketch Model of Sketch Understanding

What is human-to-human sketching like? Generally, participants are drawing and talking at the same time, producing entities to represent aspects of their thinking. We call these entities *glyphs*. Glyphs can represent physical objects (e.g., buildings), abstract entities (e.g., an earthquake), and relationships between other entities (e.g., arrows). To build a shared understanding requires both a mutual understanding of what glyphs there are, what they are intended to mean, and their implications for the shared understanding of the situation. In sketching, participants must solve three problems:

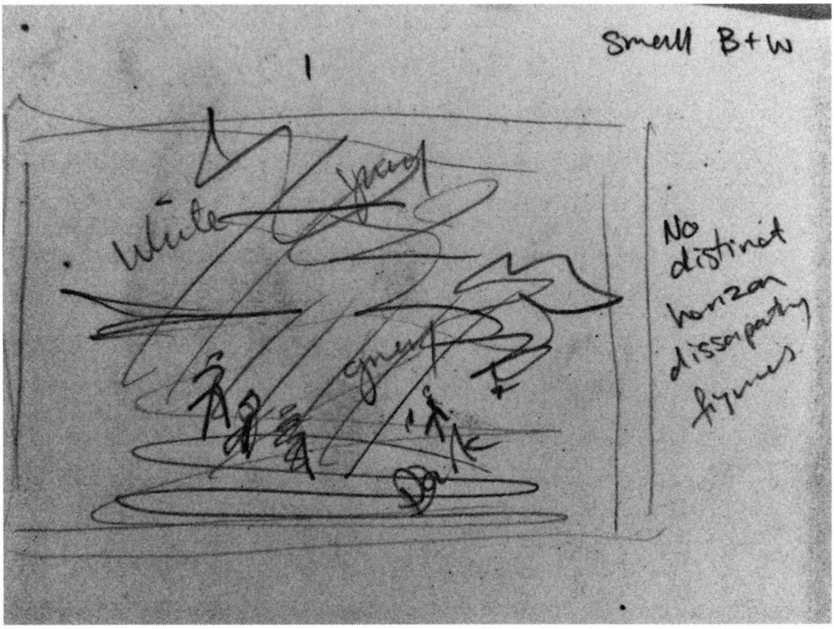

Figure 16.1
Sketching for creativity. Above is a sketch done by Shonah Trescott while in the Arctic. On page 269 is the painting that she later made from the sketch. (Used with permission of the artist.)

1. *Segmentation*. Which pieces of ink should be considered together to form a glyph?
2. *Conceptual labeling*. What is the intended conceptual meaning of a glyph?
3. *Meaning*. What does the sketch so far indicate, given the context of the interaction, about what the participants should think or do?

There are many ways to solve these problems, some more specialized than others. Sketch recognition research has focused on the first two problems, relying on a fixed context (e.g., providing input to some other software system) as the solution to the third problem. Therefore, let us focus on the first two problems initially and return to the third shortly.

In sketch recognition work, heuristics like lifting the pen from the page or long pauses are often used to automatically segment ink. Conceptual labeling in that approach derives from the ink recognizer identifying the glyph as an element of its vocabulary or combining that with the output of a speech recognizer, as outlined above. Neither of these solutions is very general, alas.

People often pause and pick up the pen when they are thinking, especially when doing creative design or learning. Most people are not artists, and even artists do not draw accurately enough to support recognition when they are sketching. Consider, for example, figure 16.1 (left), which shows the sketch that Shonah Trescott, an Australian artist, made when working on a painting, compared to the painting itself (figure 16.1, right).

Note the use of hand-drawn notes to indicate intended colors and entities. Given that most of us are not artists, recognition should best be viewed as a catalyst, a source of evidence about conceptual labels that is sometimes available but in general cannot be relied on to be sufficient.

Sketch recognition is a hard problem and has received decades less work, by many fewer people, than for example speech recognition. Speech recognition research started in the 1970s, but only in the past decade have dictation systems become good enough to be routinely used by a wide range of people. And because recognition is at best a catalyst, what do we do the rest of the time? Relying on natural language, as people do, is not practical at this time. This suggested to our group that finding alternatives for segmentation and conceptual labeling would be a useful way to make progress on understanding the meaning of sketches. Our approach is to use a simple engineering solution: we provide interfaces that enable users to tell the software how they want to segment their ink and to provide conceptual labels. For example, in CogSketch, a system built using this model, people start to draw and hit a "Finish Glyph" button when they are done with a glyph. Editing tools are provided so that they can reorganize their ink into different glyphs as they please. This manual solution to segmentation requires slightly more work by the users but in turn does not require that they unlearn many habits (like never picking up the pen or pausing to think while drawing) and does not require them to correct system errors, which can interrupt the flow of one's thinking. Similarly, an interface is provided that lets users tell CogSketch what they intend their glyph to mean, based on the concepts in a large knowledge base.[2] Thus, we sidestep the first two problems in favor of focusing on the third: how should we reason about the sketch?

The understanding of the sketch is, we believe, grounded in qualitative representations of space and shape. The representations generated by nuSketch systems are intended to be models of aspects of the representations produced by human visual systems. As described below, this includes relationships that provide distinctions needed to detect certain spatial prepositions, as well as capture important aspects of human visual thinking.

16.3 CogSketch: Representations and Processing

CogSketch is the current implementation of the nuSketch model of sketch understanding. Although its architecture and internals have aspects that are of interest to an AI audience, describing it in detail would take us too far afield here. Hence, we focus only on the representations and processing it performs, to ground the discussions of modeling in the rest of the chapter.

CogSketch starts with digital ink. Each piece of ink consists of a list of points, each containing a time stamp as well as X and Y coordinates.[3] It resamples the ink to provide an alternate rendering in terms of points that are evenly spaced along the stroke, to simplify curvature computations. CogSketch computes by default certain properties of glyphs, such as how round it is and, if not round, its major and minor axes. The RCC8 relationships (chapter 15) can be computed between pairs of glyphs to provide topological information, and positional relationships (e.g., above, left of) can also be computed.

The visual language supported by CogSketch defines three kinds of glyphs (figure 16.2):

1. *Entity glyphs* depict specific entities in the world being depicted. Examples of entity glyphs include depictions of people, trees, and houses. In

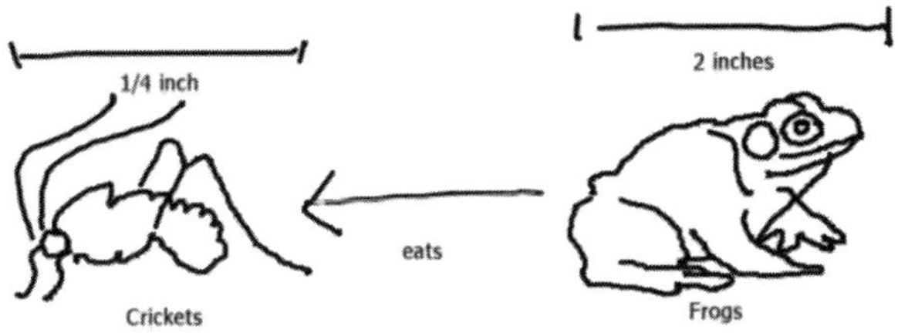

Figure 16.2
Examples of the three types of glyphs.

figure 16.2, the frog and cricket are entity glyphs. Entity glyphs can also be used to depict abstract entities, like events (e.g., running).

2. *Annotation glyphs* depict properties of entities. Some annotation glyphs are abstract (e.g., indicating the value of a physical quantity like the capacity of a container). Some annotation glyphs indicate how a visual property is to be measured (e.g., the length of the cricket and frog in figure 16.2). Other annotation glyphs indicate directions (e.g., direction of rotation, linear motion, or force application).
3. *Relation glyphs* depict relationships between the things depicted by two other glyphs. They are always drawn as arrows, and the arguments to the relationship are indicated by what the head and tail of the arrow are nearest.[4] The "eats" arrow in figure 16.2 indicates that the frogs eat crickets. Depending on the relationship chosen, any type of glyph can be an argument in a relation glyph, including annotation glyphs and other relation glyphs.

This visual language is very powerful. It is powerful enough to represent concept maps, for example, and goes beyond them to support expressing rationale for relationships. CogSketch distinguishes between visual and spatial information. Visual information consists of properties of the digital ink constituting a glyph. Spatial information consists of properties of the entities depicted by the glyphs in the situation being depicted. CogSketch is aided in understanding these intended relationships by user-specified *genre* and *pose*. The genre of a sketch frames the basic visual-spatial relationships. CogSketch currently supports three genres, which we believe correspond to the most common conventions used when people sketch:

- *Abstract*: No spatial information can be directly extracted from the visual sizes and locations of the digital ink. Examples of this genre are electronic circuit schematics and flowcharts.

- *Physical*: The visual properties of concrete entities depicted are used to represent their spatial properties. Examples include diagrams of physical systems.
- *Geospatial*: Like the physical view, but the spatial semantics for elements of maps (see below) is applied.

The pose of a sketch specifies how visual coordinates are translated into spatial coordinates. For abstract views, of course, the concept of pose does not exist. For physical views, the default looking from the side pose is that the *X* and *Y* coordinates in the visual plane of the sketch translate into *X* and *Z* coordinates in the physical world. Other poses do other translations (i.e., looking downward translates *X*,*Y* in the visual plane to *X*,*Y* in the world). For geospatial sketches, *X*,*Y* in the visual plane are translated into the east-west axis and the north-south axis, respectively. Pose also affects the computation of positional relationships (e.g., if `A` is `leftOf` and `Above` `B` in a physical sketch looking from the side, then `A` is `northEastOf B` if the same sketch were interpreted as a geospatial sketch).

When people sketch, they often divide their sketch into units depicting particular states or alternatives, as when cartoonists draw comic strips. Another important property of CogSketch's visual language is that it enables sketches to be combined into such larger structures. A sketch always contains at least one *subsketch*, with a particular genre and pose. But a sketch can contain multiple subsketches as well. Every subsketch is also a glyph in the *metalayer*, which depicts all of the subsketches in a sketch. Arrows can be drawn to depict relationships between subsketches. This enables the drawing of comic strips to show how behaviors unfold, as figure 16.3

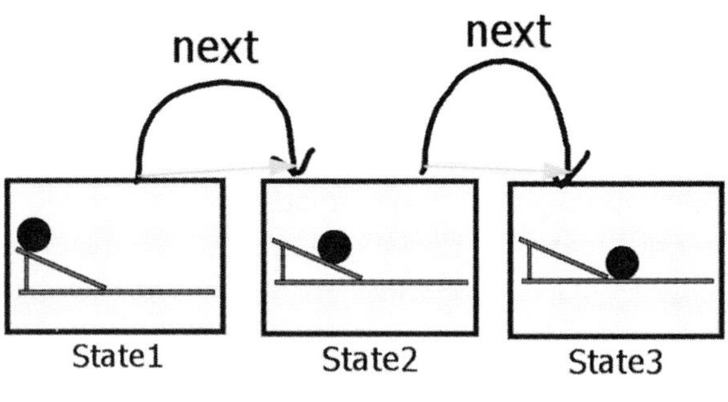

Figure 16.3
A sketch can consist of multiple subsketches. On the metalayer, subsketches are treated as glyphs, which can participate in relationships. This comic graph illustrates the behavior of a ball rolling down a ramp.

Figure 16.4
The Voronoi adjacency regions on the left are used to determine what parts to compute positional relationships between, shown on the right.

illustrates. Multiple subsketches can also be used to depict alternate points of view on a system (e.g., where a factory is located and how some key piece of equipment works) or alternate approaches to a design problem.

A hypothesis built into CogSketch is that, by default, relationships are only computed within local neighborhoods. To determine whether or not two glyphs are adjacent, a Voronoi diagram is computed based on the points constituting each stroke. (A Voronoi diagram for a set of points is a tessellation of space into cells by lines that are equidistant from a pair of points.) Two glyphs are adjacent exactly when at least two points, one from each glyph, are adjacent in the Voronoi diagram. The Voronoi diagram for a sketch of a cat is shown in figure 16.4 (left), with lines indicating the pairs of glyphs for which positional relations will be computed in figure 16.4 (right).

For instance, CogSketch will compute that the whiskers are above the body and the ears are above the whiskers, but it will not automatically derive that the eyes are above the body, because the head intervenes. Moreover, by default, positional relations are only computed between adjacent glyphs that are `RCC8-DC` (i.e., are disjoint). This is why it does not compute positional relationships between the head and anything else automatically. Models using CogSketch can make queries to derive positional relations between any pair of glyphs, of course.

CogSketch incorporates three levels of representation for its two-dimensional descriptions:

1. *Group level.* Combines glyphs into units, based on gestalt principles. Properties of groups and relationships between them are the focus of interest at this level.
2. *Object level.* Each glyph is treated as atomic, with the focus of interest being relationships between the glyphs.
3. *Edge level.* The ink constituting a glyph is decomposed into edges, which are either straight-line segments or curves, separated by corners. Qualitative descriptions of their properties (e.g., being straight or curved) and relationships between the edges in a shape (e.g., that two edges are parallel or that a corner is convex) are computed for the edges in each glyph. The focus of interest at this level is on properties of shapes.

Figure 16.5 illustrates these three levels.

On the left is an example highlighting the group level of representation, with the circles being viewed as a group below the (single-item) group consisting of the square. The middle rendition of the same stimulus at the object level treats all three entities as independent, so that one can ascertain that one circle is to the left of the other, for example. Finally, the rendition on the right zooms into the square to illustrate the edge level of representation, showing how a shape decomposes into edges connected by junctions. There is also a fourth level, pertaining to three-dimensional (3D) interpretation, of *surfaces*. CogSketch can use techniques from computer vision to divide an entity into surfaces, which it can then reason about and match in 3D (Lovett, Dehghani, & Forbus, 2008; Lovett & Forbus, 2013).

To what extent do these levels, and the representations in them, provide a model for aspects of human vision? There are many solid arguments for hierarchical representations in vision (Marr, 1982; Palmer, 1978). The particulars of the representations that CogSketch computes are motivated whenever possible by evidence from vision research. However, as with any computational model, there are always identifiability issues. Some of our representations have been developed in response to information-processing

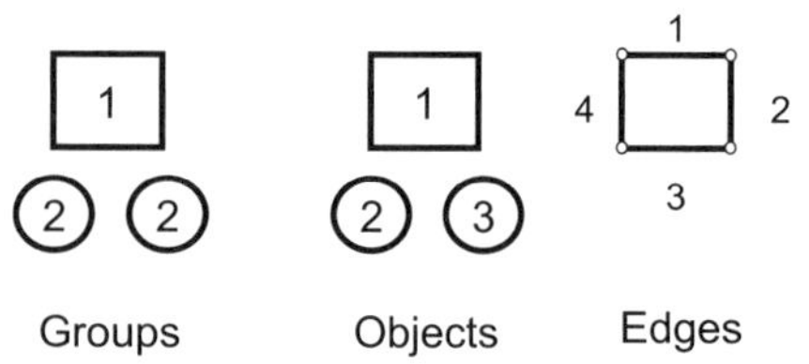

Figure 16.5
Examples of CogSketch's main three levels of representation.

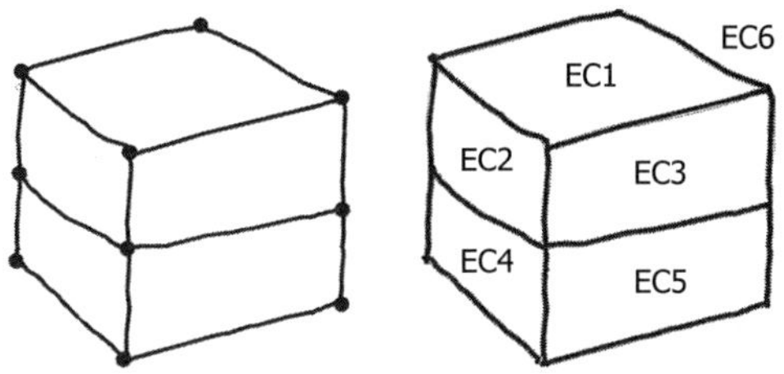

Figure 16.6
Edges and edge cycles, as found by CogSketch on a hand-drawn sketch of two stacked boxes. On the left are the fourteen edges and ten junctions that it found. On the right, the six edge cycles computed for this same sketch are labelled, with EC6 corresponding to the exterior.

constraints. For example, edge-level representations provide detailed information as to an object's shape, but they can be quite large in terms of number of entities and number of relationships between them. This increases working memory load and makes analogical matching more difficult. Consequently, one of the representations we have introduced are *edge cycles* (McLure et al., 2011), which reify connected sets of edges into contours. As figure 16.6 illustrates, this can result in a significant savings in the size of descriptions and hence the complexity of its subsequent processing.

CogSketch is being developed both as a cognitive simulation and as a platform for new kinds of intelligent sketch-based educational software. The cognitive models embodied in it play a critical role in these new kinds of educational software: for example, sketch worksheets (Forbus et al., 2017; Forbus et al., 2018; Yin, Forbus, Usher, Sageman, & Jee, 2010) uses analogy (SME, chapter 4) to compare an instructor's sketch to a student's sketch, in order to provide feedback. While I believe such software has potentially transformative implications for education, we will not discuss it further here.

16.4 Learning Spatial Prepositions

One important aspect of spatial language are spatial prepositions. Prepositions, in any language, are a relatively small set of terms (compared to the number of nouns and verbs) that concisely encapsulate aspects of the spatial world in ways important for that culture. Cognitive science research has revealed a number of important properties of spatial prepositions. First, they vary significantly across languages. For example, the sets of situations handled by the spatial prepositions of contact in English, *on* and *in*, correspond

English	Dutch	Relationship	Example
on	op	Support from below	
on	aan	Hanging attachment	
on	om	Encirclement with contact	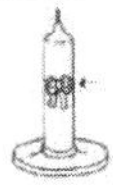
in	in	Containment	

Figure 16.7
Spatial prepositions of contact in English and Dutch. Drawings are from the original Gentner and Bowerman paper.

to four distinctions in Dutch, *op, aan, om,* and *in* (Gentner & Bowerman, 2009), illustrated in figure 16.7.

Roughly, *op* covers cases of support from below, *aan* covers cases of hanging attachment, *om* covers cases of encirclement with contact, and *in* covers cases of containment. Some languages make distinctions that other languages do not (e.g., Korean distinguishes between tight and loose fit, whereas English does not). Interestingly, children appear to shift their sensitivity to spatial relationships as they learn language; that is, children younger than twenty-four months are sensitive to tightness of fit, but as they learn a language, they become less sensitive to that distinction if it is not encoded in their language (Choi, 2006). This suggests that part of language learning may be changing one's default encoding strategies to better reflect making the distinctions that one needs to effectively communicate with others in one's linguistic community.

How might spatial prepositions be learned? Kate Lockwood (Lockwood, Lovett, & Forbus, 2008) explored this issue by using analogical generalization over sketches depicting typical situations. The idea was that a generalization pool would be created for each word that would serve as a model for that word. That is, the generalizations and outlier examples created by SAGE (chapter 4) could then be used to classify a situation as to what spatial preposition would be appropriate to it via analogical retrieval.

Children receive a lot of examples in the course of learning language. However, the simulation is not trying to learn all of the words, the grammar,

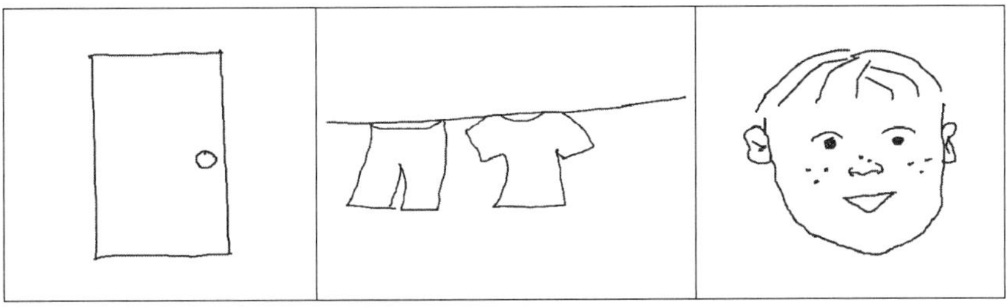

Figure 16.8
Sketched versions of three stimuli involving spatial prepositions of contact. In Dutch, the first two involve *aan* and the last involves *op*, whereas in English, all three involve *on*.

or the discourse conventions for the language. Thus, a much smaller set of example situations might suffice, given the simpler circumstances the simulation is in. The situations used were from a study by Gentner and Bowerman (2009), which examined the hypothesis that, because Dutch has more prepositions than English does for contact situations, Dutch children would be slower to learn them. That indeed was the case. The set of stimuli they used were thirty-two scenarios, eight for each preposition in Dutch, some of which are illustrated in figure 16.8.

These stimuli were used as test cases, but because they are representative, Lockwood used sketched versions of them as a training set. Because conceptual information is relevant in these judgments, the glyphs depicting objects were labeled with concepts from CogSketch's knowledge base (e.g., freckles on a face were declared to be an instance of the concept `PhysiologicalFeatureOfSurface`).

Sketches were made for thirty-two stimuli, eight for each term in Dutch (and hence eight for English *in* and twenty-four for English *on*). For learning each language, a generalization pool was set up for each word (*in* and *on* for English, *in, om, aan,* and *op* for Dutch), and each stimulus was presented with its appropriate label. Cross-fold validation was used for testing, with one stimulus held out for testing and the rest used for training. The system's answer was generated by doing analogical retrieval over the union of the generalization pools, choosing the generalization pool containing the most similar generalization (or outlier). The results are shown in table 16.1.

Even with only eight stimuli per word, the system was able to learn models that produced statistically significant results for all but English *in*. A close examination of the *in* situations revealed that two of them were particularly difficult to represent: "flower in book," meaning flower pressed inside a

Table 16.1
Results of learning spatial prepositions in Dutch and English.

	English				Dutch		
	# Correct	% Correct	*p* Value		# Correct	% Correct	*P* Value
in	6/8	75	<0.2	***in***	6/8	75	$<10^{-4}$
on	21/24	87	$<10^{-4}$	***op***	7/8	87	$<10^{-4}$
				aan	6/8	75	$<10^{-4}$
				om	8/8	100	$<10^{-4}$

book, is difficult to sketch, and "hole in towel" failed due to limitations in ResearchCyc's formalization of holes. With more examples, I expect the model would reach significance on *in* as well. It is important to note that children do not learn this rapidly. The crucial difference, we believe, is that the learning problem faced by the system is radically simpler than that faced by children. Here the situations by design are focused on a particular preposition, without the rest of the contents that would be found in naturalistic settings. Furthermore, using sketches as input provides highly refined representations compared to raw visual input. Given more naturalistic inputs, we expect that many more stimuli would be required for the correct models to emerge via analogical generalization. How that learning speed would compare with human learning speeds is, of course, a fascinating question. In any case, these results are encouraging evidence for analogical generalization over qualitative representations as a model for spatial language learning.

16.5 Reasoning about Depiction

Part of the process of understanding a sketch is interpreting it in real-world terms. This process of decoding sketches is subtle for several reasons. First, the contents of a sketch can be literally anything. Most commonly, they are spatial, of course. But there are graphical conventions, like concept maps (Novak, 1990), which enable even extremely abstract ideas to be sketches. This leads us to the second source of complexity: cultural conventions. The symbols on a map, for example, are not intended as veridical depictions of the entity that they represent: small towns on a road map, for example, may be shown as dots, whereas a map of the town itself will depict the regions within a town. Tourist maps of a town, on the other hand, only focus on particular streets and show buildings representing places of interest (or businesses that have paid to be on it), to suggest things to do for visitors.[5]

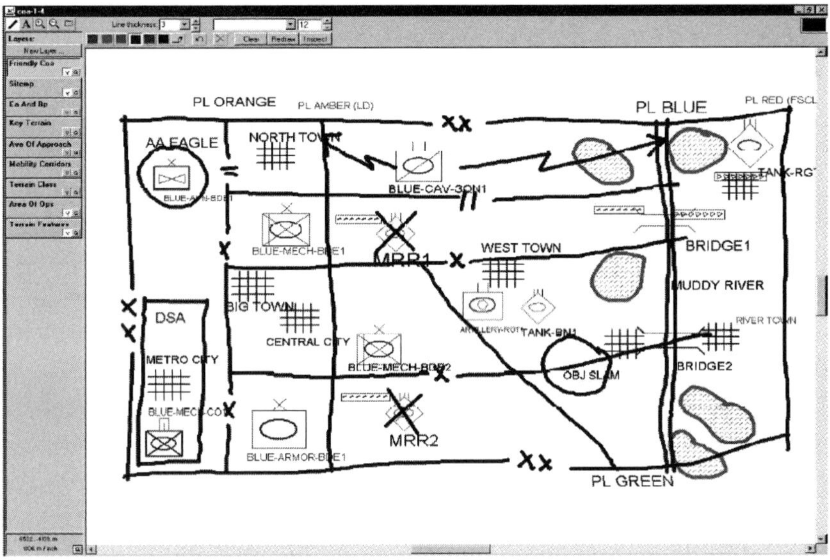

Figure 16.9
A course of action diagram.

Finally, the third source of subtlety is that relationships that are implied by the sketch need to be extracted in order to properly reason about its contents. A crucial component of sketch understanding is a *theory of depiction* that describes how people encode their ideas in sketches. Attempts to create such a theory stretch back to the 1980s (Reiter & Mackworth, 1989), but even today, existing theories are incomplete. Nevertheless, they do provide some insights into how visual, spatial, and conceptual knowledge interact in understanding sketches with cultural conventions.

Let us start with sketch maps, of the kind used in planning military operations (figure 16.9).

These maps might be drawn using markers on acetate overlays on paper maps or digital ink to communicate with others via a distributed database. The relationship between the visual properties of the glyphs and what they express about the real world is surprisingly subtle. By working with reference materials and domain experts, we found that, for the purpose of understanding the visual-spatial translations involved in them, they could be divided into five broad categories (Forbus, Usher, & Chapman, 2003):

- *Location glyphs* indicate the location of something. Military units are an example of a location glyph. Their position matters, but the size that they are drawn says nothing about their strength or the footprint that the people and equipment who make up that unit take up on the ground. The only visual property that matters for extracting location is the centroid of its bounding box.

- *Line glyphs* represent one-dimensional entities whose width of their content is not tied to the width of their ink. Roads and rivers, in most sketch maps, are examples of line glyphs. The width of what they depict is often crucial—how many vehicles at a time can move down a road helps determine how fast a unit can move, and a river that is too wide can serve as an impassable obstacle. But at the scale of the sketch map, it would be too demanding of people's drawing skills to depict it. Hence, the convention is to simply leave width out.
- *Region glyphs* indicate an area whose location and boundaries are significant. Glyphs representing terrain (e.g., mountains and lakes) and areas used in planning (e.g., assembly areas, battle positions) are examples of region glyphs.
- *Path glyphs* are like line glyphs, but their width matters, and they have a designated start and end. In military operations, planned movements need to be constrained to coordinate with other units.
- *Symbolic glyphs* serve as visual referents for abstract entities that need to be described in the sketch. Military tasks are depicted via symbolic glyphs. Sometimes there are conventions that identify role-fillers for the abstract concept, such as drawing the symbol for a defend task around a city being defended. But such conventions vary across experts, which is one reason why sketch maps are always accompanied by a narrative that provides information that is not easy to depict on a sketch.

Incorporating an understanding of these conventions enabled an ancestor of CogSketch to be used as part of a larger experimental system that military officers used to generate battle plans. In an experiment conducted by the U.S. Army, the combined system was found to be easily learned by commanders and enabled them to make plans more quickly, without impinging on their creativity (Rasch, Kott, & Forbus, 2002).

The conceptual labels that people provide when sketching with CogSketch are intended only to capture the level of information that they would provide explicitly to another person when sketching with him or her. This still leaves open the problem of making the everyday inferences that we make about the contents of sketches, based on what we know about the world. Consider, for example, the sketch of the Earth orbiting around the Sun in figure 16.10.

We think of the space inside the Sun and the Earth as part of them, whereas we don't think of the space inside the orbit as being part of the

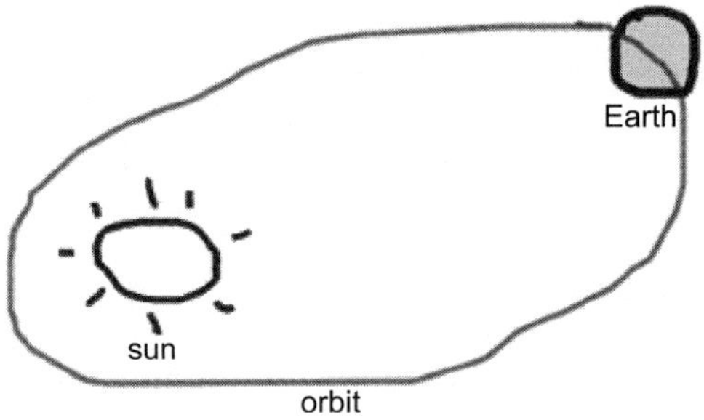

Figure 16.10
A depiction of Earth's orbit around the Sun. Using conceptual and linguistic knowledge, CogSketch is able to infer that the interior of the planet is part of the planet, whereas the interior of the orbit is not part of the orbit.

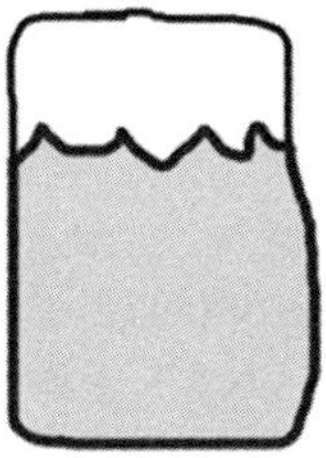

Figure 16.11
From the line indicating water in the container, CogSketch is able to infer that the spatial extent of the water is the region bounded by that line and the interior of the container.

orbit itself. This is an example of what Lockwood, Lovett, Forbus, Dehghani, and Usher (2008) call *conceptual segmentation*. Figure 16.11 illustrates another conceptual segmentation problem: the user has drawn a water tank as one glyph, with another glyph representing the water.

People usually don't draw the entire region that the water takes up; in such cases, they just draw the surface of the water and count on the other participant(s) to figure out what they mean. How can we reliably draw such inferences in a general way?

Lockwood and colleagues' (2008) solution is elegant. She observed that the conceptual and linguistic knowledge in the Cyc knowledge base provides a basis for inferring relevant information about glyphs. If the linguistic expression of an entity is a mass noun (e.g., "water" or "oil") that subclasses from the Cyc concept `TangibleStuffCompositionType`, then its depiction should be an area and, moreover, that it will fill any space

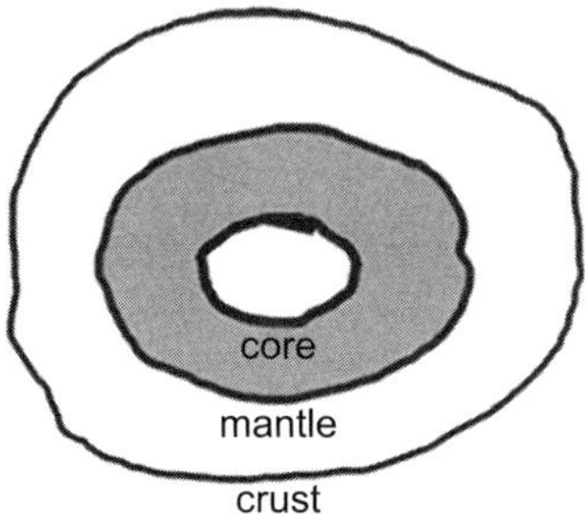

Figure 16.12
Lockwood's theory of depiction can handle interpretation of nested entities. Here, the system highlights the region corresponding to the mantle of the Earth.

below something depicting it. In such cases, CogSketch seeks a visual intersection with something that can be viewed (within the Cyc KB, given its conceptual label) as an instance of a `Container` and uses its visual capabilities to construct the rest of the depicted entity. Similarly, if the entity is an instance of `PhysicalObject`, then in most cases, it simply uses the glyph outline as the extent of the entity (as in the Sun and Earth). An exception are containers: their interior is used as their extent. Finally, if the entity is an instance of `Path-Spatial`, then the lines of the glyph itself are used as its extent. This handles not only examples like figures 16.10 and 16.11 but also nested entities (figure 16.12).

How might such conventions be learned? As you might expect by now, I believe that analogical learning is crucial. Another important problem in depiction is inferring what conceptual relationships should be inferred between a pair of entities, given the visual relationships that hold between them. Consider, for example, the wheelbarrow in figure 16.13.

Visually, there are a lot of contact relationships (partially overlapping or edge connected, in RCC8 terms) between pairs of glyphs, but as figure 16.13 shows, they are interpreted quite differently. The glyph depicting the wheel is touching the glyph depicting the ground, which we take as the wheel being above the ground and resting on it. The glyph depicting the handle is touching the glyph depicting the chassis, which we take as the handle being connected to the chassis. Similarly, the glyph representing the axle is inside the glyph representing the wheel, which we take as indicating that the wheel can rotate around the axis. But the glyph representing the rock is inside the glyph representing the bin, and we take that to mean that the rock happens to be inside the bin. These are conventions, learned through experience.

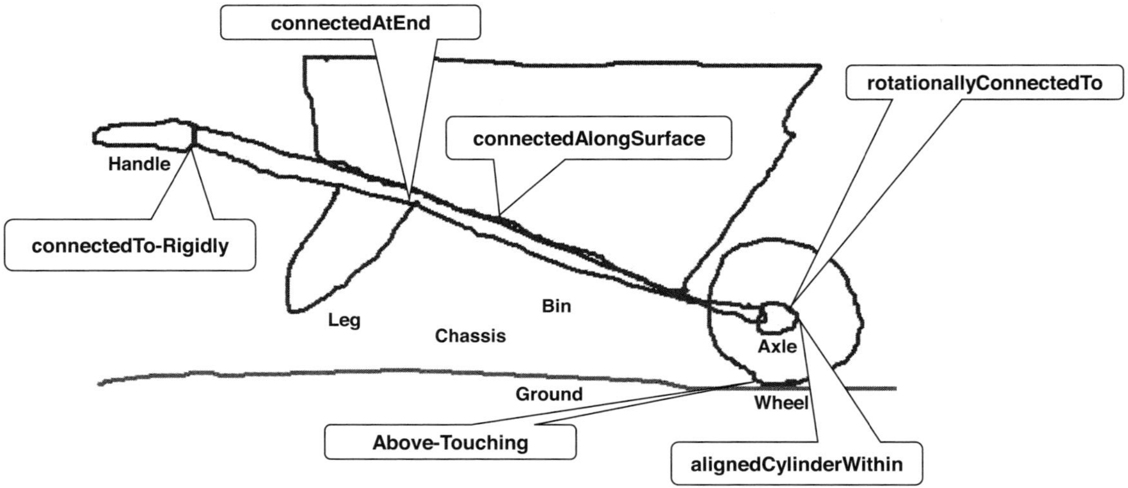

Figure 16.13

Visual relationships in a sketch can suggest conceptual relationships. This sketch of a wheelbarrow is annotated with relevant relationships between the parts.

CogSketch provides some basic links to support inferring such relationships. If a pair of glyphs is touching, then there is the hypothesis that a relationship that is a specialization of `atOrOverlapsInSketch` holds between the two entities depicted. (A similar relationship, `insideInSketch`, is conjectured to hold when one glyph is inside another.) These relationships are connected into the Cyc ontology, such that all plausible relationships that might hold in each case are specializations of these relationships (i.e., `genlPreds`). The set of potential relationships is quite large: the worst case is 204 for `atOrOverlapsInSketch` and 150 for `insideInSketch`, respectively. These numbers are often smaller because type information about the entities can be used to filter the candidates, but typically there are over 100 possible candidates on average. But when we look at a sketch, we typically jump immediately to reasonable assumptions about what relationships should hold. This kind of immediate inference, governed by experience, seems to be the sort of reasoning that can be done via within-domain analogical reasoning.

Can analogical learning be used to gather information about such conventions and apply them to new situations? Let us consider the simplest possible analogical learning model. Given a corpus of sketches representing experience, the simplest method is, when given a new sketch, to use analogical retrieval to find the most similar prior experience (via MAC/FAC) and then use the candidate inferences produced by the mapping as conjectured answers to the question, "What does this visual relationship in the sketch imply about the conceptual relationship that holds between the objects depicted?" This simple model ignores analogical generalization, rerepresentation, and iterated retrieval, all of which I suspect happen when people are learning and applying these conventions. But if this baseline does well, it suggests that these additional processes are definitely worth exploring in this context.

To gather evidence about this, a corpus of sketches depicting multipart physical objects (e.g., wheelbarrows, shopping carts, bicycles) and simple physical situations (e.g., someone balancing on a tightrope holding a pole) was created, based on situations found in the Bennett Mechanical Comprehension Test, which was originally developed for testing candidates for technician jobs but has also found a role in testing for spatial ability in cognitive psychology research. The sketches were drawn by multiple graduate students, who used the built-in facilities in CogSketch to provide conceptual relationships for pairs of glyphs by inspecting the (rather large) set of candidates it provided in each case. These fifty-four annotated sketches formed the system's experience. Using cross-fold validation, each sketch was tested by removing it from the case library, stripping it of its conceptual relationship annotations, and seeing what relationships could be inferred via analogical retrieval. This gave rise to 181 visual-conceptual relationship questions. Questions were scored by giving 1 point for a correct answer, 0.5 if the suggested relationship is one step away from the correct relation in the `genlPreds` lattice, and 0.25 if it was two steps away. The system's score with this rubric was 74.25, which is statistically significant (chance=24.2, $p<10^{-5}$). Coverage was only 54 percent, but for those questions in which an analog was found, accuracy was 87 percent (Forbus, Usher, & Tomai, 2005). A second experiment was conducted with a new set of sketches covering a larger range of phenomena (e.g., "a boat moving in water," "a bicycle"), where those doing the drawing were told to draw them in enough detail to allow someone to illustrate their principles of operation. Again, cross-fold validation was performed, using twenty sketches (ten systems, drawn by two graduate students), yielding a set of 138 questions. Coverage in this experiment was slightly smaller (46 percent), due in part to differences between sketches, but the system's performance (21.75, using the same scoring rubric) was still statistically significant. The differences in drawing style between people generating the sketches may or may not be a reasonable model for encoding variability within a single individual, but it does provide some assurance about the robustness of retrieval and matching over these sketched representations.

These experiments are very encouraging. Simple one-pass analogical retrieval provides reasonable accuracy on estimating conceptual relationships based on visual relationships plus the types of the entities involved. Coverage is lower than one might like, but there are two major differences between this experiment and what is going on when people look at these sketches. First, as mentioned above, analogical generalization, re-representation, and iterated retrieval all are likely to be occurring, and these should all increase coverage. Second, the amount of experience provided in the experiment is tiny compared to human experience. Again, one would expect an increase in coverage with more experience, although there could also be more false positives—larger-scale experimentation will be required to explore this. But nevertheless, these results seem very encouraging.

16.6 Modeling Visual Problem Solving

Psychologists have long used visual analogy problems to evaluate human intelligence. Consequently, the first computer model of analogy tackled the problem of geometric analogies (Evans, 1968). Evans only had access to an

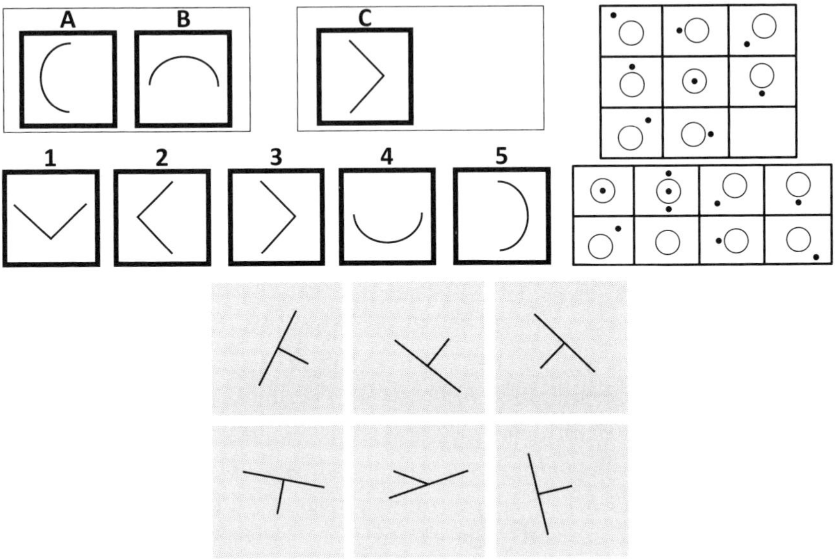

Figure 16.14
Examples of three kinds of visual problem-solving tasks. The top left is an example of the classic Evans geometric analogies task. The top right is an example of the kind of problem used in the Raven's Progressive Matrices task. The bottom is an example of a visual oddity task.

IBM mainframe that had less computing power than today's cell phones. Despite this, Evans managed to make an interesting system that was able to solve such problems.[6] Figure 16.14 shows an example of this kind of problem, along with two similar kinds of problems.

The Raven's Progressive Matrices task is like geometric analogies, albeit with the need to coordinate representations across both rows and columns. They are progressive in the sense that the difficulty of problems increases within each section of the test. The Raven's test has been shown to strongly correlate with general intelligence (Snow, Kyllonen, & Marshalek, 1984), which suggests that it goes beyond a test of simply visual processing and taps into general cognitive abilities. The visual oddity task is from Dehaene, Izard, Pica, and Spelke (2006), in which participants are asked to pick the image that doesn't belong. This test was used to compare geometric abilities of North Americans and Mundurukú, a South American indigenous group.

Visual and spatial thinking are general-purpose capabilities. The tendency for research in cognitive modeling has been to focus on modeling a single task (e.g., geometric analogies [Schwering, Kuhnberger, Krumnack, &

Gust, 2009] or Raven's [Kunda, McGreggor, & Goel, 2013]). But just as the same person is capable of doing all of these tests, a robust model should also be able to do them all. Using the same representations and processes across multiple tasks helps reduce tailorability. It can be viewed as a kind of *cognitive tomography*: tomography builds a 3D picture of what is inside something by integrating the results from multiple X-ray snapshots, each taken at a different angle. Using the same model across multiple tasks imposes stronger constraints on the representations and processes that it uses.

Andrew Lovett has developed a computational model for human visual problem solving, incorporated in CogSketch, that can solve all of these kinds of problems. It operates at human levels of performance, and what is hard for people tends to be hard for the model and vice versa. It has yielded new predictions that subsequently have been verified with behavioral experiments. The remainder of this chapter outlines his framework and some of the results obtained with it to further illustrate the power of qualitative representations and analogy.

Lovett's *spatial routines* framework (Lovett, 2012) is inspired by Ullman's (1984) visual routines idea. The idea of visual routines is that, in later stages of perception, computation starts becoming more strategic, as opposed to the highly parallel pipelined operations in earlier stages. Ullman proposed that a set of basic operations, like tracing along a curve, could be parameterized, much like a procedure call in a software library. Medium-level visual tasks are performed by routines (i.e., programs) that call these basic operations, plus other visual routines. Basic operations and perhaps some routines are part of our starting endowment, whereas most routines are learned. One source of individual differences could, in this proposal, be learning better or worse routines for a task.

Lovett's spatial routines framework applies this same idea to even higher levels of visual problem solving. It is based on three hypotheses: (1) people typically use qualitative spatial representations, resorting to quantitative representations when necessary; (2) spatial representations are hierarchical; and (3) qualitative spatial representations are compared via structure mapping. Arguments for the first two hypotheses have already been presented in this section of the book, and in fact, the particular hierarchy of representations used in CogSketch came out of Lovett's work. The hypothesis that structure-mapping operations also provide a model of visual comparison may be surprising to those who think of analogy only in conceptual terms, so let us examine it in more detail.

There is a growing body of evidence that structure-mapping principles guide visual comparison (e.g., Lovett, Gentner, et al., 2009; Sagi et al., 2012).

Recall that structure mapping provides several kinds of useful information: it identifies what goes with what when comparing two situations, thus highlighting commonalities. It produces conjectures about the target and base via projecting unmapped structure between them. These inferences also serve to highlight differences, what are called *alignable differences*. Alignable differences are psychologically very salient compared to nonalignable differences. An example of an alignable difference is a difference in color, whereas an example of a nonalignable difference is extra entities in one situation versus the other. One prediction of structure mapping that has been found to hold in both conceptual and visual stimuli is that, when two items are very dissimilar, it is faster to say that they are dissimilar than it is to name a difference, but when two items are very similar, it is faster to name a difference (Sagi et al., 2012). SME is the only model of analogy that predicts this.

Structure mapping, Lovett (2012) found, provides an elegant way of modeling mental rotation. The phenomenon of mental rotation has been heavily studied. In the simplest form of the task, two figures are shown and participants are asked to say whether they are the same thing. In some cases, one figure is the same as the other figure but rotated, in two or three dimensions, depending on the experiment. In other cases, the two figures are actually different but placed at different orientations as well so that it is not easy to see that they are different. A robust finding is that, in many circumstances, the time it takes to confirm that two stimuli are the same is a function of the angular distance between them (Shepard & Cooper, 1982). This is interesting because, unless one postulates massive parallelism, it assumes that people already know the direction of rotation that should be used when they compare the two stimuli. Lovett argues that mental rotation occurs as a two-step process. The first step uses SME on orientation-independent qualitative representations. The best mapping from this step provides a basis for quickly ruling out pairs that are very different and hypotheses for which elements correspond qualitatively. The second step estimates an angle of rotation for pairs of corresponding parts, computes a transformed representation, and does a quantitative comparison to see if they do indeed match.

The spatial routines for sketches framework incorporates three types of basic operations:

1. *Visual perception* generates a qualitative, symbolic representation of a sketch (or portion thereof). This step produces humanlike representations, but it does not necessarily attempt to do the processing in a humanlike way. It has parameters to control which level of the hierarchy it should operate at (e.g., edges, objects, groups), and the symbolic

representations it produces are always grounded in quantitative representations (i.e., CogSketch serves as a metric diagram).

2. *Visual comparison* describes the commonalities and differences between a pair of representations, using SME. Because SME is domain independent, these operations can be performed on the symbolic representations of sketches, but also to generalize across a set of differences or to find the differences between generalizations. In other words, it is capable of doing higher-order mappings.
3. *Visual inference* applies differences found by visual comparison to a portion of the sketch to infer a new representation (e.g., a candidate answer to a problem). It does this by adding or removing qualitative relations and applying quantitative shape transformations.

These operations have been used to build spatial routines for each of the three tasks illustrated in figure 16.14. Because psychologists typically use PowerPoint to render their stimuli, Lovett used CogSketch's ability to copy/paste from PowerPoint to avoid hand-generation of the stimuli in all of these experiments. Let us examine his results for each class of problem in turn.

16.6.1 Geometric Analogies

In this task, participants are asked the classic proportional analogy question, "A is to B as C is to ?" Psychologists have proposed two competing models for how people perform it. In terms of the spatial routines framework, we can think of them as follows:

1. Visual inference (Sternberg, 1977) compares A and B to find the differences and then compares A and C to find their corresponding elements. The A/B differences are then applied to the corresponding elements in C to infer an answer. This inferred answer is compared to the choices offered, with the most similar being chosen.
2. Second-order comparison (Mulholland, Pellegrino, & Glaser, 1980) also compares A and B to find the differences. But then it compares each answer choice *AC* to C to get the difference between *C/ac*. It selects the answer choice whose differences are most similar to the A/B differences.

Evans's (1968) model also used the second-order comparison strategy, albeit with different domain-specific matching processes for the two types of comparisons. Lovett (2012) observed that actually both strategies are found in people. There are good reasons for this. Consider the number of comparisons required in each strategy. Visual inference requires fewer comparisons (i.e., two to construct the answer, plus four to compare the inferred answer with the proffered choices, so it is more efficient). However, it is not always possible to do, because the answer cannot be inferred or the inferred answer doesn't match any of the choices. In such cases, people may revert to second-order comparison, even though it is less efficient (i.e., five comparisons of images, followed by four second-order comparisons of the differences). Lovett's model predicts that such problems will take longer for people.

To evaluate the model, the twenty problems from Evans (1968) were drawn in PowerPoint and given to both human participants and the model. Overall, the model chose the answer preferred by people on all twenty problems. This is an easy task, so error rates were not considered, but reaction times were recorded to check the prediction of backtracking. A linear regression was calculated, using as factors the number of objects, whether the model reverted to second-order comparison, and whether the model made additional strategy shifts during problem solving. These factors accounted for 0.95 of the variance in human response times (i.e., $R^2=0.95$). The best match was achieved by using the number of objects in the A/B differences, rather than the number of objects in the answer image, as the measure of working memory load (Lovett & Forbus, 2012). Lovett conjectures that people may have more problems keeping track of abstract differences in working memory compared to concrete image representations. This could be explained by people accessing special-purpose perceptual processing capabilities for doing the concrete image comparisons and using more general-purpose comparison capabilities for comparing the differences.

16.6.2 Raven's Matrices

Raven's Progressive Matrices (Raven, Raven, & Court, 2000) is a visual intelligence test. There are three versions of the test; what we focus on here is the Standard version.[7] It starts with a section incorporating simple texture comparisons, followed by some 2×2 matrix problems, and ends with 3×3 matric problems, like those shown in figure 16.14 (top right). Lovett's model for this task uses the same basic strategy as for geometric analogies. It starts by trying to use visual inference, applying differences in the upper rows to the bottom row to attempt to infer an answer. It reverts to second-order comparison if needed, plugging in each possible choice and comparing the representation it gets for the bottom row to the others.

Raven's problems are much harder than Evans's geometric analogy problems. Consider figure 16.15 (top).

As per the hypothesis that people start with the highest-level representations that they can, the representations for the top two rows each include

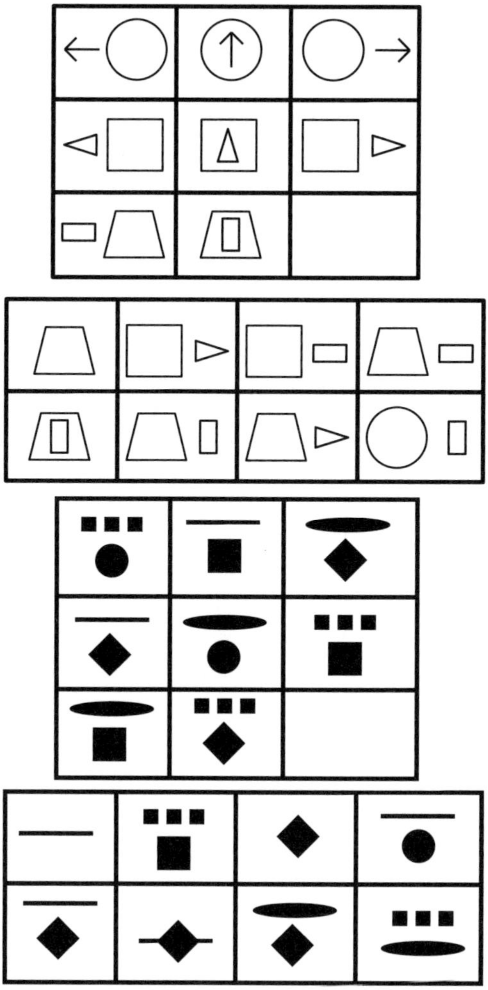

Figure 16.15
Problems in the style of Raven's Progressive Matrices. Actual Raven's problems are not shown to protect the security of the test.

two objects, one whose shape and orientation are constant across the rows (circle in the top row, square in the second row) and an object with changing orientation (arrow in the top row, triangle in the second row). Comparing the corresponding elements enables the model to discover that the orientation of the changing objects is the same in each column and corresponds to a 90-degree rotation and a translation between each column. This enables it to project what the missing element in the third row should be qualitatively, which allows it to match this answer with the qualitative descriptions for the candidate answers. The model is capable of perceptual reorganization, a form of re-representation. For example, it can break up groups into objects and objects into edges when doing so is necessary to form a consistent pattern describing changes across a row. There are several types of such reorganization. For example, the model by default represents a row via differences between adjacent images.

Consider, however, figure 16.15 (bottom). Here the important properties are the common elements in each row rather than the differences. The model compares the top two rows to determine whether they are best represented in terms of differences versus common elements, picking the representation that leads to them being viewed as most similar.

The model does quite well on the Standard test, which consists of sixty problems. It correctly solved fifty-six problems, which, according to 1993 norms, places it in the 75th percentile for American adults. Importantly, the four problems it failed on were among the five hardest for human participants. A parallel behavioral study was conducted to look at how well the model could account for human reaction times. A subset of thirty-one problems that students did well on (leaving out texture comparison problems and problems that people did badly on) were selected for further analysis. These problems were coded for number of elements and strategy shifts required to solve them. The coding was done by ablating the model, selectively removing the ability to perform an operation and seeing which problems could no longer be solved. The model explained 80 percent of the variance in human accuracy ($R^2=0.8$), as found via linear regression (Lovett, Forbus, & Usher, 2010; Lovett & Forbus, 2017). As expected, perceptual reorganization imposed a significant cost. Moreover, consistent with the results of the geometric analogies experiment, the working memory load cost was higher for problems involving a difference strategy, relative to a common-elements strategy. This again suggests that remembering abstract differences is harder than remembering image contents.

16.6.3 Oddity Task

In an oddity task, participants are asked to find the image that "doesn't fit" or "doesn't belong." Figure 16.14 (bottom) shows one of the forty-five problems that Dehaene et al. (2006) used to explore geometric processing across cultures. Each problem was intended to measure whether a particular spatial concept was understood, such as perpendicular lines, orientation, concavity, containment, and symmetry. One culture, the English-speaking North Americans, has extensive schooling and English has a strong vocabulary for geometric concepts. The other, the Mundurukú, an indigenous South American culture, does not conduct formal schooling, and their language does not have words for most of the concepts involved. The Mundurukú performed above average on most problems, like the North Americans, which they took as evidence for an innate geometry module, available to all humans.

A close look at their data, however, suggests that some of the capabilities needed for this task are learned. For example, the North American adults outperformed the North American children, whereas the performance of children and adults for the Mundurukú was the same (Newcombe & Uttal, 2006). Lovett and Forbus (2011) proposed that these differences may lie in differences between representations between the two groups. That is, the visual processes (i.e., structure mapping across qualitative representations) are universal, but groups may vary how well they encode particular spatial relations or how well they use particular levels in the spatial hierarchy. To test this hypothesis, Lovett built a spatial routine model for oddity tasks. It operates by performing analogical generalization over half of the images (i.e., the top or bottom row) and then comparing that generalization to each remaining image. If one image is noticeably less similar, it is taken to be the answer. Like the other models, it starts with the top (group) level of the hierarchy, moving downward until it finds a good solution. This leads to two predictions: (1) people will identify the odd image more easily if it varies qualitatively from other images, and (2) people need to identify the level for which the salient difference is found.

Lovett compared the model against four groups (Lovett & Forbus, 2011). The Mundurukú were treated as one group, because their performance did not depend on age. The North Americans were divided into three groups, ages four to eight, eight to twelve, and adults (age eighteen to fifty-two). The model correctly solved thirty-nine of the forty-five problems. This is comparable to the highest-performing group, the North American adults, who were at 83 percent compared to the model's 87 percent. Moreover, the model's error patterns correlated well with human errors on these problems. This was found by assigning for each problem 1 if correct and

0 if incorrect. The lowest correlation was with the Mundurukú ($r=0.49$), and the highest was with American adults ($r=0.77$). Thus, the model is doing a reasonable job of capturing the overall behavior of North American adults.

To examine the differences in performance more deeply, the thirty-nine solved problems were analyzed further. As with the other models, each problem was coded for what operations were needed to solve it (via ablation) and for the number of elements required, and linear regression used to compare the ablated models with each group's performance. Some factors were common across cultures. For example, both groups had difficulties with problems involving shape comparison. Such problems require that participants first encode the internal structure of an image before comparing it, which might explain this difference.

Importantly, there were differences between the two cultures. North Americans had more difficulty with edge-level representations. Moreover, this factor was significant for North American children but marginally significant for North American adults, suggesting that education matters. The Mundurukú, by contrast, had more difficulty with problems requiring group-level representations. Because the basic shapes were geometric elements for which English has names, one possible reason the Mundurukú might be worse at these problems is that learning geometric categories (e.g., circle, square) provides an encoding advantage. Because this task did not require encoding abstract differences, number of elements was not a factor in the correlations, consistent with the previous models.

16.6.4 What Makes an Effective Visual Problem Solver?

Lovett (2012) observes that these results, taken together, make some interesting suggestions about what it takes for someone to be an effective visual problem solver. First, as always, working memory capacity matters. But more specifically, the capability for remembering more abstract representations is particularly important. Second, flexible re-representation capabilities, here in the form of perceptual reorganization, matter. Whether or not working memory capacity for particular types of information is malleable is still an open question. It seems likely that perceptual reorganization strategies can be improved with practice (Uttal et al., 2013), leading to the possibility of improving people's visual thinking.

Another factor, of course, is the set of problem-solving strategies that individuals have. The spatial routines approach provides two exciting new opportunities. The first is to model individual differences by creating sets of routines that can explain the performance of particular individuals, as measured by close comparison on their performance in multiple tasks. The second is to model the process of learning spatial routines. This could have several important benefits, such as better understanding the developmental trajectory of visual thinking and, to the extent that the basic operations and lower-level routines are malleable, finding new ways of training people to become better visual thinkers. Given existing evidence for the rapidity of analogical learning, it is probably important to pursue these opportunities in tandem, because there is probably learning occurring even in the number of trials used in many behavioral experiment paradigms.

16.7 Summary

Sketching is a sweet spot for exploring human cognition, as others in cognitive science have found (e.g., Gagnier, Atit, Ormand, & Shipley, 2017; Scheiter, Schleinschock, & Ainsworth, 2017; Sheredos & Bechtel, 2017) and this chapter further illustrates. Equating sketch understanding with recognition, as many do, is an oversimplification. The deepest problems concern meaning, what the sketch is depicting, which relies heavily on world knowledge and context. Fluently constructing representations, at multiple levels, and re-representing as needed appear to be central in understanding sketches. Moreover, combining a rich set of automatically constructed visual representations with analogical reasoning provides a surprisingly powerful explanation for multiple visual problem-solving tasks, including mental rotation, geometric analogies, oddity tasks, and Raven's Progressive Matrices.

Ashok Goel's group has also explored the use of analogy and visual reasoning, integrating it with their structure-behavior-function models to provide a geometric layer for reasoning about designs (Yanner & Goel, 2008). Their work provides another line of evidence for the utility of analogical reasoning in visuospatial problems, especially for problems that integrate spatial reasoning with conceptual knowledge. They use a domain-specific matching algorithm, but I suspect that SME, given its ability to incrementally extend matches, would work on their representations as well.

Although the progress discussed in this chapter is exciting, much remains to be done. For example, the 3D representations in CogSketch are somewhat rudimentary. This may not be incompatible with accounts arguing that human vision is mostly viewpoint specific but would be with accounts that argue for rich 3D models. Another line of investigation is better understanding the nature of depiction by building a broad catalog of visual concepts and their depiction conventions to uncover patterns. Finally, the hypothesis that the visual representations constructed by CogSketch can provide a useful model for human high-level vision can be further tested by integrating it with models of lower-level vision (e.g., edge finding, stereopsis, lightness).

IV Learning and Reasoning

So far, we have explored the panoply of qualitative representations and how they can be used in reasoning and communication in a wide variety of tasks, providing evidence for their utility and for their psychological plausibility. We have already seen some examples of how such representations tie into other forms of conceptual structure, as in the discussion of how qualitative representations play a role in natural-language semantics in chapter 13 and the discussion of how qualitative representations play a role in high-level vision in chapters 14 and 16. Here we go further in explicating the roles for qualitative representations in human conceptual structure by looking three important topics:

- Chapter 17 describes how qualitative representations and analogical reasoning have been used to model learning and conceptual change. A framework for mental models in physical domains is presented, which motivates a number of computational explorations of important problems. Using analogical generalization to learn prototypical behaviors (*protohistories*), we argue, provides a starting point for learning about the world, which are then refined to construct first-principles knowledge. Friedman's *assembled coherence theory* of conceptual change shows how to bring together two competing views of conceptual change, where the fragmentation proposed by the *knowledge in pieces* view is explained by composable model fragments that are used to locally model phenomena, with the within-explanation consistency proposed by the *theory theory* view arising from qualitative reasoning within local examples. Cross-domain analogical learning is also examined, showing how persistent mappings can be accumulated to support incremental learning.
- Chapter 18 explores commonsense reasoning further. It contrasts some commonly proposed views of commonsense in AI research with the view proposed here and draws upon results from cognitive development

research to further support the idea that human commonsense reasoning is heavily based on qualitative representations and analogical reasoning. An important aspect of commonsense reasoning is that it often includes quantitative estimates, not just qualitative information. How qualitative representations provide a framing mechanism for analogical estimation is described, along with Paritosh's general theory of back-of-the-envelope reasoning. Most of the examples in the book so far have been about the physical world, because physical domains have seen the most attention in the qualitative reasoning community. But they are more widely applicable than that, as the rest of chapter 18 demonstrates. It examines how qualitative representations and analogy can be used to account for conceptual metaphors. Finally, it examines how qualitative representations can be used to model problems in social reasoning, that is, aspects of emotions, blame assignment, and moral decision making.

- Chapter 19 explores how qualitative representations are used in systems that perform expert reasoning. The ability to perform at human levels on engineering and scientific tasks provides strong evidence for the importance of qualitative representations in human expert reasoning. This chapter looks at the key problems in engineering reasoning (i.e., analysis, monitoring, control, diagnosis, design, and system identification), showing how qualitative representations and reasoning have been used to build systems that solve them. It also looks at how qualitative representations have been used in systems that perform scientific modeling, including dynamical systems, genetic regulatory networks, and ecosystem modeling.

17 Learning and Conceptual Change

We have examined how people use qualitative models to reason about the continuous world. In exploring how analogy interacts with qualitative representations to simulate based on experience (chapter 12), we have seen the beginnings of how such knowledge might be learned. This chapter delves into learning and conceptual change in more detail. I start by laying out the broad theory that Dedre Gentner and I have been building about the range of types of human representations that get constructed in the course of learning in physical domains. Then I will drill down into particular aspects of this framework, beginning with how prototypical behaviors are constructed and used to build initial qualitative models. Because many of our concepts come from our culture, rather than discovered on our own, a model for conceptual change is described that accounts for how models are changed in response to new information. The use of analogy to learn expert models, both within and across domains, is also discussed.

17.1 A Framework for Mental Models in Physical Domains

Figure 17.1 illustrates a framework that Dedre Gentner and I have been developing since the 1980s (Forbus & Gentner, 1986b). People start with experiences, which they remember and use by analogy to make predictions about new situations. Explanations that they are given or construct about those particular behaviors are applied by analogy to new situations, thereby providing a form of analogical abduction. These behaviors are also used to construct generalizations, using the SAGE (chapter 4), which provides more abstracted descriptions that can be transferred to a wider range of new situations. These abstracted behaviors are called *protohistories*, because they represent prototypical behaviors in terms of what happens rather than what

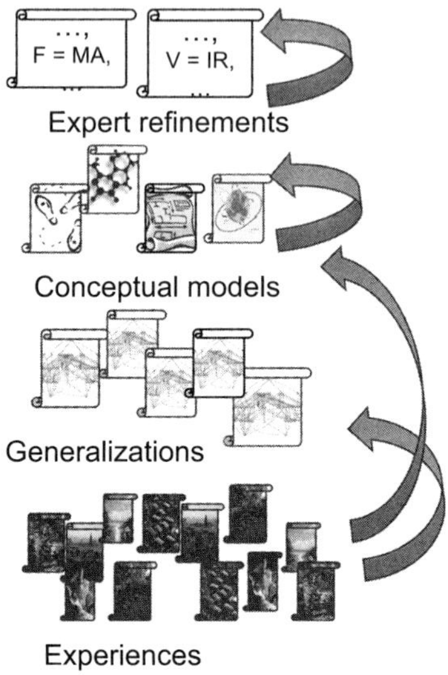

Figure 17.1
A framework for learning mental models in physical domains.

generates them. Many conceptual models are grounded in such experience (e.g., that something initially moves faster when you push it harder). Such conceptual models are added to protohistories by introducing causal, explanatory relationships (such as qualitative proportionalities), based on statistical properties and comparisons.[1] Many conceptual models come from our culture (e.g., our notions of thermodynamic properties like specific heat and entropy). Part of a learner's job is to tie these culturally provided models to their experience appropriately. (As any educator knows, this is difficult and often works only partially.) Expert refinements to conceptual models are almost all culturally derived, and although still linked to protohistories and conceptual models, their internal structure of rich, nested relationships (i.e., the explanations and proofs of formal fields of study) means that people can operate solely within them for long periods, deriving powerful and subtle consequences that would not be possible with less precise models. But even expert refinements exploit qualitative representations as part of model formulation and evaluation (as discussed in chapter 11).

Let us begin at the beginning, with how protohistories might be formed and what can be done with them.

17.2 Learning Protohistories

Protohistories are generalizations constructed from specific behaviors. We assume that specific behaviors are described in qualitative terms, because such representations are relatively easy to extract from sensory data. For example, our visual systems compute direction of motion, which directly corresponds to the sign of derivative of position. We also assume that some estimates of quantitative values are constructed when feasible. The possible nature of these estimates is discussed in chapter 18. Behaviors over time are separated into qualitative states, within which the qualitative properties perceived remain constant. For example, when something stops or starts moving, there is a change in sign of derivative and hence a change in qualitative state. Other perceptible properties are used to divide behavior into states as well. For example, if a rock is placed on the surface of a lake, it quickly goes into the lake and moves downward until it reaches the bottom. Being on the surface versus going into the lake versus being totally immersed in the lake correspond to different RCC8 relationships (chapter 15). That these relationships are important are also signaled by language (i.e., the rock is on the surface of the water, the rock is in the water, the rock is on the bottom of the lake). Changes in linguistic description thus also serve as clues for how the world should be divided into meaningful pieces.

As experience accumulates, predictions can be made. Rocks placed on water sink. Trying to apply this prediction to a leaf placed on water leads to a failure of expectations: the leaf remains on the surface of the water. This expectation failure, especially if accompanied by an observer noting that "the leaf is floating," suggests that there is a distinction here worth attending to. Recall that SAGE organizes behaviors into *generalization pools*, which have an entry pattern that indicates whether or not something should be included in it. (Any new example can be added to multiple generalization pools, because it could include examples of multiple new concepts.) At this point, generalization pools for something floating and something sinking might be set up to accumulate behaviors whose generalizations can be compared and contrasted to come up with a more accurate theory of why these different outcomes occur. More attention will be paid to behaviors relevant to those generalization pools, leading to a tendency to encode and/or retain more quantitative information when possible. The use of QP theory to represent mental models provides a strong inductive bias: presumably, some parameter has a limit point that determines whether or not something will float or sink. As experience accumulates in these generalization pools, the data needed to make reasonable hypotheses are collected.

"The woman *bodyInLiquid0* floats in water *liquid0* in a pond *container0*. The mass of the woman *bodyInLiquid0* is 60 kilograms. The volume of the woman *bodyInLiquid0* is 62039 cubic centimeters. The woman *bodyInLiquid0* is moving but the water *liquid0* is standing still."

NLU

```
(isa bodyInLiquid0 AdultFemaleHuman)
(isa container0 Pond)
(isa liquid0 (LiquidFn Water))
(in-UnderspecifiedContainer liquid0 container0)
(massOfObject bodyInLiquid0 (Kilogram 60))
(volumeOfObject bodyInLiquid0
  (CubicCentimeter 62039))
(isa gliding0 MovementEvent)
(primaryObjectMoving gliding0 bodyInLiquid0)
(isa stillLiquid0 StandingStill)
(doneBy stillLiquid0 liquid0)
(in-Floating bodyInLiquid0 liquid0)
```

Figure 17.2
An example of a stimulus for learning about floating versus sinking. The version on the left is the simplified English given to a natural-language understanding system, which produced the predicate calculus assertions on the right. The use of quantitative values and units is a simplification.

Scott Friedman (Friedman & Forbus, 2008) developed a computational model that demonstrates the feasibility of this account. Sixteen descriptions of floating and fourteen descriptions of sinking were given to the simulation in simplified English, as illustrated in figure 17.2.

The rules he used in constructing the stimuli were that objects float when they move autonomously, when the water has a current, or when their density is less than 0.001 kg/cc. Density, however, was not encoded explicitly, although weight and volume were. (Precise quantitative values were used for convenience; the same algorithms would work if they were cruder estimates.) Given two alternate generalization pools G_a and G_b, the strategy used to conjecture quantity conditions was to look for ordinal relationships between parameters in their protohistories that were uniformly true for protohistories in G_a and false for those in G_b or vice versa. Sometimes, the limit point to be used in the comparison is straightforward (e.g., zero). However, sometimes the limit point must be derived from examining quantitative estimates in the protohistories. In the floating protohistories, for example, the values for density will always be lower than the values of density for the sinking protohistories. Thus, the limit point for density would lie in the interval defined by the minimum of density for the sinking protohistories and the maximum of density for the floating protohistories. As noted above, density was not explicitly encoded. Following Langley's (1981) BACON, rules are used to suggest new, possibly relevant quantities based on combinations of existing quantities (i.e., using sums, differences, products, and quotients of existing quantities). Such derived quantities are only sought when the strategies for finding quantity conditions and limit points fail. The nature of the combined quantity is also used to derive qualitative proportionalities, thereby providing a causal model for the new parameter. The new quantity introduced to explain the difference between these stimuli is

```
Q=(/ (mass body) (volume body))
Q<[0.001, 0.00102] kg/cc
(qprop Q (mass body))
(qprop- Q (volume body))
```

This model does not address a number of issues needed in a full account of how people construct such conceptual models. For example, it does not provide criteria for when to give up or when to change encoding strategies based on a failure to find a good model with the information currently being gathered. Nor does it describe where behaviors should be stored when questions have not yet been identified. (It seems likely that there are default generalization pools used for this purpose.) It also does not specify how a complex experience describing multiple behaviors should be further subdivided into pieces for best focusing analogical generalization. Nevertheless, it seems to be a promising start.

The issue of carving up complex behaviors has been examined in modeling how intuitive notions of force and motion are learned. There is a sizable body of research on misconceptions that students have concerning force and motion (e.g., diSessa, 1993; McClosky, 1983). Interestingly, these misconceptions can persist after physics instruction, even in students who do well in traditional quantitative physics problems (Clement, 1983; Halloun & Hestenes, 1985). The distributed model of qualitative reasoning described here provides a natural explanation for this. Given an everyday situation, the intuitive, causal models formed by experience are retrieved and used. In a classroom setting, such models might initially be retrieved, but because they do not prove effective in solving problems posed there, new experiences are remembered that incorporate the explanations from textbooks and include the operationalization of equations (at whatever level that the student has been able to achieve). Subsequent attempts to solve similar problems retrieve (one hopes) prior examples of problems from physics class or generalizations constructed from them. The speed of analogical generalization, in terms of small number of examples, enables people to construct more broadly transferrable descriptions from even a handful of worked examples. However, these new generalizations do not replace the generalizations formed by experience, which presumably are quite strong (as measured by the number of situations they have been successfully used in), and hence experience-based generalizations are quite likely to be retrieved when thinking about the everyday world. This explains the repeated finding that people's misconceptions often persist despite instruction. And yet, in many

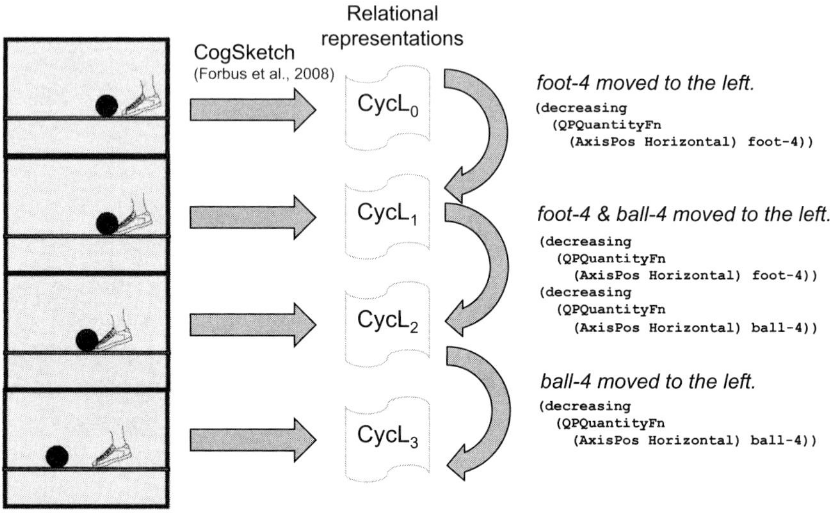

Figure 17.3
Using sketched comic strips to produce stimuli for modeling conceptual change.

people, such misconceptions do seem to get uprooted and replaced. How this might happen, we return to shortly.

How might intuitive models be learned from experience? To model the experience of motion, Friedman and Forbus (2010) used the Companion cognitive architecture, which integrates qualitative reasoning, analogy, and CogSketch. Friedman sketched *comic strips* that indicate successive frames of motion (figure 17.3).

CogSketch (chapter 16) was used to sketch these strips. Its visual processing capabilities were used to construct qualitative spatial representations of the visual properties in each panel. CogSketch's conceptual labeling facility was used to identify each type of object, serving as an approximation for recognition (here, foot and ball; see figure 17.3). Changes between panels were automatically computed by comparing corresponding visual quantities between frames (e.g., the foot moving to the left, followed by the ball and foot moving to the left, and then the ball moving to the left).

In Friedman's *assembled coherence* theory of conceptual change (Friedman, Forbus, & Sherin, 2018), all changes in quantities need to be explained. In this example behavior, the motion of the ball needs to be explained. Because initially the Companion has no process for describing motion, one is introduced. Because the ball is moving horizontally, it introduces a process with a direct influence that affects its horizontal velocity (see figure 17.4).

```
(defModelFragment M1
 :participants ((?e :type Entity))
 :conditions nil
 :consequences ((Quantity (rate ?e))
                (> (rate ?e) zero)
                (I+ (AxisPos Horizontal ?e) (rate ?e))))
```

Figure 17.4
An initial process model for motion.

```
(defModelFragment M2
 :participants ((?e :type Entity))
 :conditions ((> (q ?e) zero)
 :consequences ((Quantity (rate ?e))
                (> (rate ?e) zero)
                (qprop+ (rate ?e) (q ?e)
                (I+ (AxisPos Horizontal ?e) (rate ?e))))
```

Figure 17.5
Heuristics modify process descriptions based on comparisons of behavior across time. Here, *q* is an arbitrary conceptual quantity suggested by the heuristic.

Friedman's theory specifies a set of heuristics for constructing and modifying continuous process descriptions. For example, because the ball stops, a new conceptual quantity is postulated to represent what is different about a moving object from a nonmoving object. The initial specification of this quantity is abstract, with additional constraints layered upon it as more reasoning is performed. Because it is intended to determine when something is or is not moving, a heuristic adds an ordinal constraint to the newly created process that conditions its activation on this new quantity being greater than zero (figure 17.5). Newly explained behaviors are stored, along with their explanations, in the Companion's memory.

Given a new behavior, MAC/FAC is used to retrieve similar behaviors whose explanations might be reused in this new situation. There are two ways this might be performed. The first is to attempt to apply the explanation of the prior situation via analogy to the new situation. This is the simplest way to proceed, but it does have limitations. If there are more entities in the new behavior, then multiple mappings must be used to provide explanations for all of it (chapter 12). If there are fewer entities in the new behavior, then either new entities must be postulated, based on the candidate inferences from the prior explanation, or those inferences must be filtered and perhaps modified to reflect the differences between the new and old situations. (In the case-based reasoning literature, these kinds of

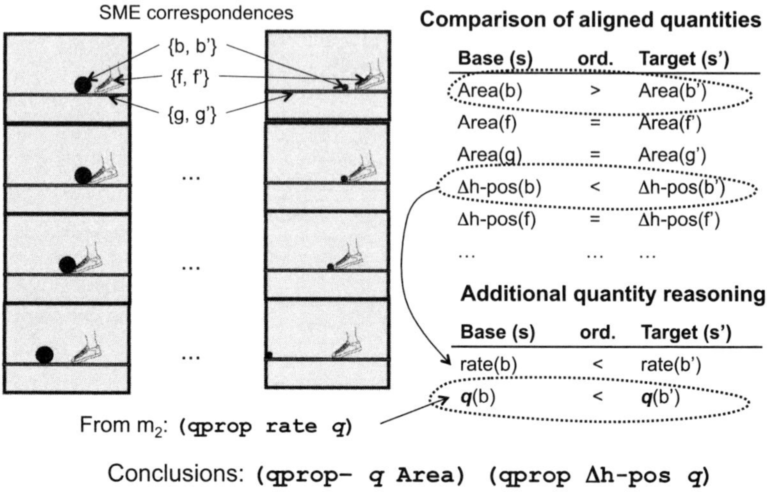

Figure 17.6
Close comparison suggests additional causal relationships.

operations are called *adaptation*.) The second alternative is to construct a new explanation via abduction but limit the selection of model fragments that can be used to exactly those used in explaining the previous behavior. This second alternative is what Friedman's conceptual change theory uses to avoid the adaptation problem and to provide more transferrable knowledge.

In addition to explaining the new behavior, Friedman's theory proposes that people sometimes do *interscenario explanation* (i.e., compare a retrieved behavior in more detail to postulate new causal relationships based on differences between the descriptions). Figure 17.6 shows how qualitative proportionalities between the new quantity q and visual properties of motion can be introduced by comparing two descriptions.

Finally, Friedman's theory postulates that a force for regularization in our conceptual structure is *retrospective explanation*. Suppose we discover in reasoning about one situation that a particular model fragment or quantity representation is inadequate and use one of the conceptual change heuristics to produce an improved version. A link is added between the new and old versions, so that the system can know that it is using an outdated version of a concept. Thus, it is able to tell that it should look for a better explanation of the retrieved behavior, using the upgraded concepts.

If linguistic inputs are also available, the simulation can do more. Each verb is used as a signal to add that portion of the behavior to a generalization

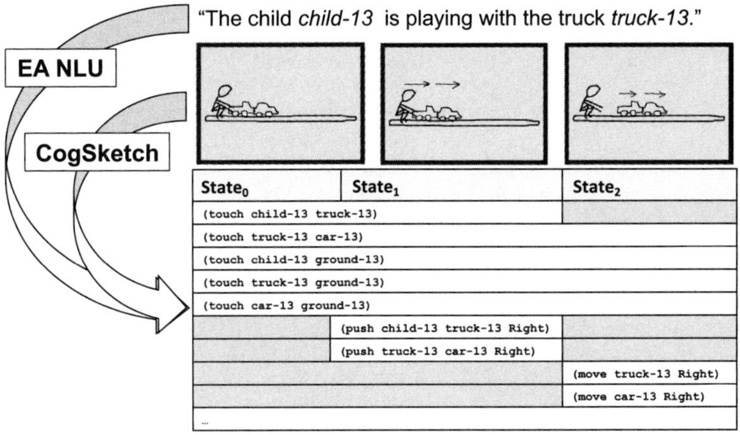

Figure 17.7
Natural language can label behaviors (e.g., pushing, touching, moving), providing more information about the behavior in comic strips.

pool. Temporal analysis of the actions indicated by arrows is combined with language to produce more information. Thus, the three panels of behavior in figure 17.7 give rise to multiple behaviors used for learning.

In the original experiment (Friedman & Forbus, 2009), seventeen comic strips were used, which gave rise to fifty examples of moving, pushing, and blocking. These fifty examples yielded ten generalizations across the three concepts, with twelve of them being unassimilated. Blocking had one generalization constructed out of five examples, with one additional unassimilated outlier; moving had three generalizations, built out of ten examples, with five outliers; and pushing had six generalizations, built out of twenty-one examples and six outliers.[2] This illustrates that SAGE operates conservatively, building disjunctive concepts in a way consistent with pastiche models found in human mental models (Collins & Gentner, 1987).

17.3 Constructing First-Principles Knowledge via Protohistory Statistics

Although these generalizations can be used directly for analogical reasoning, they can also be used to construct logically quantified domain knowledge. Recall that SAGE constructs probabilities for each statement in a generalization by keeping track of how frequently something matching it appeared in assimilated examples. High-entropy generalizations are filtered out, because they are uninformative. Within informative generalizations,

```
(defEncapsulatedHistory Push05
 :participants ((?p1 :type Entity)
                (?p2 :type Entity)
                (?p3 :type PushingAnObject)
                (?dir1 :type Direction)
                (?dir2 :type Direction))
 :conditions ((providerOfMotiveForce ?p3 ?p1)
              (objectActedOn ?p3 ?p2)
              (dir-Pointing ?p3 ?dir1)
              (touches ?p1 ?p2)
              (dirBetween ?p1 ?p2 ?dir1)
              (dirBetween ?p2 ?p1 ?dir2))
 :consequences ((normal-Usual
                  (and (PushingAnObject ?p3)
                       (providerOfMotiveForce ?p3 ?p1)
                       (objectActedOn ?p3 ?p2)))
                (causes-SitProp Push05
                  (exists ?m1
                   (and (MovementEvent ?m1)
                        (objectMoving ?m1 ?p2)
                        (motionPathway ?m1 ?dir1))))))
```

Figure 17.8

An encapsulated history learned for pushing. This adds an additional layer of causal explanation over preexisting concepts in the Cyc ontology (e.g., `MovementEvent`, `PushingAnObject`).

facts whose probabilities are below a particular threshold are also filtered, because they are very likely accidental. The facts remaining are examined to ascertain their temporal relationship with the instance of the concept *c* holding. If a fact *f* starts with or before *c*, it might cause *c*. Conversely, if *f* starts with or after *c*, *c* might cause *f*. If *f* temporally subsumes *c*, then *f* could be a condition for *c*. This information is used to hypothesize the contents of an encapsulated history (chapter 7). For example, an encapsulated history for pushing derived by this simulation is shown in figure 17.8.

Because the motion event continues after the pushing in the generalization, it is hypothesized that the pushing causes the motion event. The conjunction of statements identified as `normal-Usual` are things that must be true during the situation.

The intuitive models of force generated by this simulation were tested by using tasks from the mental models literature. For example, in one problem used by Brown (1994), students were asked whether or not a book resting on a table pushed up against the book. Less than half of the seventy-three high school students correctly answered yes; the other explanations are shown in figure 17.9.

# Students	Answer	Explanation
33	Yes	It must, to counteract the downward force of the book.
19	No	Gravity pushes the book, and the book exerts a force on the table. The table supports the book.
7	No	The table requires energy to push.
5	No	The table is not pushing or pulling.
4	No	The table is just blocking the book.
4	No	The book would move if the table were exerting a force.

Figure 17.9
A challenging problem for students in high school. Does the table push up on the book?

The system's explanation of this situation correctly includes that the book pushes the table and that the table pushes the ground, and that both are blocked from moving. However, the system also agreed with the four students who thought that, if the table exerted a force, the book would move, because the encapsulated history that would be active in that case (figure 17.8) predicts that the pushing will cause motion.

Similarly, one of the questions from the Force Concept Inventory (Hestenes et al., 1992) asks about the motion of a puck moving with constant velocity on a frictionless surface. Students had to choose which picture best describes its path when the puck is given an instantaneous kick. Figure 17.10 shows the situation and the alternatives. The system answered this question by examining each picture (provided via CogSketch sketches) to see if it could instantiate encapsulated histories that would explain the behavior. The only option for which it was able to do so is the most popular misconception: that the moving object would move straight upward after the kick, in exactly the direction of the kick.

These results provide evidence that people might indeed be learning these concepts this way, because the answers they provide correspond to misconceptions exhibited by people. But conceptual change research also indicates that people go through sequences of models over time. Can this account capture sequences of models that people go through (i.e., the trajectories of conceptual change)? The simulation of learning mental models

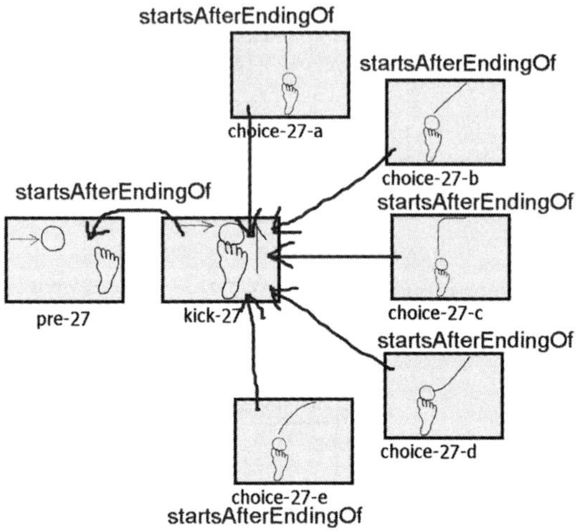

Figure 17.10
A problem from the Force Concept Inventory, represented in terms of a forced choice between sketches.

of force suggests that the answer is yes. In simulating conceptual change in learning about intuitive notions of force, a ten-problem questionnaire developed and used by other researchers (diSessa, Gillespie, & Esterly, 2004; Ioannides & Vosniadou, 2002) was periodically administered to the Companion. The questionnaire used drawings, so we used CogSketch to create a machine-understandable version of the test, as shown in figure 17.11.

Table 17.1 illustrates the distribution of models found by Ioannides and Vosniadou (2002) in students by grade. Although this table does not directly indicate trajectories, it does indicate shifts in the kinds of models held by this population. Figure 17.12 illustrates the trajectories of models that the simulation went through in terms of these force concepts. Because SAGE is order dependent, we ran ten trials with the order of stimuli varying for each trial. Notice that the sequence of models that the simulation goes through is compatible with the human data.

One significant difference between the simulation and the students is that the simulation's movement through the space of models is far more rapid than that of the students. We believe there are three reasons for this. First, the model's only response to an anomaly is to revise its models. Human students often ignore conflicting information (Feltovich, Coulson, & Spiro, 2001), and thus it may take many more examples to convince them to change their models. Second, the model's inputs, although automatically constructed, are highly refined and noise free. This makes the statistical tests in generalization

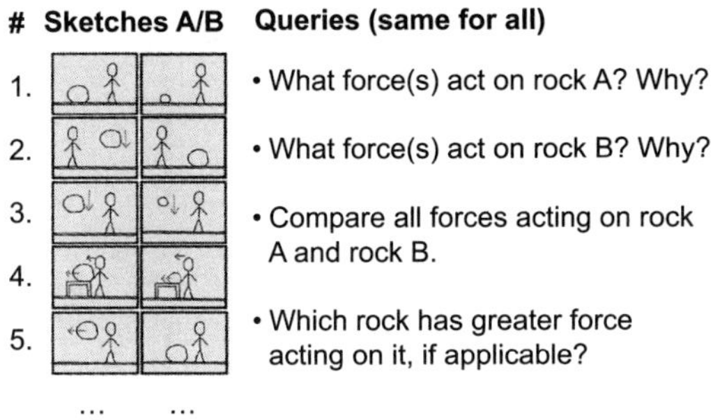

Figure 17.11
CogSketch version of a test used by mental models researchers.

Table 17.1
Mental models of force by grade.

Concept of Force	Kindergarten	Fourth	Sixth	Ninth	Total
Internal	7	4			11
Internal/movement	2	2			4
Internal/acquired	4	10	9	1	24
Acquired		5	11	2	18
Acquired/push-pull			5	10	15
Push-pull				1	1
Gravity/other		3	1	16	20
Mixed	2	6	4		12

become satisfied with many fewer examples than might be needed in a world with more realistic levels of noise and inattention. Finally, the stimuli provided are highly alignable, making comparison easier and making it easier to extract more from each comparison. There is evidence suggesting that, when circumstances provide highly alignable stimuli, learning can be drastically sped up (Kotovsky & Gentner, 1996). Thus, we believe that this rapidity is not unreasonable, given the differences between the strategies available to the simulation and the environment that it is operating in.

17.4 Distributed Knowledge, Explanation Structure, and Conceptual Change

Storing causal laws and other domain theory information with specific experiences, or generalizations over them, provides a radically different way to look at the organization of conceptual structure. It offers an explanation

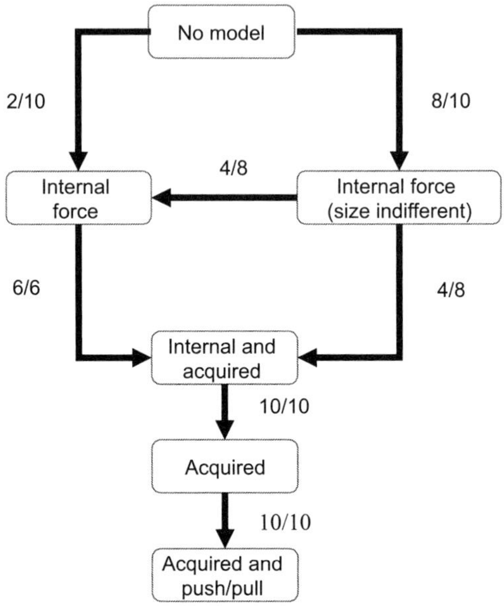

Figure 17.12
Trajectories of mental models of force exhibited by Companions in ten cross-fold validation runs.

of a number of mysteries. Why do mental models look so fragmented (e.g., the *pastiche models* of Collins & Gentner, 1987)? Because they are distributed. Different situations can lead to different retrievals, each of which can have somewhat incompatible domain knowledge, because they have been formulated in distinct parts of the underlying distribution of behaviors. Repeated experience, especially experiences that cause different combinations of protohistories to be retrieved together, leads to regularization of our domain knowledge. This can be catalyzed via language, by using the same term for a process in situations that look quite different to a novice.

A long-standing controversy in conceptual change research concerns the nature of conceptual structure itself. The *theory theory* community (e.g., Ioannides & Vosniadou, 2002) holds that everyday conceptual structure is essentially the same as scientific theories (i.e., systematic, general, organized, and highly structured). The *knowledge in pieces* community (e.g., diSessa, 1988, 2008) holds that everyday conceptual structure is fragmented, consisting of small pieces in isolation that are assembled on an ad hoc basis. The account that Friedman (2012) developed represents a synthesis of these seemingly incommensurate views. We have seen how analogical processing with some simple statistical reasoning can provide a computational model for accumulating experience into protohistories and annotating them with causal hypotheses. This captures many of the intuitions of the knowledge in pieces community, albeit via a different computational mechanism (analogy) and representation (QP theory) than has been proposed before. Now let us see how the intuitions of the theory theory community play out in this account, in terms of the structure of explanations.

Friedman (2012) assumes that when people are constructing explanations, they record the justification for each belief, in terms of other steps in the explanation or beliefs about the world. These beliefs include model fragments as well as facts about the world.[3] Explanations themselves are reified, so that statements can be made about them (e.g., that one explanation is preferred over another). It is helpful to visualize explanations graphically, as shown in figure 17.13. The upper tier consists of tokens denoting explanations. The lowest tier consists of tokens denoting conceptual knowledge (e.g., contingent facts about the world and model fragments). The middle tier consists of justification structure, introducing intermediate steps in the explanation and linking each step with its antecedents.

Explanations are constructed by trying to prove, abductively, the fact (or facts) to be explained. This is done in a backward-chaining process (i.e., starting from a fact to be explained, finding model fragments and/or rules that would enable it to be derived, and then seeking their antecedents in

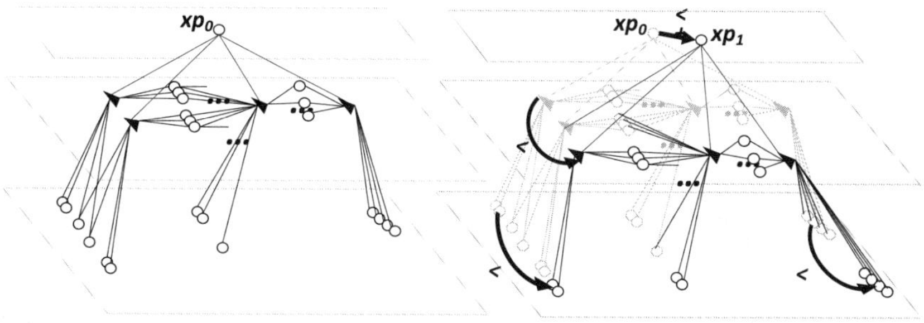

Figure 17.13
Friedman's tiered explanation structure. The bottom layer consists of domain concepts and information about the world. The middle layer justifies some elements from the bottom layer in terms of others. The top layer consists of nodes representing entire explanations at the justification (middle) level. The ordinal relationships here express preference (i.e., particular concepts, model fragments, and explanations can be preferred over one another).

turn). If an antecedent is not known, under some circumstances, it can be assumed to hold. This power must be carefully restricted, because otherwise anything can be explained by simply assuming it. As noted above, the only model fragments that are considered are those retrieved from prior explanations. (If no prior explanations are retrieved, a broader search is conducted.) This normally significantly limits the scope of the search process and should provide scalability, even as knowledge grows to human-level amounts.

Explanations are evaluated via a preference scoring system. Each fact, justification, and model fragment used incur some small cost. Assumptions incur a larger cost, thereby enforcing a preference for avoiding them. For example, uncaused quantity changes may be needed as a scaffolding in the intermediate stages of thinking about a phenomenon, but extending one's theories to explain them and thereby discharge that assumption should then become a high priority. Contradictions are extremely expensive and hence are rarely tolerated.

To see how this works, let us consider a simulation of Sherin et al.'s (2012) study of how middle school children reason about why there are seasons. Sherin and his colleagues interviewed twenty-one middle school students to find out their explanations for why the seasons changed. Quite detailed protocols were taken to examine their mental models. A common misconception, as found by others, is that summer occurs when the Earth is closer to the Sun, and winter occurs when the Earth is far away. If a student's model, like this one, did not account for different seasons in the Northern and Southern Hemispheres, then the interviewer pointed this out (i.e., when it is summer in Chicago, it is winter in Australia). Some students changed their explanation in response to this. The detailed interview information was used to construct formal representations of beliefs and explanation structures for five of the students interviewed. Once the initial explanation was in place, the simulation was presented with the contradictory fact about different seasons in different hemispheres. Once this contradiction was detected, the high cost of the current explanation caused the system to look for alternatives. The simulation was capable of simulating the behaviors of the five students who were interviewed in depth by assuming different background knowledge. The simulation was also pushed far enough to produce the scientifically correct model (i.e., that the difference in incident angle of sunlight causes the seasons, although the interviewers did not push the students that far, because it is not clear that they knew enough about the inverse square law at that point to construct that explanation).

Differences in background knowledge are one source of individual differences. Another can be different explanation preferences. Friedman's (2012) explanation scoring system incorporates four dimensions of preference:

1. *Specificity*. Prefer more specific knowledge to more abstract knowledge (e.g., prefer "The heart has four chambers" to "The heart has chambers").
2. *Instruction*. Prefer information learned via instruction over prior beliefs.
3. *Prior knowledge*. Prefer prior knowledge over new information.
4. *Completeness*. Prefer model fragment instances with more complete bindings and involving preferred entities. (Incomplete bindings are resolved via abduction, requiring new assumptions, and hence should be more expensive.)

These factors are combined via an *explanation strategy*, consisting of an ordering between these factors. For example, one student might prefer instructional knowledge, whereas another might prefer his or her own prior knowledge (i.e., Feltovich et al.'s [2001] notion of *mental shields*).

To see how well explanation strategies could account for individual differences in conceptual change and to test the theory on a third domain, Friedman (2012) set up a simulation based on an experiment on self-explanation conducted by Chi, de Leeuw, Chiu, and LaVancher (1994). In that experiment, twenty-one eighth graders read a passage about the circulatory system, with a test administered before and after. The control group read each sentence twice. The self-explanation group was prompted to explain each sentence after reading it. Sixty-six percent of the self-explanation group were correct at posttest, compared to only 33 percent of the control group, thereby showing that self-explanation resulted in greater learning gains. Moreover, the test was designed to elicit which mental model of the human circulatory system that students had. The transitions between the two groups are show in figure 17.14. Each arrow indicates a transition between models, with numbers indicating the number of students who made that particular transition.

Friedman (2012) simulated this experiment by encoding the contents of the passage, as well as the set of initial models that were found in the pretest, to provide starting points compatible with each of the students in the experiment. For each trial, two factors were varied: (1) the initial domain theory and (2) the explanation strategy used for computing preferences over explanations. By varying these two factors, the simulation was able to model over 90 percent of the transitions between models found in the original experiment. Most of the students in the control group were best modeled via a strategy of preferring prior knowledge. Most of the students in the

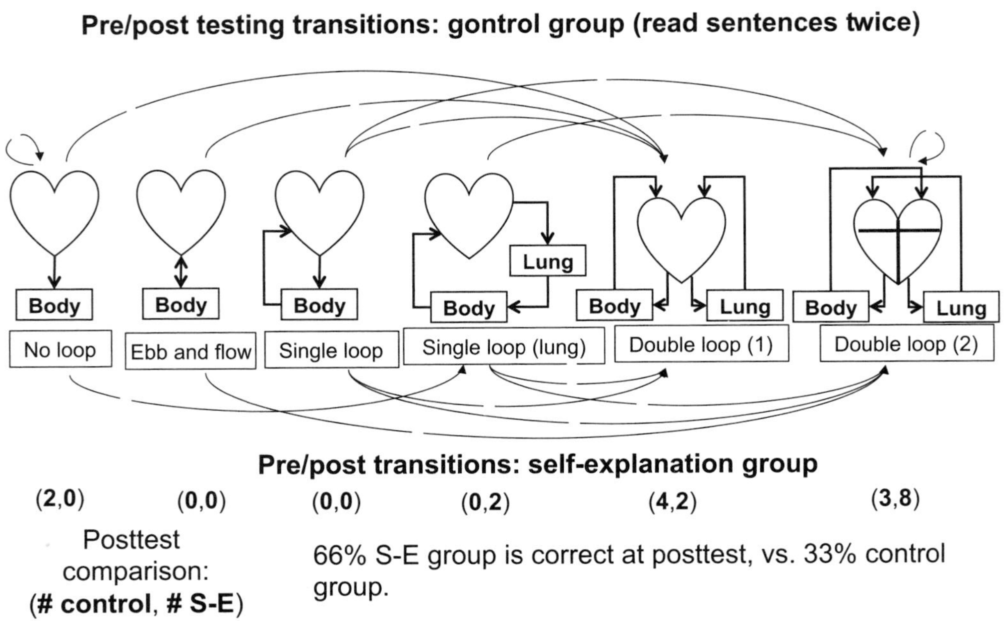

Figure 17.14

Human subject results from Chi et al. (1994). Participants either read a passage about the heart twice or read it once with self-explanation. Arrows indicate transitions between mental models, as measured by pretesting/posttesting, for both conditions.

self-explanation group were best modeled via a strategy of preferring more specific knowledge, followed by preferring knowledge from instruction. Thus, Friedman's theory provides an explanation for the self-explanation effect and can simulate conceptual change as driven via instruction.

17.5 Learning via Cross-Domain Analogies

As knowledge accumulates, the potential to reuse rich conceptual structures learned in one area in another grows. Historically, analogies have been a fertile source of scientific discovery. For example, heat and electricity were both initially understood via analogies to liquid flow, which enabled important properties to be predicted. For example, both have an identifiable intensive parameter (temperature and voltage, respectively) whose difference drives flow. The differences between the domains discovered by seriously testing the predictions of an analogy can also lead to surprising and important insights. For example, heat was at one time modeled as *caloric*, a fluid that permeates objects like water permeates a sponge, such that the more caloric something had, the hotter it was. This model suggests that an object could be emptied of caloric, because there could only be a finite amount of it inside. Careful experimentation by Count Rumford ascertained that this was not so—boring the barrel of a cannon could produce an arbitrary amount of heat. Thus, cross-domain analogies, although rare, can be very insightful.

Great cross-domain analogies are rare for two reasons. The first is that not everything is alike: pick two random explanatory theories, and they very likely have little in common. The second is that human memory is optimized toward literal similarity matches (chapter 4), where both surface information and structural information overlap. This makes sense in terms of the function of a memory, because most of the time, we want to retrieve explanations of similar behaviors, as Friedman's (2012) model of conceptual change suggests. Cross-domain analogies are facilitated by two factors. The first is being told about them, from one's culture. The second is the detection of some abstract commonality between two domains that serves as a bridge between them. For example, if the qualitative patterns of behavior for two phenomena are similar, then it may make sense to look for similarities involving the domain theories that causally explain those behaviors.

The second factor was the insight behind Falkenhainer's (1987, 1990) PHINEAS model of cross-domain analogical learning. PHINEAS used similarities between behaviors of observable quantities to construct correspondences between domains. Given a new behavior to explain in a domain in which it had no theory, PHINEAS looked through its memory for an analogous behavior. (This was before MAC/FAC, so the search process was serial and task specific. Although people do such searches when consciously working through a problem, we suspect that even then, the process involves reformulating a probe and examining what analogical retrieval provides.) Once it found a behavior, PHINEAS used SME to try to construct an analogy between them. If successful, this analogy provided a set of correspondences between quantities across the two domains. These correspondences were then used with SME again, but this time with the base being the previously understood domain theory and the target being an empty domain theory, seeded only with the entities from the prior match. The QP model fragments were thus candidate inferences, which PHINEAS then used as a model for the new domain. This new domain theory was used to qualitatively simulate the new behavior to see if it was in fact satisfactory.

PHINEAS only dealt with qualitative models. What about learning to solve physics problems, where quantitative answers are expected? As discussed in chapter 7, encapsulated histories are typically used to model how equations are represented in domain theories, because they enable time to be used as an explicit variable, unlike model fragments. Therefore, we already have the machinery needed for representing equations. One feature of such problems is that learners are provided with examples that contain worked solutions. These worked solutions are explanations used for communication, not the more detailed form of explanations used above that represent what is happening in the learner's mind. Klenk (Klenk & Forbus, 2013) showed that analogies between worked solutions could prime the pump for cross-domain analogies. Consider two analogous problems, one from linear kinematics and the other from rotational kinematics, as shown in figure 17.15. Comparing their worked solutions can build a set of correspondences between the two domains, as figure 17.16 illustrates.

Recall that tiered identicality in structure mapping allows two nonidentical relations to match if they have a close common superordinate. These correspondences can include quantities, modeling abstractions, domain relationships, events, and encapsulated history types.

Like in PHINEAS, these initial correspondences can be used to initialize a theory for the new domain by projecting the encapsulated histories in the base domain into the new domain. This new domain theory is tested by retrying to solve the new problem that originally failed. If it succeeds, the new domain theory is kept; otherwise, it is discarded. Importantly, unlike PHINEAS, Klenk's model (Klenk & Forbus, 2009b, 2013) accumulates correspondences between domains via *persistent mappings* that accumulate

Base: linear mechanics	Target: electricity
Estimate the net force needed to accelerate a 1,500 kg race car at -5 m/s^2.	A 75 volt EMF is induced in a 0.3 Henry coil by a current that rises uniformly from zero to I in 2 milliseconds. What is the value of I?

Figure 17.15
Example of a distant cross-domain analogy, linear mechanics to electricity.

Linear mechanics	Electricity
`PointMass`	`Inductor-Idealized`
`objectTranslating`	`objectActedOn`
`ForceQuantity`	`VoltageAcross`
`Distance`	`Charge`
`Compliance-Linear`	`Capacitance`
`DefOfNetForce`	`DefOfSelfInduction`

Figure 17.16
Persistent mappings can be built between domains via solving multiple problems.

over time. This enables it to handle large, complex domains incrementally while maintaining consistency. This model was tested with multiple target domains: rotational kinematics, electricity, and thermal problems, with the base always being linear mechanics. It was able to do correct cross-domain transfer 87 percent of the time when an analogous worked solution was provided. When it had to find an analogous worked solution itself, it fared far worse, which is what one would expect, given how rare such retrievals are in daily life. In fact, the simulation's retrieval rate of 40 percent for cross-domain analogies is something that we view as a bit too high and very likely due to the small size of the memory set.

17.6 Summary

This chapter has examined the claim that learning and conceptual change in reasoning about continuous phenomena can be explained in terms of analogical processing over qualitative representations. The simulations described here are not comprehensive: As a practical matter, it is not yet feasible to bootstrap a software organism through the entire range of models, across all domains that people learn in. However, they do supply strong evidence that these representations and mechanisms can in principle explain the phenomena.

18 Commonsense Reasoning

One of the deepest problems in cognitive science is understanding how commonsense reasoning works. It has received a great deal of attention in AI (e.g., Davis, 1990; Mueller, 2014) but surprisingly little in other areas of cognitive science. We reason about the everyday physical, social, and mental worlds rapidly, with reasonable accuracy, across a breathtakingly broad range of situations. Just think about cooking for a moment. Cooking requires making plans and predictions about the physical world, using causal reasoning and perception. Where can I set this bowl down on a crowded countertop? Could I add another cup of water to its contents? Is it safe to go check something in the next room while the soup is cooking? The picture I have painted so far argues that qualitative representations are central in such reasoning. The ontologies developed by qualitative dynamics provide the causal theories that enable us to reason about the processes and events that will happen as a consequence of our actions over situations framed in part by qualitative spatial representations derived via perception. Much of this reasoning, I have argued, occurs via within-domain analogical reasoning over qualitative representations. Experience thus plays a central role in common sense. Some first-principles reasoning does occur: a situation might require us to combine influences in novel ways, for example, or to reason about situations for which we have no close analog. Some things are better thought through than experienced—for example, what happens if a running blender falls into the sink or someone tries cooking a hard-boiled egg in a microwave oven?

This chapter steps back and looks at the commonsense problem again. We start by looking at alternate accounts that have been proposed and arguing that qualitative reasoning plus analogy provides a better account. Then we look at how everyday quantitative reasoning might work by outlining models of *back-of-the-envelope reasoning* and *analogical estimation*. Common sense is not only about the physical world, of course. The rest of the chapter examines initial forays into applying these ideas to social reasoning and decision making, which provides intriguing evidence of their potential generality in these realms. We discuss the use of qualitative representations in conceptual metaphors, particularly in politics. Qualitative models of emotions and blame attribution are described next. The chapter closes by showing how qualitative representations and analogy have been used in a computational model of moral decision making, which incorporated sacred as well as utilitarian values.

18.1 How Common Sense Doesn't Work

Artificial intelligence researchers have made three families of proposals for how commonsense reasoning works. Each of them has some merits, but I believe that none of them are sufficient to explain it.

The first family of proposals equates common sense with the use of a particular logical formalism or theory. Commonsense reasoning was one of the major motivations for research on *nonmonotonic logic* (Strasser & Aldo, 2015), for example. Think about the set of beliefs an agent might have, both explicitly and derivable from what it explicitly knows. In traditional logic, adding a new belief can never decrease the size of the set of beliefs. The new belief can increase the size of the set of beliefs, both by itself plus all of the new inferences that follow from it. (If you find out that a new friend plays a musical instrument, for example, you can infer that he or she has practiced with that instrument.) In nonmonotonic logics, the size of this set can decrease (i.e., adding a new fact can eliminate others from being believed, resulting in a net decrease in beliefs). A classic example of this is default reasoning. If you hear about a bird, you might assume it can fly. If you later discover that it is a penguin, stuffed or baked, you will change your mind. Nonmonotonic reasoning clearly occurs in commonsense reasoning. Unfortunately, current nonmonotonic logics have problems that make them poor candidates for models of psychological reasoning. These problems all boil down to what should happen when a new belief is contradictory. Strong constraints need to be imposed to not violate our notions of causality (see the Yale Shooting Problem; Hanks & McDermott, 1987). Often these take the form of theories of action (Chou & Winslett, 1994; Giunchiglia, Lee, Lifschitz, McCain, & Turner, 2004). But to date, these action models have not been robust enough to successfully encode the range of domains that research in qualitative reasoning has been able to do. (One reason is that they have focused more on discrete actions, ignoring continuous change altogether.)

Let us think through how contradictions should be handled, returning to our abstract agent as a foil. If the new belief is contradictory with respect to its existing beliefs, then the entire set becomes contradictory. Several distinct possibilities arise, depending on the agent's reasoning mechanism. If it uses one large logical theory and traditional theorem proving, then it can believe anything, because any statement would follow via an indirect proof, given the existence of a contradiction in the theory. This is one reason why the approach of having a single-context logical knowledge base with just first-principles theorem proving is too simplistic to be a model for human commonsense reasoning. As noted in chapter 2, modern agents have knowledge bases that are contextualized, so that the effects of a contradiction are isolated and the rest of the system carries on without problem. What to do about the contradiction can be reasoned about to diagnose the problem and figure out how best to resolve it. In the simple case of a single default rule and new reliable information, resolution is simple: retract the default assumption, and the problem is resolved. Alas, there are many more complex cases, especially if default rules are used in deriving facts that are then in turn used in other default rules. Suppose the interpretation of the sentence that our bird was a stuffed animal ultimately, via a long chain of reasoning, depended on the default assumption that it could fly. (The bird remained on its perch for multiple days, when we looked in the open window, unmoving and never leaving.) Then retracting the default assumption also loses our new belief. Such loops in logical dependencies can and do occur in real reasoning systems. It gets worse: if there are an odd number of defaults in such loops, there is no stable solution to that set of beliefs. This is the sort of problem that designers of nonmonotonic reasoners and nonmonotonic logics must deal with. A number of practical solutions have been worked out to avoid such cases, and so it is usually not a problem in practice. But the formalizations of nonmonotonic logic currently lag behind practice.

Given the uncertainty of the everyday world, it seems sensible to incorporate some notion of probabilities in commonsense reasoning. Another proposal is that Bayes nets (Pearl, 2009) provide a sufficient formalism for commonsense reasoning. I find this dubious, because Bayes nets are propositional. The approach covers what to do when one has a Bayes net, not how to construct a Bayes net for a given situation, and thus it provides at best only a piece of the puzzle. Various integrations of logical reasoning and Bayesian reasoning have been proposed and are actively being explored (e.g., Markov logic nets [Richardson & Domingos, 2006], BLOG [Milch, Marthi, & Russell, 2004]), which could, in the long run, overcome this limitation. Unfortunately, they all rely on propositionalization to do their reasoning and thus require huge amounts of computation to do even simple problems. Moreover, none of these accounts provide the distinctions needed for causal reasoning that qualitative process (QP) theory and other representations developed by the qualitative reasoning community provide (see chapter 9).

The second family of commonsense proposals are that common sense will just emerge once we have vast amounts of knowledge. For example, Cycorp's approach to common sense has been to hand-engineer a knowledge base by hand initially and transition to having the system itself extend the knowledge base via various forms of learning (e.g., learning by reading, knowledge capture via web-based games). Having large amounts of knowledge is certainly part of the solution. The Cyc knowledge base can be viewed as a model of the semantic memory that a person might have. Unfortunately, the episodic memory is currently missing: formalizing a mass of everyday experience in terms of concrete situations by hand is too daunting for anyone to undertake. There are attempts to remedy this, for example, crowdsourced data collection (e.g., Cyc's FACTory, OpenMind, and others). However, the kind of knowledge involved is crucial. Most of these efforts have not tried to use richer modalities than text, in contrast with, for example, Friedman's use of comic strips to provide input for modeling conceptual change. But knowledge is not enough. Our arguments presented in chapters 4, 12, 16, and 17 suggest that analogical reasoning provides a missing piece of the puzzle.

Finally, the third class of proposals is that common sense will emerge from embodied cognition. That is, common sense arises from the experience of being an agent in the physical world, able to perceive and manipulate the world. I believe that using physical agents could be useful in accumulating the knowledge needed for common sense, but they are not a necessity. We learn about social reasoning, for example, from stories as well as from interactions with others. Even for learning about the physical world, our culture provides critical constraints that help us organize and refine our experience (e.g., language). It should also be observed that this approach has been tried many times, with little success. (The number of "robot baby" projects that have been run over the past twenty years is surprising.) Historically, such projects have not gone beyond learning a handful of simple distinctions and then quietly die off. Part of the problem is that today's robotic platforms are so impoverished compared to biological creatures, which significantly limits what can be done with them, no matter how well intentioned and creative researchers are. Hopefully, changes in manipulation and sensor technology will make more interesting experiments feasible.

Stepping back, the account proposed here has features from each of these three families. It proposes a particular form of theory—qualitative representations—as the currency for common sense, the representations that centrally organize such knowledge. But it also makes analogical processing central to the account, unlike previous proposals. It proposes that large bodies of experience are necessary, using analogy to both reason directly with it and to construct more portable knowledge via analogical generalization. Finally, it proposes that qualitative representations provide a bridge between perception and cognition. Even for embodied agents, I propose that qualitative representations must play a central role.

18.2 Some Psychological Considerations Concerning Common Sense

Cognitive science research, particularly research on cognitive development, has provided some interesting insights into the nature of commonsense reasoning. To summarize,

- It is learned, often very early.
- It is localized, often within specific types of situations within a domain.
- Generalizations arise partially and slowly.
- Language provides a catalyst for learning and generalization.

Let us examine these insights in more detail. Studies of causal reasoning in infants typically use a violation of expectations method, based on measuring looking time. That is, infants look longer at novel events than they do at familiar events. By using tricks borrowed from stage magicians, experimenters set up seemingly impossible events for infants to view. Looking longer at an impossible event can be taken as evidence that the event violates their expectations. (Experimenters control for novelty of objects and other factors, of course.) From this work, an interesting pattern occurs.

Some examples inspired by the work of Baillargeon, Hespos, and others are shown in figure 18.1.

Figure 18.1a shows the before and after scenes of a physically possible event, which serves as a baseline. Figure 18.1b shows something that is impossible, and when infants see such an event, even at three months, they exhibit longer looking times. This is taken as some notion of contact being connected to some notion of support is available even to infants at three months. That these notions are complicated and evolving can be seen in figure 18.1c, which shows a situation that infants do not look at longer until they reach 4.5 to 5.5 months, even though (absent glue or one object painted to look like two) it is impossible. This suggests that, before that time, contact of any kind may be viewed as sufficient for support.

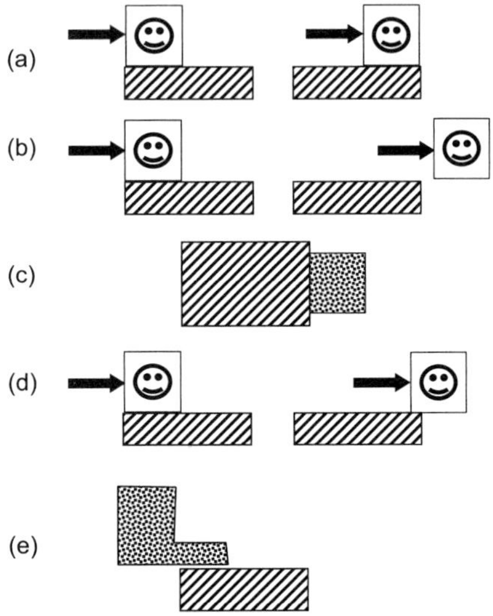

Figure 18.1
Examples of situations used to probe the evolution of infants' models of support. The arrow indicates where, in the experiment, a hand was pushing the block.

Figure 18.1d illustrates a situation that does not lead to longer looking times until around age 6.5 months. It appears that now the amount of contact becomes relevant to them. Around 12.5 months, situations like figure 18.1e start leading to longer looking times. This suggests that, by then, infants are sensitive to what we would call, using adult concepts, center of mass, as estimated by area using assumptions of uniform density.

Interpreting such results requires caution. Baillargeon (1994, 2002) interprets these changes as a movement from binary features being sufficient (e.g., contact versus not contact), followed by the introduction of a continuous parameter (e.g., the amount of overhang). Causal relationships constraining this continuous parameter provide more sophisticated judgments (e.g., how much overhang is required?).

Baillargeon and others have found similar patterns in other everyday phenomena. Examples include the following:

- Can moving a tall object behind a short object cause the object to not be seen?
- Can a moving screen rotate through a solid barrier?
- Will a moving object colliding with a stationary object cause it to move and, if so, by how much?

The basic findings supporting this pattern replicate, although there can be differences in details, such as how early a transition occurs. There are major differences in how they are interpreted (e.g., which belief is being violated). A controversial question is whether or not such beliefs are "hardwired" (i.e., the nativist perspective [e.g., Spelke & Kinzler, 2007]) versus learned (e.g., Baillargeon, 1994; Gentner, 2010).

One possible nativist explanation for these phenomena is that infants' brains are rapidly growing during this period, so they might simply not have the capability yet for finer-grained distinctions. There are two reasons to doubt this explanation. The first is that, even within what might be reasonably viewed as a capability, improvement in performance is not global. The ability to judge sizes, for example, seems a not unreasonable candidate for a hardwired ability that might come "online" at some point in development, so to speak. But infants only become able to judge sizes at different ages for different phenomena. For example, by 4.5 months, infants are able to estimate heights in occlusion events, whereas other reasoning about height does not seem to happen until around 7.5 months (Hespos & Baillargeon, 2001). If there were a single, simple principle that came online, one would expect these both to happen at the same age, so either there is learning involved, or the physical principles have to be subdivided at a much finer granularity than previous analyses.[1] Spelke's *core knowledge* theory (Spelke & Kinzler, 2007) can be viewed as such an approach. The second reason is that Baillargeon found that she could get infants to show a particular piece of performance much earlier by using closely spaced trials. This is more compatible with an analogical learning account, being an example of progressive alignment as described below, rather than with maturation.

A second line of evidence for analogical learning as being core to common sense comes from research on situated cognition. Transfer is notoriously difficult—this is often called the *inert knowledge problem* (Brown et al., 1989) in the education literature. Students memorize facts, but they do not know how to apply them in their daily life. A particularly dramatic experiment was carried out with children who worked after school as street vendors in Brazil (Carraher, Carraher, & Schliemann, 1985). Experimenters kept track of arithmetic problems that students failed to solve in the classroom and then had confederates pose the equivalent arithmetic problem, but in the context of making change, which they solved successfully. Our interpretation of this phenomena (Gentner et al., 1993) is that analogical retrieval is the bottleneck. One cannot use what is not retrieved in a situation. Retrieval tends to rely on surface features. Analogical generalization can help, in that accidental surface features are suppressed or even removed as the number of

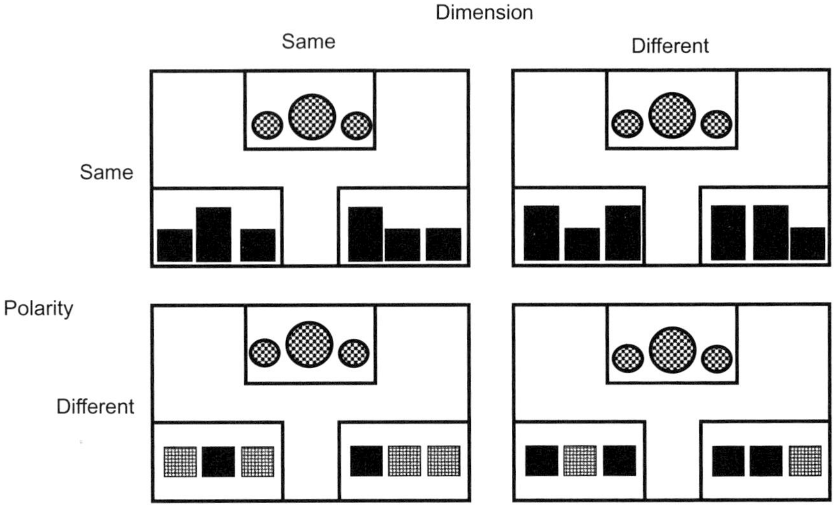

Figure 18.2
Examples of stimulus triads for a forced-choice task. Four-year-old children find cross-dimensional comparisons quite difficult.

examples contributing to a generalization grows. But another way to help is via language. As mentioned in chapter 12, Kotovsky and Gentner (1996) showed children, age four or eight, stimuli like those in figure 18.2.

Children were asked to say, "Which is the top pattern more like?" For cross-dimensional stimuli like the bottom two of figure 18.2, where the top varies in size and the bottom stimuli vary in shading, eight-year-olds can do it, but four-year-old children find it very hard. They discovered two ways to speed learning, so that four-year-old children performed, on this task, as well as the eight-year-olds. The first is *progressive alignment* (i.e., blocked trials). The second is to use language (e.g., "even" for the left pattern and "more and more" for the right).

These experiments provide evidence for our account of common sense. Qualitative representations are the heart of commonsense reasoning. They provide a level of abstraction that supports effective within-domain analogical reasoning based on experience and learning via analogical generalization. Schematized knowledge does emerge, but slowly via analogical generalization or more quickly via instruction from our culture (e.g., language, formal schooling). First-principles reasoning plays a limited role in constraint-checking the results of analogical reasoning or in reasoning about very novel situations, but it is always buttressed by analogical reasoning.

18.3 Quantitative Aspects of Common Sense

Most of our explorations of commonsense physical reasoning about quantities have involved purely qualitative representations. But in many situations, practical quantitative knowledge is also crucial. When cooking, for example, we have to add specific amounts of an ingredient. Often this is specified, but if one has modified the recipe or the instructions are to add an ingredient "to taste," we are forced to come up with some estimate. Even in professional situations, rough and ready estimates are often crucial. When the Russian space station *MIR* had a rupture in 1997, a crucial question was determining how much oxygen was left for the cosmonauts to breathe. When the USS *Cole* was damaged by a terrorist attack, a key question that arose was how long it would take to repair it. These and many other questions require generating a reasonable estimate in the face of considerable uncertainty. We will start with such *back-of-the-envelope* questions because they provide a setting where evaluation is straightforward.

Back-of-the-envelope reasoning has long been an interesting topic in science and engineering because of the need to make feasibility estimates. In physics, these are often called "Fermi problems," but they can be found everywhere (e.g., Bundy, Sasnauskas, & Chan, 2015; Harte, 1988). They are also important to commonsense reasoning to quickly ascertain if our reasoning has "gone off the rails" or if we have misunderstood something that we had read. In a TREC question-answering competition, for example, one system's answer to the question of how much folic acid a pregnant woman should consume was 16 tons per day. Presumably, no awake and alert adult human being would ever give this particular answer. How much money is spent on newspapers every year in the United States? The business prognosticators say this number is dropping, but from what? We can estimate it using logic like this:

1. Total money spent = money spent per buyer * number of buyers.
2. Money spent per buyer = units bought per year * cost per unit.
3. Newspapers cost $0.75.
4. There are 365 days per year.
5. Annual expense per buyer = $250 (from steps 2, 3, and 4).
6. Assume one newspaper per four-person family.
7. Assume 300 million people are in the United States.
8. Number of buyers = 75 million (from steps 6 and 7).
9. Total spent = $20 billion (from steps 1, 5, and 8).

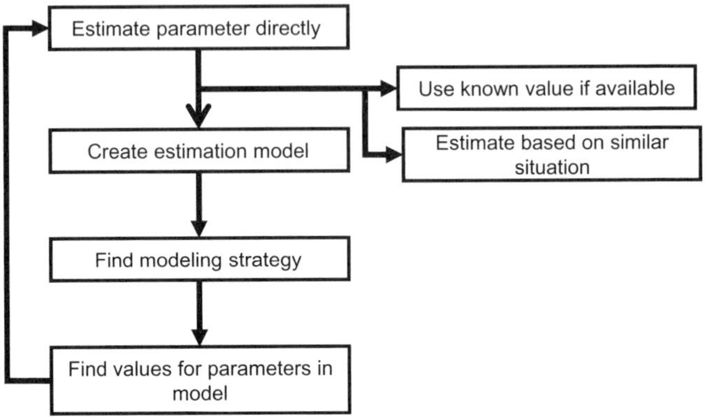

Figure 18.3
Flow of reasoning in Paritosh's model of solving back-of-the-envelope problems.

According to the Statistical Abstracts of the United States for 2003, the amount spent was $26 billion, so this estimate is not bad, especially given how few resources we put into deriving it. How can we formalize such reasoning?

Praveen Paritosh (2007; Paritosh & Forbus, 2005) proposed that reasoning like this can be captured by the model shown in figure 18.3.

The first step is to try to estimate the parameter directly. One might already know a reasonable numerical value, as we did for days of the year and population of the United States. Alternately, one might need to estimate based on a similar situation. If we wanted to know the number of papers sold in 2004 and knew how many were sold in 2003, our first estimate might be to use the same number, and our second estimate might be to discount it, based on a guess about how rapidly newspaper consumption is falling. When we cannot come up with a parameter value directly, we try to construct a model of the situation from which we can derive a value. That means first finding a reasonable model, then solving for the other parameters in that model. Sometimes we know general models, like we did in the newspaper case (i.e., steps 1 and 2 are facts of economics). Other times, we might use models that were useful in similar cases (e.g., decompose *MIR* into its major components, estimate the volume of each component). As we will see, both aspects of this process involve an interplay of analogy and qualitative representations.

18.3.1 Analogical Estimation of Numerical Values

The question of how people form estimates of numerical values has received considerable attention in cognitive psychology. The seminal *anchoring and adjustment* model (Tversky & Kahneman, 1974) argues that if people are given some initial number, the anchor, their subsequent estimates are biased by the value of the anchor. In other words, the anchor serves as a starting point, and their estimates of specific values are adjusted based on the particulars of the situation. For example, a real estate agent might know that apartments that are larger and in desirable locations tend to have higher rents than those that are smaller or are in less desirable neighborhoods. Anchors are often generated by people themselves during estimation (Epley & Gilovich, 2005), based on their experience. If people are given some values for quantities in a set (e.g., populations of European countries), they can exploit such *seeds* to improve their estimates (Brown & Siegler, 2001).

These findings and others led Paritosh and Klenk (2006) to propose a model of *analogical estimation.* Anchor values, they argued, are found via analogical retrieval. The causal knowledge used for adjustment can be modeled via qualitative proportionalities:

```
(qprop (price apartment) (size apartment))
(qprop (price apartment)
       (desirability (neighborhood apartment)))
```

Pilot analyses of protocol data from two experts (a car salesman and a real estate broker) indicate that the linguistic equivalents of such statements (see chapter 13) do indeed appear in what they say when thinking aloud about estimation problems. Qualitative laws tell us what factors are relevant, but they don't provide the metric information to perform quantitative adjustments. They propose that people solve this problem by retrieving more examples for which the same causal laws hold and use those data points to generate an approximate quantitative function to use for estimation. (Kareev, Lieberman, and Lev [1997] provide evidence that people can estimate correlations with a handful of samples.) Differences in each independent quantity (e.g., size, neighborhood) are used to compute adjustments to the dependent parameter, using a linear approximation. They showed that using analogical adjustment yielded estimates that were significantly more accurate for four parameters out of six in a database of basketball statistics.

18.3.2 Qualitative Representations Can Enhance Similarity

If you think back to the description of our models of analogical processing in chapter 4, you may notice a problem with the estimation account provided above. One of the current limitations of SME (and hence of MAC/FAC and SAGE) is that it currently does not factor in any quantitative aspects of similarity at all. Numerical values can be used in analogical matching, of course, but they are treated simply as entities, like Fred. So how can SME and MAC/FAC be more sensitive to quantitative values to support processes like analogical estimation?

Paritosh (2004, 2007) proposed that cases should be augmented with qualitative information about quantitative values, so that structural alignment processes could be more sensitive to them. We have already seen (chapters 5 to 10) how limit points can be used to impose qualitative structure on numerical values by breaking them up into meaningful regions, where a change in value corresponds to some change in the qualitative structure of the situation. Therefore, although the set of ordinal relationships that define qualitative values might not supply sufficient discriminability—in the MAC stage of retrieval, only the relative number of ordinal relations compared to the rest of the relations, functions, and attributes matters—the changes in a set of active model fragments should provide some utility. But what about domains where we don't yet have a solid grip on their causal laws? The literature in support of fuzzy logic and other vague representations argues that some numerical values appear to not have crisp boundaries—what does it mean for a person to be "tall" or a nation to be "underdeveloped"? Paritosh argues that qualitative structure can be imposed on such properties via clustering. Suppose we have a set of linearly ordered symbols (e.g., Small/Medium/Large, or underdeveloped/developing/developed). There is ample evidence that such qualitative descriptions of quantities are highly contextual (e.g., a tall person may not be tall when considered as a basketball player). Thus, the statements

```
(isa LarryBird
          (MediumValueContextualizedFn Height
                                             BasketballPlayer))
(isa LarryBird
          (HighValueContextualizedFn Height Person))
```

say that Larry Bird (who stands 6′ 9″) is considered as having medium height, in the context of being a basketball player, whereas with regard to people in general, he is considered to have a high value of height (i.e., tall). Paritosh proposed automatically constructing these *distributional qualitative representations* via k-means clustering,[2] where k = the number of symbols in the linear ordering. A pilot study (Paritosh, 2004) suggests that, for a task involving classifying the sizes of unfamiliar countries, the partitions generated by k-means provide a reasonable overall match for the representations that participants produced. (Other clustering algorithms may be more appropriate for different distributions, but the principle is the same.)

Notice the interesting complementarity between structural and distributional qualitative representations. Structural qualitative representations have crisp boundaries (i.e., the limit points that delineate regions). But membership in the regions is delineated incrementally, because the set of ordinals constraining a quantity can be partial. Distributional qualitative representations have crisp regions (i.e., something that has `Large` attributed to it relative to a class membership is indeed to be considered large). On the other hand, the boundaries are delineated incrementally by those members who are considered to be, or not to be, members of it.

In the analogical estimation experiment described above, distributional qualitative representations were constructed for the set of examples in the database as a preprocessing step and indeed led to better performance than simply retrieving the closest match.

18.3.3 Strategies for Back-of-the-Envelope Reasoning

One of the aspects of estimation reasoning that makes it seem more art than science is the sheer range of types of knowledge that can be brought to bear. We have already seen that knowing some numerical values, especially knowing closely allied values that are causally related, is needed. When we do not have a reasonable estimate for a quantity, we have to find some model we can use to estimate it. Imposing some structure on the form of possible models is important, because then we can turn the problem from being completely open-ended into one where there are a finite number of families of solutions to explore. Paritosh (2007) observes that the basic form of an estimation problem can be cast as

```
(Q O ?v)
```

where Q is the quantity to be found, `O` is the object whose property it is, and `?v` is the value to be found. He observes that there are only three families of strategies, and these can be further decomposed into just seven heuristics that capture all back-of-the-envelope reasoning strategies:

Object-based strategies involve transforming the object `O` into parts, solving the parts, and then combining them appropriately to form the total solution.

1. *Mereology heuristic*: If `Q` is an extensive quantity, then for a nonoverlapping partition of `O`, $\{O_i\}$, `?v` is simply the sum of the solutions to all of the O_is. For example, if you want to find out the weight of the contents of a shopping cart, add up the weights for each of the objects in it. If `Q` is an intensive quantity, then the value for the total is still found by combining the values of the parts, but each contribution has to be

weighted according to its size. For example, if a bead of cold water drips into a bathtub filled with hot water, we know we can basically ignore the change in temperature that it causes.

2. *Similarity heuristic*: If a value V′ can be found for a similar object O′, then use V′ for V. For example, the price of a car might be approximated by the price of another similar car.
3. *Ontology heuristic*: If we have category information about O, check that information to find possible values and estimation models.

Quantity-based strategies use laws of the domain to decompose the problem into a simpler problem. There are three quantity-based heuristics:

1. *Density heuristic*: Many estimation problems involve ratios, so finding the numerator and denominator separately provides a useful method of decomposition. For example, in the newspaper argument above, assuming that there is one newspaper per family and four people per family are examples of this heuristic.
2. *Standard domain laws*: Laws of a domain provide a model for one parameter in terms of the others (e.g., $F=MA$).
3. *Approximation assumptions*: For example, in ascertaining how many kernels of popcorn could fit into the room you are reading this in, it is reasonable to approximate the rather complex shape of a popped kernel as a cube, for simplicity.

System-based strategies exploit constraints that govern whole systems. There is one system-based heuristic (because other compositional heuristics are listed under object-based heuristics):

1. *Conservation heuristic*: Use conservation laws to solve for more subtle properties based on easier to measure properties. For example, to estimate the rate of leakage from a gas tank, measure its level before and after a known period of time, subtracting out the intended flow over that period from the difference.

Finding relevant decompositions and similar examples, of course, relies on analogical retrieval. For example, Linder (1991) asked participants (in this case, mechanical engineering students) to estimate the energy stored in a 9-volt battery. None of the participants tackled the problem by using first-principles knowledge of chemistry. Instead, they thought about everyday systems they have worked with that use batteries (e.g., music players, clocks, flashlights). They also adapted estimates from 1.5-volt batteries (which are more common) or car batteries.

18.3.4 How Well Does This Model Do?

Are these seven heuristics really enough to solve all back of the envelope reasoning questions? Two sources of evidence support this claim. The first is a corpus analysis of Swartz's (2003) *Back-of-the-Envelope Physics* book. Paritosh (2007) analyzed all forty-four problems from four sections of the book (Force and Pressure, Rotation and Mechanics, Heat, Astronomy). All but five of the seventy-nine transformation steps used were identifiable as one of the seven heuristics. Four of the five that were not covered were designing experiments to estimate a quantity, and the fifth was a complex problem from statistical mechanics.

For another test of his model, Paritosh (2007) tested it against a corpus of questions from the Science Olympics. The Science Olympics is a set of science and math competitions at all grade levels. For U.S. high school students, one event (out of twenty, typically) includes Fermi Questions. This event typically consists of about thirty questions, and scoring 90 out of a possible 150 points will get the team a medal. Only a few out of hundreds of participating teams achieve this level of performance. Paritosh tested his BotE-Solver on thirty-five practice problems from the University of Western Ontario Science Olympics website. Here are some examples of the questions:

- How much energy does a horse consume in its lifetime?
- How many bricks are there in London?
- What is the mass of all the automobiles scrapped in North America this month?

Having all of the specific world knowledge needed to answer these questions is something that lies outside the back-of-the-envelope estimation theory. The key test is whether or not the seven heuristics, proposed in print before this evaluation, sufficed for solving all of these questions, given appropriate world knowledge. They were indeed sufficient, providing additional evidence for their completeness.

18.4 Qualitative Representations in Conceptual Metaphors

It has long been argued that metaphors are common in human languages and everyday life (e.g., Lakoff & Johnson, 1980). Many of these metaphors involve continuous phenomena (e.g., someone getting so angry that he or she started steaming). Open a newspaper, and you can easily find stories like this:

> Mr. Koizumi says he understands the growing calls for sanctions against North Korea, but he believes that a combination of dialogue and pressure is the best way to succeed.

Pressure, here, does not refer to the physical quantity, naturally, but to some application of a force of some sort to the situation. Such everyday reasoning is naturally expressed via qualitative representations. It involves continuous parameters, but nobody knows the specific quantitative values (or even the units!). Ordinal information is often available (e.g., broadening the range of trade goods in an embargo applies more pressure, engaging in more trading reduces the pressure). There might be implied limit points (e.g., someone steaming with anger or melting down in a crisis), but again, no specific numerical information about such values is available. Nevertheless, it is useful to reason about them qualitatively because even direction of change can be valuable for predictions.

One of the hotly debated questions about such metaphors is the degree to which they are computed online. One extreme is that whenever such language is used, a new comparison is made to the physical phenomena referenced, and that comparison is used to draw inferences. The opposite extreme is that such metaphors become "frozen" in language, becoming alternate meanings for the terms that are simply used by whatever processes are normally constructing semantic representations. Evidence suggests that novel metaphors are indeed processed via online comparisons (i.e., comparisons during the process of understanding the sentence itself; Gentner et al., 2001) and that very familiar metaphors do not require such processing, suggesting that there is a learning process that moves common metaphors into lexical knowledge (Bowdle & Gentner, 2005). Although this is a fascinating issue, we won't consider it more here, because it isn't relevant to the point we are trying to make. The key point is that the qualitative, causal representations of QP theory are equally applicable to decidedly nonphysical phenomena.

Consider reasoning about politics of the sort one does when faced with newspaper stories that include arguments like those above. We seem to use a notion of political agents, which have many of the same properties as people. Most likely, we form models of people that include continuous parameters describing their social properties, like how patient they tend to be, and then analogize from those models to political entities ranging from clubs to political action groups to nations and transnational organizations. In modeling the physical world, a small set of types of physical processes tends to be sufficient (e.g., motion, flows, transformations, creation/destruction). In the political realm, the numbers and kinds of interactions are much broader. There are diplomatic interactions, such as fact finding, negotiations, and exchanges. There are economic interactions, such as trade, embargos, and tariffs. And of course, there are military interactions, such as staging exercises, invasion, and insurgency. Each of these has many aspects that aren't continuous, so it is probably better to think of the continuous information as simply part of whatever representations are used for such interactions. Because people are continually finding new ways to interact (e.g., outsourcing financial services, using hijacked airplanes as suicide weapons), this requires the ability to expand such representations as needed. A final source of complexity is that many of these actions or events can be viewed as discrete. Kim's (1993) discrete process representation provides a conceptual tool for integrating continuous change with discrete events.

Although promising, this line of research is still in its early stages. Part of what is required is a set of qualitative models for the kinds of phenomena commonly studied in social sciences. Thus, we turn to social reasoning next.

18.5 Social Reasoning

One of the interesting frontiers of qualitative reasoning is exploring its roles in social reasoning. Many aspects of our models of other people, ourselves, and social relationships have continuous aspects: we gauge someone as a good friend but might feel distanced from that person after a slight but warmer after a positive shared experience. Qualitative representations, I argue, provide a useful representation for the continuous aspects of such models and for causal models that underlie our social reasoning. Moreover, qualitative reasoning combined with analogical processing looks like a promising model for human social reasoning. Work in this area is much less well developed than work in modeling human reasoning and learning about physical domains, but I think it is very promising. This section outlines some initial investigations.

18.5.1 Modeling Aspects of Emotions

Emotions are of great interest to cognitive science for several reasons. Understanding the ways that they help guide behavior is a central problem. But an understanding of emotions is also vital for social reasoning, because understanding the emotions of others is needed to help predict what they will do under various circumstances and how one might interact with them successfully. We will focus on modeling emotions for social reasoning here, because exploring the roles of emotions in cognitive architectures would take us too far afield.[3]

An early computational model of emotions was the Ortony, Clore, and Collins (1988) model of emotions, hereafter OCC. OCC is an appraisal theory of emotion, focusing on how emotions about events, agents, and objects are generated. Events are appraised with respect to one's goals (e.g., one feels joy when winning a lottery and distress when injured, because these events are desirable or undesirable with regard to standing goals of maintaining well-being). The actions taken by an agent are appraised by comparing them to one's standards (i.e., the expectations about how an agent should behave). The agent can be yourself (e.g., pride at writing a particularly clever program and shame when it turns out to have a serious design flaw). Objects are appraised relative to one's standards: one might find Dijon mustard a bit dull for one's taste but think that a Sharf German mustard is just right.

OCC has been implemented in a variety of ways. The original specification is in terms of rules, numerical values, and assignment statements. The basic ideas can be recast in QP theory, and doing so has two advantages. First, it avoids having to choose ad hoc numerical values for the large number of parameters implied by the theory. Second, the same kinds of causal reasoning that is done on physical systems, by using model fragments to instantiate causal networks of influences, can be done to support reasoning about emotions. This approach is described in more detail in (Forbus & Kuehne, 2005), but summarizing the basic ideas here illustrates its potential.

Let us look at the emotion of joy, which in OCC concerns the feeling one has about an event. Two model fragments replace a number of rules more concisely, as figure 18.4 illustrates.

OCC uses a threshold to trigger joy when a combination of factors rises to a certain level. In the QP model, we simply introduce a limit point, joy threshold, for a person and an event, indicating that its intensity depends on how high the potential is over the threshold. Another model fragment defines the constraints on the potential for a desirable event. Following OCC, there are a number of parameters that it depends on. Some of these parameters are probably themselves directly influenced (e.g., arousal, with a decay process that reduces it over time, thereby attenuating joy and ultimately removing it as a feeling).

Suppose we did a complete implementation of OCC or a more recent model, such as EMA (Gratch & Marsella, 2004). How could it be used? Given partial information about a situation, of the kind one might find in a story, it could be used to make predictions about the emotions of the characters in the story. This might be done via qualitative simulation, either via first principles or analogically. The latter seems more psychologically plausible for prediction, whereas reasoning about perturbations in the causal structure and how they might change a person's actions might be better for planning how to influence someone in novel situations.

```
(defModelFragment PossibleJoy
 :participants ((?p :type Person)
                (?e :type Event))
 :conditions ((> (Desires+ ?p ?e) zero))
 :consequences ((qprop+ (JoyPotential ?p ?e)
                        (Desires+ ?p ?e))
                (qprop+ (JoyPotential ?p ?e)
                        (SenseOfReality ?p ?e))
                (qprop+ (JoyPotential ?p ?e)
                        (Proximity ?p ?e))
                (qprop+ (JoyPotential ?p ?e)
                        (Unexpectedness ?p ?e))
                (qprop+ (JoyPotential ?p ?e)
                        (Arousal ?p))))

(defModelFragment Joy
 :participants ((?p :type Person)
                (?e :type Event))
 :conditions ((> (JoyPotential ?p ?e)
                 (JoyThreshold ?p))
 :consequences ((= (JoyIntensity ?p ?e)
                   (- (JoyPotential ?p ?e)
                      (JoyThreshold ?p)))))
```

Figure 18.4

Expressing the OCC model of joy using QP theory. These model fragments more concisely express the contents of a number of OCC rules.

18.5.2 Blame Assignment

Figuring out responsibility for events is an important aspect of social reasoning. Social psychologists like Shaver (1985) have developed theories of attributing blame for negative events that are described in terms of abstract continuous parameters. These include the following:

- *Intentionality*. The degree to which the person intended for the action to occur.
- *Coercion*. The degree to which the person was forced to act in the way he or she did, as opposed to being free to choose a different course of action.
- *Foreknowledge*. The degree to which the person understood the implications of his or her action.

As Tomai and Forbus (2008) describe, this makes Shaver's theory amenable to formalization via qualitative representations. Shaver's theory of blame attribution involves four distinct modes of judgment, with strictly increasing responsibility in this order:

1. Causal without foreknowledge
2. Causal without intent

3. Intentional but coerced
4. Intentional in the absence of coercion

These four modes can be expressed via six model fragments to capture the distinction between someone intending to do an action directly versus causing it to occur via coercion. Knowing which model fragment is active to describe a person's role in a situation provides the relative magnitude of blame associated with them, compared to someone whose role is captured by a different model fragment. For two agents in the same role, relative blame between them depends on the causal factors that constrain responsibility in that mode.

An experiment by Mao (2006) used simple scenarios to find out how people assigned relative blame. Table 18.1 shows the human data from Mao's experiment, using the scenarios like the two shown in figure 18.5.

The blame figures are averages across subjects. In scenario 3, people assign more blame to the chair, because they ordered the environmentally harmful choice. However, it's important to note that the vice president (VP) still gets some blame. In scenario 4, the chair still gets some blame, because people have abdicated their responsibility as good corporate citizens, but more blame is assigned to the VP, because they could have made either choice freely and yet chose the option that harmed the environment.

The QP model produces ordinal results that are consistent with most of the human data for the cases in Mao's experiment. For the two scenarios shown here, table 18.2 shows the results both within and between scenarios. Notice that we can infer that the least blame across these scenarios is garnered by the VP in scenario 3, because they had no choice. The qualitative model correctly shows that the chair in scenario 4 gets less blame than either the chair in scenario 3 or the VP in scenario 4. However, the qualitative model has no basis on which to predict the relative blame

Table 18.1
Human participant results for blame experiment.

	Human Data		Mao Model		
	Chair	VP	Chair	VP	Degree
Scenario 1	3.00	3.73		Y	Low
Scenario 2	5.63	3.77	Y		Low
Scenario 3	5.63	3.23	Y		Low
Scenario 4	4.13	5.20		Y	High

Scenario 3. The chairman of Beta Corporation is discussing a new program with the vice president of the corporation. The vice president says, "The new program will help us increase profits, but according to our investigation report, it will also harm the environment. Instead, we should run an alternative program, that will gain us fewer profits than this new program, but it has no harm to the environment." The chairman answers, "I only want to make as much profit as I can. Start the new program!" The vice president says, "Ok," and executes the new program. The environment is harmed by the new program.

Scenario 4. The chairman of Beta Corporation is discussing a new program with the vice president of the corporation. The vice president says, "There are two ways to run this new program, a simple way and a complex way. Both will equally help us increase profits, but according to our investigation report, the simple way will also harm the environment." The chairman answers, "I only want to make as much profit as I can. Start the new program either way!" The vice president says, "Ok," and chooses the simple way to execute the new program. The environment is harmed.

Figure 18.5
Two blame scenarios used by Mao (2006).

Table 18.2
Results from QP model of blame assignment.

Chair1 VP1	VP3	VP2	Chair4	VP4 Chair2 Chair3

Results from QP model of blame assignment. Degree of blame increases going left to right, with characters in the same column being unordered within the model. The prefix indicates the character (i.e., the chair versus the vice president [VP]), and the suffix indicates the scenario number.

assigned to the chair of scenario 3 versus the VP in scenario 4. These results are better than Mao's original model, which could only assigned blame to a single person in a scenario. There were twenty-eight possible comparisons, of which the model correctly inferred twenty-one. The remaining seven cases all arise from the lack of knowledge of the functions implied by the qualitative proportionalities. Note that this analysis is inferring the ordinal relations that people would give across scenarios based on taking the ratings they give as being on an absolute scale. Whether or not this is an accurate way to estimate human judgments is an open question.

18.5.3 Moral Decision Making

Economic theories often focus on utilities as driving human decision making. But there is considerable psychological evidence that people are not solely driven by utility. For example, under some circumstances, people will make decisions that they view as more consistent with their moral principles, even if that leads to worse outcomes, based on utility. Consider this scenario (Ritov & Baron, 1999):

> A convoy of food trucks is on its way to a refugee camp during a famine in Africa. (Airplanes cannot be used.) You find that a second camp has even more refugees. If you tell the convoy to go the second camp instead of the first, you will save 1,000 people from death, but 100 people in the first camp will die as a result. Would you send the convoy to the second camp?

In utilitarian terms, the choice is obvious: 100 deaths is bad, but 1,000 deaths is worse. But by changing the destination of the convoy, it is your order that is causing the 100 deaths, and hence you are responsible for them. Participants were indeed more likely to not divert the convoy! An explanation for this is that people have *sacred* or *protected* values, which are not allowed to be traded off, no matter what (Baron & Spranca, 1997). Multiple subsequent studies have found similar results and refined the phenomena considerably. For example, the degree to which people are insensitive to outcomes varies with contextual factors, such as whether the action will influence the patient of harm rather than the agent (Waldmann & Dieterich, 2007), and the effect varies across cultures (Lim & Baron, 1997).

How can qualitative representations be used to model this phenomenon? Morteza Dehghani et al. (2008) proposed that qualitative order-of-magnitude representations provide an elegant solution. He defined a variation of the ROM(R) model (chapter 5), which imposes three potential qualitative distinctions between a pair of numerical values: greater than, equivalent, and equal. A single parameter, k, provided a means of making the model more or less sensitive to situational factors. One prediction of this model is that, for a big enough difference in utility, a sacred value could be overwhelmed. There is some evidence that this can occur, as can be seen by the variation in percentage of people willing to divert as the number saved at the original camp gets smaller and smaller.

This representation was used in MoralDM, the first computational model of moral decision making that incorporated sacred values. The architecture of MoralDM is illustrated in figure 18.6.

Given a new moral dilemma, expressed in simplified English, it extracts the essence of the problem. Essentially, it reads the story by looking for a decision to be made, the alternatives it has to choose from, and what the consequences are for each alternative. It uses rules to determine if an impact is positive or negative, including identifying any numbers involved (e.g., number of people at risk). The scenario is also analyzed to see if any sacred

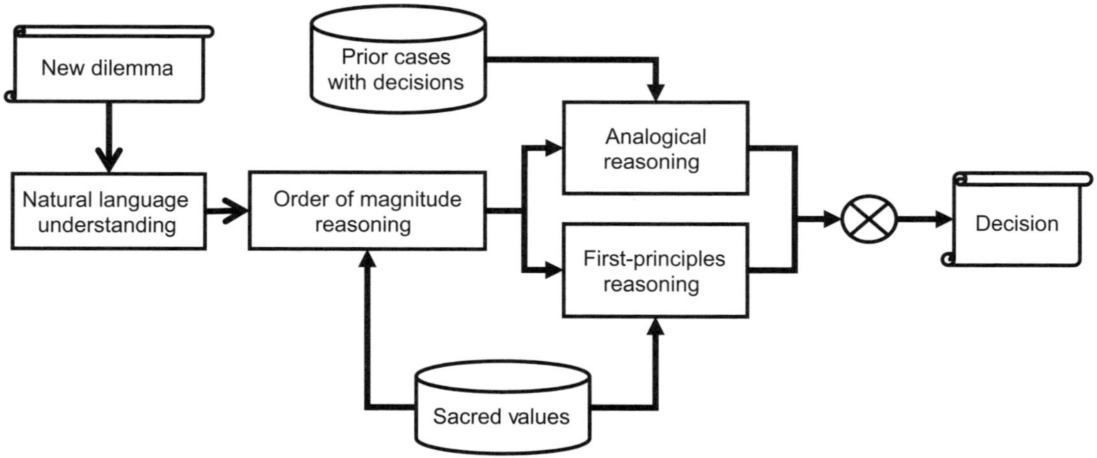

Figure 18.6
Results from the QP model of blame assignment. The suffixes indicate which scenario, and the prefix indicates which character (i.e., the vice president versus the chair).

values are applicable. The utilities, plus the relationship between them, are computed by the order-of-magnitude reasoning module. This module uses information from the story to ascertain an appropriate value for *k* and determines what relationship holds between the utilities for the two choices. If a sacred value is involved, the relationship computed will be close to, rather than greater than, making the system less sensitive to the numeric utility of outcomes. But if no sacred values are involved, then *k* is adjusted downward, making it more sensitive to utilities. It also uses rules to adjust *k* based on other factors found to be relevant in the literature, including whether or not the agent (in this case, the model) must intervene.

For example, consider again the starvation scenario. The model ascertains that there are two choices, an ordering event and an inaction event. The ordering event has two consequences, 1,000 people saved in the second camp and 100 people dying in the first camp. It knows that saving people has positive utility and people dying has negative utility, so the utility for the ordering event is 900. Similarly, for the inaction event, 100 people will be saved, but 1,000 will die, leading to a utility of –900. Because there is a sacred value involved, the setting of *k* is such that the relationship computed between them is `closeTo`. The three utilities and the relationship are then provided to the two decision-making modules. We start with the first-principles reasoning module and return to the analogical reasoning module below. The first-principles reasoning module uses utilitarian reasoning, picking the option with the highest utility, unless a sacred value is involved. If there is a sacred value at stake, the system looks at the relationship computed between the two utilities. If it is `closeTo`, the system chooses the alternative that avoids violating the sacred value because it views the difference in utilities as unimportant. Otherwise, it selects the alternative with the higher utility, capturing the intuition that, if the stakes are high enough, one might actually violate a sacred value.

Knowledge about various types of utilities (e.g., money, lives) and sacred values (e.g., not taking actions that will directly cause deaths) is part of the system's knowledge base. It is assumed that the knowledge of utilities is reasonably complete but that the system's knowledge of sacred values is not. There is evidence that people use analogy in decision making (Markman & Medin, 2002), including moral decision making (Dehghani, Sachdeva, Ekhtiari, Gentner, & Forbus, 2009). Consequently, MoralDM also includes an analogical reasoning capability, which compares the incoming story with scenarios that it has previously solved. It examines the decisions made for scenarios that are sufficiently similar (as determined by a threshold on a normalized version of SME's structural evaluation score) and that have the same order-of-magnitude relationship between their utilities. The analogical reasoning module's selection is the choice that is recommended by the most number of relevant analogies. If there are no relevant analogies, this module indicates that it cannot provide an answer. For example, in one run, MoralDM solved the starvation problem by analogy with a situation involving reallocating funds for a program to reduce deaths from traffic accidents from one area to another, involving saving 50 lives versus 200 lives.

The outputs of the first-principles and analogical modules are used to calculate the system's choice. If both modules suggest the same alternative, that is presented as the system's choice. If either fails to suggest an alternative, the answer from the other module is used. If neither suggests an alternative, the model fails to provide a choice. If the suggestions from the two modules are different, the answer from the first-principles reasoning module is chosen. Whenever it solves a decision problem, its understanding of that decision problem, including its analysis, is stored in the system's case library, so that it can be used in subsequent reasoning. Thus, first-principles reasoning bootstraps the model's processing, but as its range of scenarios grows, it can expand its range via analogies.

MoralDM was able to solve eight scenarios from the psychological literature (i.e., Ritov & Baron, 1999; Waldmann & Dieterich, 2007). Moreover, as the number of cases in the case library grows, the system's performance improves, in the sense that it is able to respond more often and to respond correctly more often. Figure 18.7 shows the results for a cross-fold validation using all possible case libraries of sizes 1 to 7, testing it on each of the scenarios not in the library. In essence, the model is learning how to do better moral reasoning by accumulating experience in the form of cases.

MoralDM has several important limitations. For example, all utilities within a problem must be of the same type—it cannot handle cross-dimensional trade-offs (e.g., money vs. lives). There is little psychological literature to draw upon at this point concerning such trade-offs, so this is perhaps a question best left until more evidence about what people do is available. The system's grasp of natural language is also limited—this is one of many problem areas where a deeper understanding of language is needed than simple statistical techniques can provide. Fortunately, there is more research occurring in this direction as well. A particularly interesting facet of the model is the potential for modeling cross-cultural issues via using a case library of moral stories from other cultures. In other words, the methodology of translating a culture's stories into simplified English, which can then be automatically understood, provides a more transparent workflow for building models than, for example, trying to set numerical parameters

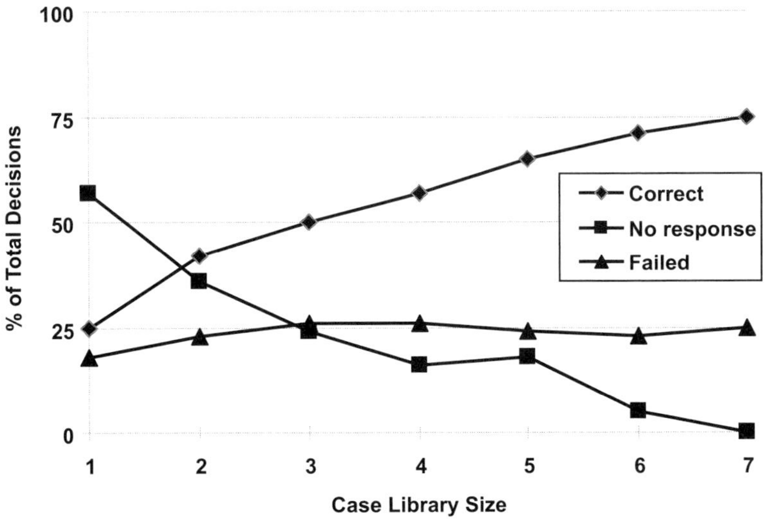

Figure 18.7
Moral decision making improves as analogs accumulate.

in a traditional mathematical model. Thus, this approach has the potential for providing a way to build models of cultural moral thinking in a more principled way than other computational modeling approaches.

18.6 Summary

The examples presented here provide evidence for the hypothesis that qualitative representations are central to human commonsense reasoning. Numerical estimation is something that we all do, and the use of quantitative representations to provide distributional qualitative representations provides a means of bootstrapping models of quantities via analogical generalization. The work on back-of-the-envelope reasoning indicates that even highly sophisticated commonsense reasoning can be facilitated by the use of qualitative representations. Qualitative representations provide an appropriate level of detail for describing many of the causal entailments of conceptual metaphors, whether they are being processed online or are frozen into our language. Qualitative representations appear to be able to capture important aspects of appraisal theories of emotions and to provide a formalization for a theory of blame assessment that provides results that are consistent with human data. Finally, qualitative order-of-magnitude representations provide a means for modeling the influence of sacred values in moral decision making. Thus, for both the physical and social realms, qualitative representations appear to play a key role in commonsense representation and reasoning. What about the mental realm, that is, the models we build of ourselves? To the extent that the *like me* hypothesis (Meltzoff, 2007) holds, we may first build our social models via analogies with qualitative models we have built of ourselves, honed by social interactions with others. But at this writing, the application of qualitative representations to commonsense reasoning about mental states and events remains an open frontier.

19 Expert Reasoning

One of the roles of qualitative representations, I am arguing, is to provide a foundation for expert knowledge used in professional reasoning. Many of the examples in part II involve phenomena of interest to experts, such as thermodynamics, fluid flow, and motion. But the reasoning there was more of the kind that any of us can do, without much professional training. Is expert reasoning completely different, isolated, and insulated from our everyday intuitions? Or is it tightly integrated, with everyday knowledge interacting with the kind of professional knowledge acquired through formal education?

The answer seems to be, unfortunately, is that it depends. As noted in chapter 17, many students seem to treat everyday knowledge and professional knowledge as completely different, as studies of physics misconceptions have illustrated (Clement, 1983; McCloskey, 1983). However, it also seems that the best students aggressively seek to integrate what they are learning in school with their everyday experience. Both domain experts and learning scientists agree that deeper conceptual understanding is an important component of expertise (Hestenes et al., 1992).

To gain insight on this question, we look at it here from a different perspective. Many efforts in qualitative reasoning have been focused on creating software systems that achieve expert-level performance in some area of scientific or engineering reasoning. Although these systems were not developed as cognitive models per se, they still provide important clues as to the nature of expert reasoning. They provide evidence about the constraints imposed by the tasks themselves. Task-level constraints are harsh masters. Someone, or something, attempting to do a task must first be judged as to whether or not they can do it. How well they do it, over what range of circumstances, and with what resources (both in terms of inputs and processing/materials needed during performance) are important questions. For example: Does an approach require detailed numerical data about the situation? Is the construction and/or use of external diagrams or three-dimensional models necessary? There is often more than one way to do something, but alternate methods can be judged based on several criteria. Even in an application setting, gathering information has costs, so methods that require less information to derive an equally good answer are preferred. Similarly, methods that require less computation tend to be preferred over methods that require more, so the same constraint of parsimony that one expects in evaluating cognitive models is also commonly used in evaluating application-oriented research.

As noted in chapter 12, although evidence suggests that the qualitative representations developed in qualitative reasoning research are often quite good approximations of human representations, there are reasons to doubt that human reasoning is entirely based on the kind of first-principles reasoning commonly used in qualitative reasoning research. These differences would show up as differences in the complexity of reasoning (e.g., different amounts of time required, which would be true in any case, given the quite different information-processing architectures involved) and in error patterns (e.g., an envisioner can catch possibilities that a less complete method can miss). But the kinds of information required as inputs and the forms of representations needed for the tasks are both properties that hold for any information-processing system attempting those tasks.

There are several reasons that practitioners, as well as qualitative reasoning researchers who work with them, give for the importance of qualitative models in science and engineering. The first is that they are viewed by practitioners as providing conceptual descriptions that are more in line with their intuitions. As discussed below, the insights often wrung from visualizations and data mining are often qualitative descriptions. In other words, qualitative representations are the level at which insights are expressed. A second reason is that, for many purposes, the more abstract level of detail that they operate at is valuable for reasoning and for framing more quantitative problems. Finally, the goings-on within complex artifacts often must be communicated in more intuitive terms to those who aren't their designers. Such people include those who repair cars, photocopiers, and aircraft, as well as those who operate and maintain factories, chemical plants, and power stations.

We start by looking at engineering reasoning. Textbook problems are intended to be practice for the kinds of analyses that engineers do, so we examine how qualitative representations have been used to model human performance on them. Monitoring and diagnosis are examined next. The role of qualitative representations in design is also considered, showing that qualitative representations provide a natural way to explore alternatives. System identification, where experiments are conducted to reverse-engineer how a device or system should be modeled, is also examined, because it provides a bridge to the next topic, scientific reasoning. That section explores how qualitative representations are being used by practitioners in biology, ecology, and geoscience—clear evidence that practicing scientists find qualitative representations compatible with how they think about their domains.

19.1 Engineering Reasoning

Engineering covers a broad range of human activities. It is the harnessing of natural laws and phenomena in service of human needs and goals.[1] There are many types of engineering, with radically different cultures and concerns. (When someone talks about "real engineers," look out—it's about as accurate a phrase as "real people" when describing a cultural practice.) Nevertheless, there are some broad commonalities. A hallmark of engineering practice in most fields is the use of mathematical models, and a core practice is moving from everyday terms into formal models that provide quantitative predictions. Systems break, so it is important to monitor them, look for problems, and then diagnose and repair them. Design is at the heart of engineering: how does one put together systems—often structured hierarchically, using components and subsystems involving multiple physical domains—that perform needed services? Finally, system identification is a form of what is commonly called *reverse engineering* (i.e., taking a finished artifact and ascertaining, by experiments on it, how it works). This section examines the role of qualitative representations for each of these tasks in turn.

19.1.1 Analysis

Textbook problems are designed as a model for engineering analysis. Let us look at the content of a typical first course in thermodynamics for mechanical engineers as an example. Figure 19.1 illustrates a typical sequence of types of problems. Students start out first learning about the properties of fluids, such as air and steam. (These are called *working fluids* in engineering thermodynamics.) They learn about approximations like the *idea gas law*, which holds for air and a variety of other substances. But they also learn that many substances, like steam (i.e., water in its vapor state), cannot be modeled this way and that numerical tables are needed instead. Interactions between pressure, temperature, and force are all explored to help them cement their understanding of working fluids. The problems given in this part of the course typically involve a single qualitative state that is missing some piece of information.

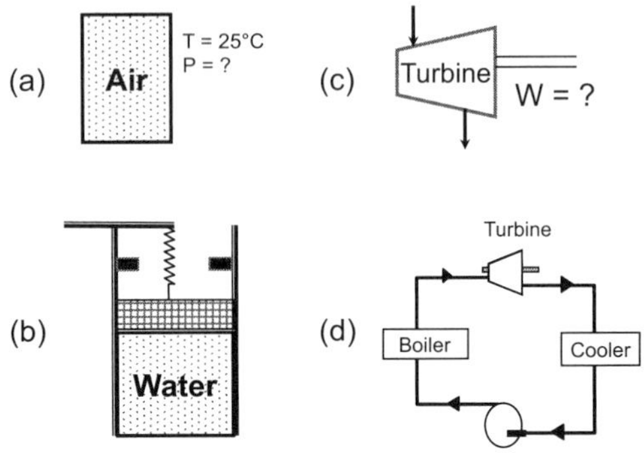

Figure 19.1
Four kinds of problems from engineering thermodynamics courses for mechanical engineers.

For instance, they might be asked about the pressure of a piece of air, given that it is at a particular temperature (figure 19.1a).

Next they move on to problems where there are two qualitative states, where they have to derive how properties change over time. Consider figure 19.1b. Heat is applied to the working fluid in the cylinder until it reaches a specific, given temperature. How high is the piston in the cylinder at that time? Even solving this simple type of problem requires a number of qualitative insights. First, because there are no holes or pipes, we know there are no mass flows in or out of the system. At some point, the cylinder may rise up far enough to touch the spring, which will apply a force against further expansion. The two stops in the cylinder will put an upper bound on how high the cylinder will go. Applying heat to a fluid generally causes its temperature to rise, unless the fluid is in the saturated state (think boiling, which is a special highly turbulent case). When a fluid is saturated, adding heat changes more of it from liquid to gas. Thus, the qualitative state of the liquid determines which mathematical models are applicable. All of these are conclusions that must be drawn by reasoning about the system before a mathematical model can even be formulated. And they are all, in essence, qualitative.

This particular problem has a feature that, based on instructor observations, students hate. It has a qualitative ambiguity: will the piston reach the stops or not? The only way to solve the problem is to make an assumption about the outcome, formulate that model, solve it, and look for a contradiction. This is quite different from the "plug-and-chug" approach that most students take (and too many courses encourage) of recognizing the right equation and plugging in numerical values to produce the desired numerical answer.

Returning to figure 19.1, the last two kinds of problems involve what are called steady-state analyses (i.e., the working fluid is flowing through a system, undergoing various changes in different parts of it, to do useful work). Although the properties of the working fluid are changing as they go through the system, at each physical location in the system, its properties can be treated as constant over time. (We have seen this idea before, in chapter 11, and the representations and techniques discussed there are used here, too.) They start with simple problems involving particular components and steady flow (i.e., given the properties of the working fluid going into and out of a turbine, how much work is being produced by it, in figure 19.1c). Finally, they analyze entire thermodynamic cycles, which are abstract descriptions of power plants, engines, refrigerators, and other useful systems (figure 19.1d). As we have seen, even though most answers in textbook problems involve numbers, getting to those numbers can require qualitative reasoning.

We have seen one example of how qualitative knowledge is needed to identify relevant phenomena and modeling assumptions that need to be made to produce quantitative knowledge. That is not an isolated example. CyclePad (Forbus et al., 1999), an intelligent learning environment for engineering thermodynamics, uses a combination of quantitative and qualitative representations and reasoning to help students learn how to design cycles. It includes simple qualitative models linked to the quantitative models to help detect implausible assumptions students can make when doing designs. For example, some students come up with parameters that would have the pump in their system producing, rather than consuming, work, which is impossible. Thus, qualitative models provide a form of reality check on mathematical models.

Yusuf Pisan (1996, 1998) found another role for qualitative representations: they provide a means of classifying equations in ways that enable systems to set up mathematical models and solve them in more human-like ways. His Thermodynamics Problem Solver (TPS) solved over 150 thermodynamics problems, representing 75 percent of the types of problems found in the first four chapters of thermodynamics textbooks for mechanical engineers. The other 25 percent were mostly purely conceptual questions, involving essays for answers, with a handful involving counterfactuals involving world knowledge beyond what the system had (e.g., a hypothetical alien fungus). It is well known that novices tend to thrash on such problems, doing much more searching than experts do. It is easy to say that experts know more, but what form does this knowledge take? Let us look at prior work on a simpler problem that provided part of the motivation for Pisan's approach.

Consider the simpler problem of solving an algebraic equation with a single unknown, such as

$$3X = 5X - 2$$

Such equations are solved by applying laws of algebra to rewrite the equation, until it is in the form of the unknown on the left-hand side of the equation and an expression not involving the unknown on the right-hand side of the equation—in this case,

$$X = 1$$

There are a lot of algebraic laws. Knowing which one to apply when is a problem of control, often called *metaknowledge*. Bundy (1983) argued that the difference between novices and experts was a particular kind of metaknowledge: experts classify laws of algebra into functional roles. Specifically,

- *Attraction laws* bring occurrences of the unknown closer together in an expression. An example of an attraction law is $WU + WV \rightarrow W(U + V)$, where U, V contain the unknown and W does not.
- *Collection laws* reduce the number of occurrences of the unknown in an equation. An example of a collection law is $(U + V)(U - V) \rightarrow U^2 - V^2$.
- *Isolation laws* reduce the depth of the occurrences of the unknown in the equation. An example of an isolation law is $U - W = Y \rightarrow U = Y + W$, where again, U contains the unknown, and W and Y do not.

If you think about what it means to solve an equation, these make sense. Attracting occurrences of the unknown together makes it easier for collection laws to reduce the number of occurrences, and because the goal is to have a single occurrence of the unknown, these bring one closer to the goal. Having just the unknown on one side of the equation is another constraint on solutions, so bringing it up to the top (i.e., reducing its depth until there is nothing above it) is useful. Bundy (1983) showed how a set of rules could analyze algebraic laws to determine their functional role automatically and that using these functional roles enabled a system to solve equations with far less effort than a system that tried applying them more blindly.

Pisan (1996, 1998) found an analogous phenomenon in solving thermodynamics problems. The three categories of equations he identified were as follows:

- *Internal equations* consist of variables pertaining to just one object. For thermodynamics, the objects can be quite complex, with pieces of working fluid modeled with up to twenty parameters, resulting in over thirty internal equations relating them. Internal equations provide a means for calculating the values of unspecified properties based on properties that are known. Which set of internal equations is appropriate depends on qualitative properties of the objects (e.g., a fluid being an ideal gas or not and, if nonideal, saturated or not).

- *Bridge equations* involve the parameters of two or more objects. Bridge equations express interactions between objects and provide a means for inferring information about one object given information about the other. Bridge equations in thermodynamic cycles, for example, are tied to physical processes that describe the transformation of the working fluid from one place to another in the cycle.
- *Frame equations* are central laws in a domain that serve as a starting point for analysis. The first law of thermodynamics is an example of a frame equation.

Qualitative analysis is needed to set up each of these types of equations. The general form of the first law of thermodynamics is quite complicated, involving mass, velocity, gravity, height, enthalpy, heat, and work. But for any particular problem, some of these parameters are irrelevant, allowing it to be considerably simplified. For example, in most engineering cycle analyses, height (and hence gravity) is factored out. TPS used this functional decomposition to control how its equations were used, resulting in more expert-like solutions compared to a problem-solving system that did not use this decomposition (i.e., CyclePad, which used exhaustive constraint propagation to analyze student designs).

Another role for qualitative representations in solving textbook problems is in facilitating the use of analogy to solve problems. Evidence suggests that one aspect of expertise is having learned better encoding processes (e.g., Chi et al., 1981). Part of the encoding process for textbook problems, we hypothesize, is a qualitative analysis of the problem to identify what processes and conceptual distinctions (i.e., model fragments) are applicable in it. To test this hypothesis, Ouyang and Forbus (2006) used a modified version of Pisan's TPS, which incorporated MAC/FAC for analogical retrieval and SME for applying solution methods used in prior problems to new problems. Using a corpus of eighty-two problems, we found that doing a qualitative analysis of the problem led to a noticeable improvement (9 percent) of the overall performance, as measured by a reduction in the number of nodes in the search space. The impact was stronger on more difficult problems.

Thus, we see qualitative representations play three key roles in textbook problem solving, which is itself a surrogate for engineering analysis:

- They identify what phenomena are relevant and what modeling assumptions are appropriate for the domain.
- They provide a means of classifying equations functionally, providing control knowledge that facilitates more expert-like problem solving.
- They facilitate analogical retrieval, leading to better solutions.

19.1.2 Monitoring, Control, and Diagnosis

Consider a complex engineered artifact, like an airliner. Such artifacts have many systems that must continue to work reasonably well for the artifact to operate as intended. These systems must be monitored to ensure that they are operating correctly. Modern airliners are so complex that a combination of people and software is needed to fly them. Some of their subsystems are so complex (e.g., engines) that they have their own subsystems for controlling their operation, based on commands from the rest of the aircraft and their own monitoring systems to inform the rest of the aircraft if the engine is starting to malfunction. When there is a problem, it needs to be diagnosed and an appropriate course of action determined. An engine drifting slightly out of its usual settings may not require any extraordinary steps during flight but might lead to a more detailed maintenance check upon landing.

The problems of monitoring, control, and diagnosis are deeply intertwined. Based on the work of researchers building AI systems to perform these tasks, there is now a body of evidence suggesting that qualitative representations often play an important and sometimes essential role in them. Monitoring involves summarizing the behavior at a system that is useful for taking action. These summaries are typically qualitative. An engine being out, for example, means that the processes usually taking place in it aren't. A pilot cares that the thrust being produced is low and whether it is below some critical threshold—there is not a different strategy for every 2 percent deviation in thrust value. Similar problems occur in other types of artifacts run by people, including factories, chemical plants, and power stations. These problems also occur in artifacts that are expected to run autonomously: we want our appliances and equipment to explain to us what is wrong, when possible, if they fail to work. (As of this writing, most of them, sadly, do not.)

Feedback controllers rely heavily on quantitative knowledge: sensors measure numerical information, the control strategy calculates quantitative responses, and the actuators apply those responses to physical properties of the artifact to correct its behavior. But what sets the control strategy?

For some simple systems, such as the flushing mechanism of toilets, a fixed control strategy will do. But for others, where a system undergoes distinct qualitative states, different control strategies may be needed for different qualitative states. This is what is known as *supervisory control*. Qualitative representations have proven quite useful for automated reasoning for supervisory process control. The first industrial application of qualitative reasoning was in curing epoxy-cast parts (LeClair, Abrams, & Matejka, 1989). Some complex parts are manufactured by putting epoxy over molds and then baking them in an oven to cure the resin. If the heat is too high, bubbles form from gases trapped in the epoxy, ruining the part. If the heat is too low, the process is inefficient, because it takes longer to cure each part and more energy to do so. Le Clair and his colleagues (1989) figured out that by using a simple qualitative model of the processes involved, they could create a controller that detected when it was safe to crank the heat up. This technique was quite successful and was commercialized.

Another example of supervisory control are the strategies used by machine operators, for instance, using a crane to unload cargo from a ship. Move too quickly, and the cargo starts to swing, which could damage the crane and the surroundings. Move too slowly, and the whole process of unloading becomes less efficient (and hence less profitable). Suc and Bratko (2002) showed that, by recording the strategies of crane operators, they could use qualitative representations combined with machine-learning techniques to reverse-engineer efficient strategies for unloading cargo.

One of the major drivers of the cost and complexity of modern electromechanical systems is the cost of the control software to run them. The software can only be written after the machine is designed, a process that typically requires significant manual labor and testing, leading to increased development expenses and increased time to market. Perhaps the most dramatic use of qualitative models is to enable complex artifacts to automatically do their own planning and scheduling, eliminating the need for manual construction of control software (Crawford et al., 2013). This work was done in the context of high-end printing machines, the kinds of printing machines that can produce a full hardcover book (including binding) in less than a minute. Such machines are enormous, with multiple print engines, paper sources and paths, and processing units. By creating qualitative models of the kinds of components that can be used to assemble such machines, AI planning and scheduling techniques can be used to dynamically reconfigure the flow of paper through the components, in real time, as each page moves through the machine. The models contain quantitative information, as well as qualitative information, to provide timing information needed for scheduling purposes. Moreover, new parts can be added to the machine while it is powered off, and when it is restarted, it detects the new parts, updates its self-model, and proceeds to generate new plans that best incorporate the new capabilities. Crawford and colleagues (2013) point out that another purpose of the models is to facilitate communication between teams of engineers—more evidence that qualitative models provide a language that seems conceptually natural to them.

Automated diagnosis has been an area of much research, because many billions are lost each year to systems not working and to the cost of maintaining and repairing them. Qualitative representations are heavily used in diagnosis, because most minor quantitative differences in component parameters don't matter. Every component's parameters have some range of tolerances in their specification, and a well-crafted system builds in a bit of extra resilience against out-of-tolerance components. However, detecting when a system has drifted too far out of tolerance is important. Moreover, there are hard, sudden failures (e.g., a capacitor burning, a resistor shorting out, a pump failing), which can in turn lead to a cascade of other failures. This makes diagnosis complex, because there can be several explanations for the symptoms that indicate a failure. One of the surprises has been how often extremely simple models suffice for diagnosis problems: for photocopier diagnosis, for example, a simple okay/not okay distinction is often enough (Fromherz et al., 2003). A similar lesson came out of NASA's Remote Agent project (Muscettola, Nayak, Pell, & Williams, 1998), where the onboard controller for the *Deep Space One* spacecraft was built using qualitative models described in terms of deviations from norms. They found that qualitative models provided robustness in the face of design changes—if hardware designers changed a thruster valve to produce more thrust, the qualitative model remains the same. Moreover, qualitative models enabled the system to use propositional reasoning for fast inference, compared to numerical models or simulations, to guarantee real-time response during the mission.

Understanding the consequences of failures also often requires qualitative models, because quantitative failure models often don't exist. There are too many ways for physical systems to fail. For example, a fan blade from an engine on an airliner can slice through various parts of the aircraft and sometimes even enter another engine on the other side of the airplane. This makes constructing failure models extremely expensive, because one has to consider multiple levels of interactions between different systems. It can also be dangerous: to create quantitative failure models for in-flight airline engine failures would require test pilots, aircraft, and the ability to "inject" faults into heavily instrumented versions of systems at different altitudes.

Consequently, for most artifacts across many industries, such quantitative failure models simply don't exist.

In the automobile industry, onboard diagnostics for monitoring pollution produced by cars is now a legal requirement. There is plenty of computation available inside cars these days (some have up to 100 different computers in them at this writing), but automatic generation of diagnostics is complex, for many reasons. One reason is that cars can be ordered with lots of different options, and these options can affect the diagnostics (e.g., whether or not particular electricity-using accessories are installed). Another reason is that the electrical environment inside a car is very noisy: qualitative representations provide a useful way to abstract away from sensor noise (Sachenbacher, Struss, & Weber, 2000).

There is an interesting way in which most of today's diagnosis systems are very different than people. Most fielded systems use a technique known as *consistency-based diagnosis*. Suppose you have a model of a system, described by composing models for its components. Given a set of observations of that system, consistency-based diagnosis involves using discrepancies between the observations and the model's predictions to hypothesize which combinations of component models cannot be operating normally. This technique works well in many circumstances. It is popular because it has a major advantage: explicit models of how systems can fail are not needed. Fault models can be very hard to develop, in part because there are so many ways that something can fail. However, there are two problems with this approach, in terms of capturing the full range of human diagnostic reasoning. The first is that sometimes the details of the failure mode truly matter: if the level of cooling water in a nuclear reactor is dropping unexpectedly, it is extremely important to know where that water is going! The second is that consistency-based diagnosis can, with some models, produce counterintuitive results. Consider a simple electrical circuit consisting of a switch, a battery, and a light bulb, as might be found in a flashlight (figure 19.2).

Initially, the switch is closed, and the light bulb is on. Now we open the switch, and unexpectedly, the light bulb does not go off. If we ignore the possibility that there are problems with the connections and just focus on the named parts, most of us would assume that the switch has a short circuit (i.e., it is still acting as if it were closed). Consistency-based reasoners will find one additional hypothesis: that the light bulb has failed in the ON state! It is easy to build into the model that this particular possibility cannot happen, but that does seem somewhat ad hoc. A more elegant solution was suggested by Collins (1993), who observed that the components of a system are always bound to follow the laws of Nature, and hence abductively

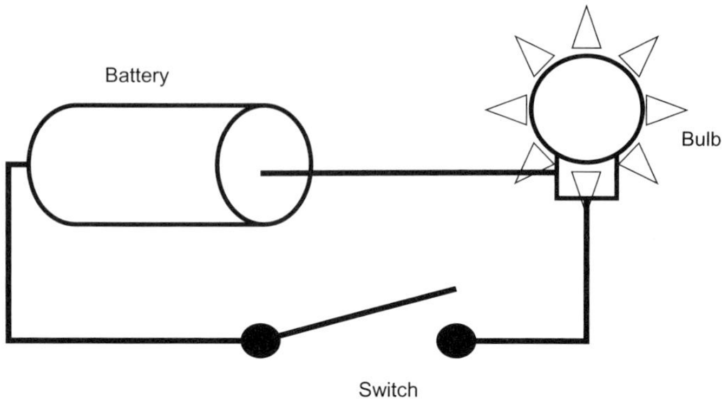

Figure 19.2
A simple flashlight, exhibiting a failure—light bulb is on, but the switch is open.

constructing a model of what the component is doing instead could be a useful way to do diagnosis in some circumstances (e.g., where you need to understand the possible indirect effects of a problem more clearly).

19.1.3 Design

Design is the creative center of engineering. Design is often divided into two phases. *Conceptual design* involves the basic idea of a design, only specifying enough detail to allow it to be worked through to see if it is plausible. *Detailed design* is when all the parameter choices are made, the exact shapes are determined, and enough information is provided so that numerical simulation is possible.[2] Most software tools for design focus on detailed design. Qualitative representations have enabled the construction of tools that help in conceptual design as well (Klenk et al., 2012; Shimomura, Tanigawa, Umeda, & Tomiyama, 1995). For example, the first product to market that used qualitative representations in its design was the Mita Corporation's DC-6090 photocopier. It was the first fielded *self-maintenance machine*, capable of reconfiguring itself dynamically when it detected certain kinds of faults. It did this by keeping track of which qualitative state it was in, so that it produces the best-quality copy that it can, given its current health. This was made possible by a conceptual design system that used qualitative simulation to construct an envisionment, including fault models, at design time. This envisionment was used to construct the copier's control software, to monitor and reconfigure itself, at design time. (The PARC fully automatic, real-time composition of photocopier/printer control software described above represents the latest developments in this area.)

In most fields, design strategies and methods are described in natural language. Qualitative representations, which seem to be used in natural-language semantics (see chapter 13), can provide formalizations that allow such strategies to be automated. For example, in chemical engineering, several methods for designing distillation plants have been so formalized, with automatic methods producing designs for binary distillation plants comparable to those found in the research literature (Sgouros, 1998).

One important problem in designing complex systems is *failure modes and effects analysis* (FMEA). That is, what sorts of things can go wrong with a system, and for each thing that can go wrong, what are the possible outcomes, especially the worst-case outcomes? (The designers of the Fukushima nuclear power plant did not anticipate a tsunami, for example, leading to flooding that damaged critical systems.) FMEA basically requires a form of envisionment. That is, it requires looking at all of the possible behaviors of the system, both when it is operating normally and when various faults in components of the system (or extreme operating conditions, e.g., a river so warm that a power plant cannot transfer waste heat away quickly enough) occur. Qualitative models, because they carve up the infinite number of numerical states a system may go through into a finite set of meaningful regions, provide a natural representation for FMEA. And indeed they have been successfully applied to this task. The most famous example concerns designing the electrical systems of cars. AutoSteve (Price, 2000) has been in daily use since the end of the twentieth century by designers at Ford Motor Company, and now by other car manufacturers as well, to perform FMEA for the design of their electrical systems.

19.1.4 System Identification

Sometimes engineers have to figure out how a new component or system works. This is the problem of *system identification*. It is, in essence, the engineering culture's version of scientific modeling, so it serves as a perfect bridge to our next topic. Typically, the goal is to construct a differential equation model of the dynamics of the system, using experiments to gather numerical data about how the system responds to perturbations. Even though the goal is a quantitative, numerical model, qualitative representations turn out to play a useful role in system identification as well. Systems can be understood at multiple levels, and understanding the qualitative behavior of a system requires less data and serves as a guide to constructing the more detailed, quantitative models. For example, the PRET system (Bradley, Easley, & Stolle, 2001) takes as input a physical system to understand and, using sensors and actuators to experiment with it, constructs a differential equation model describing its behavior. The kinds of systems it operates on include networks of electrical components, mechanical systems, and more complex devices, such as a radio-controlled car. Such systems often require nonlinear differential equations to understand, making them quite complex. PRET starts with extracting qualitative properties of the system's behavior. For example, if a system is oscillating, that immediately rules out the underlying equation being lower than second order. Thus, qualitative representations are used to guide the search for quantitative models.

Similarly, Pade (Bratko & Suc, 2003) can be used to learn qualitative models from examples, including active learning (Zabkar, Janez, Mozina, & Bratko, 2010), where statistics are used to guide the next examples to be examined.

19.2 Scientific Modeling

One of the hypotheses of those doing qualitative reasoning research is that qualitative representations provide a natural level of description for the intuitive models used by scientists in their professional reasoning. Several lines of qualitative reasoning research lend support to this hypothesis. The first is the work on producing qualitative summaries of spatial numerical data. The insights that scientists prize are often qualitative in nature. For example, Yip's KAM system produced qualitative descriptions of the behavior of fluid dynamics models that led to journal-quality publications (Yip, 1991). The formalization of the *Spatial Aggregation Language* (Bailey-Kellogg & Zhao, 2003; Zhao et al., 2007) has led to techniques for structure discovery that are useful with spatial data sets more broadly. For example, Ironi and Tentoni (2011) use these techniques to gain insight to cardiac phenomena.

Second, model building in science often has a character similar to system identification in engineering. In some fields, one can experiment with the subject of study (e.g., cell biology), but in other fields, one must be content to gather data by making measurements (e.g., cosmology). Nevertheless, the same ontology of processes is applicable across a broad range of fields. Biologists and ecologists, for example, both informally talk about processes, and this talk of processes guides their construction of quantitative models. For example, from the work of Langley and his collaborators on *inductive process modeling* (IPM),

> In contrast [to empirical models], mechanistic models contain unobserved relationships, explain system dynamics, and emphasize the physical, chemical, and biological processes that generate system behavior. Ecologists use mechanistic models to understand how system behavior may change in response to changing

> environmental conditions. A central problem of building models with more realistic structures is determining which entities and processes to include and which mathematical representation is most appropriate. (Borrett, Bridewell, Langley, & Arrigo, 2007, 1)

In other words, processes are used to frame what is happening in a situation. The processes are not identical to the mathematical formulation, because multiple mathematical formulations can be used for the same process. This is compatible with QP theory's notion of continuous process, even though the model fragments that IPM systems use focus on quantitative models. These systems start with a domain theory, expressed as a library of such processes, and a collection of data to explain. The system is then tasked with coming up with a process model, or improving an existing model, to fit the data. Sometimes it works autonomously; in other situations, it works in tandem with scientists. The model constructed consists of a set of instantiated processes from the domain theory, with the evaluation criteria being the root mean square of deviations of the simulations produced using the model to the data set. When the system sets out to build (or improve) a model, it searches through the space of possible instantiations of the processes to descriptions of the entities in the current situation. It also uses ideas from compositional modeling to express constraints between processes to help guide the search. For example, it uses the equivalent of assumption classes (chapter 11) to ensure that, if grazing is relevant, exactly one of the ways of modeling grazing is included in the model under construction. The system has been shown to be capable of building models for several data sets, including the Ross Sea ecosystem. The formalism has been extended to handle spatially extended models as well, what would normally be modeled using partial differential equations (Park, Bridewell, & Langley, 2010).

Sometimes qualitative models provide the right level of representation for scientific reasoning. If your model suggests that a balloon filled with mercury vapor will rise, then it doesn't take curve fitting to see that if it falls, your model is wrong. Qualitative models have long been used in economics (e.g., Simon, 1953), although not originally recognized as such. Another set of examples of qualitative models being directly used for reasoning through hypotheses comes from biology (e.g., Karp, 1993; King, Garrett, & Coghill, 2005; Langley, Shiran, Shrager, Todorovski, & Pohorille, 2006; Trelease & Park, 1996). For example, in understanding bacterial regulatory networks, experiments often provide very little data, and much of them are noisy, making quantitative modeling difficult (Batt et al., 2012). On the other hand, a special-purpose qualitative simulator has been built that enables scientists to automatically see the implications of their hypotheses about a network and, using AI model-checking techniques, automatically test those hypotheses against laboratory data (de Jong, 2008).

Qualitative representations have also been used as a means of collecting intuitive models from people and refining them to represent the understanding of a group. For example, farmers living in an area often have a deep understanding of the factors that affect plant growth in their areas. This knowledge is often localized, because different flora and fauna have different ranges, and thus there are subtle differences between local ecosystems. Agricultural scientists conduct field studies, interviewing people and using that information to build models for particular areas to better understand ecosystems and to help improve agriculture. These models are qualitative in nature, describing influences between various factors affecting plants. The Agroecological Knowledge Toolkit (Sinclair & Walker, 1998) provides a set of language-based ways of describing qualitative models for coding interviews. It is broadly compatible with the ideas discussed in chapter 13 but developed independently, providing independent evidence for the attractiveness of qualitative representations. This toolkit has been successfully used in a variety of studies to capture and refine the local ecological knowledge of multiple groups. Similarly, Dehghani, Unsworth, Lovett, and Forbus (2007) analyzed protocol data from two groups to formalize the qualitative models of individuals about aspects of ecosystems and used analogical generalization to construct group-level models from that data.

Scientists, being human, do not always agree. The level of discussion in these disagreements typically concerns mechanisms, not quantitative details. The exact form of a quantitative approximation rarely inflames the journals, whereas whether or not a kind of process is occurring can and does (e.g., plate tectonics). The CALVIN system (Rassbach, Bradley, & Anderson, 2011) uses qualitative representations of evidence to help geoscientists construct arguments about the interpretation of cosmogenic isotope data in rocks. This is a difficult interpretation problem because it involves reconstructing past occurrences of processes from very indirect data, complicated by the sparsity of available samples, combining laboratory analyses (subject to their own errors) with qualitative field observations and a variety of alternative explanations for most situations. Experts in these tasks build arguments for and against competing hypotheses. CALVIN contained just over 108 rules that combine qualitative, quantitative, and certainty information. In an experiment using published arguments from the literature, the system was able to closely reproduce expert arguments 62 percent of the time and produce similar arguments a further 26 percent of the time, as well as occasionally producing novel plausible arguments, as judged by experts.

Finally, qualitative representations are being explored in scientific modeling as a means of helping nonscientists understand complex systems. For example, the GARP3 environment (Bredeweg, Linnebank, Bouwer, & Liem, 2009) has been used in a variety of ecological modeling projects. One such project concerns the Danube Delta Biosphere Reserve, in Romania, which is one of the great wetlands of the world. How can sustainable development, which provides economic benefits but does not wreck the environment, be carried out? GARP models were built as a way to improve communication, that is, to

> explain and educate environment agency representatives, decision-makers, and stakeholders about the working of processes within the Danube River and how these can be managed. (Bredeweg et al., 2008, 5)

The models focus on biodiversity conservation and protection measures for flora and fauna, such as habitat preservation. Avoiding algal bloom is an example of one such problem (Cioaca, Linnebank, Bredeweg, & Salles, 2009). The models are also intended to support student learning about sustainability.

19.3 Summary

This chapter provides examples of how qualitative representations have been used to produce human-level performance in multiple aspects of engineering and scientific reasoning. This provides evidence that qualitative representations can provide an explanation for how people are doing such reasoning. To date, because most of the research has been by AI scientists, the measures used for evaluation have been ability, although this includes the ability to provide explanations that human experts find plausible, so compatibility with human ways of operating is an important criterion even here. Future studies of the psychology of scientists and engineers will hopefully help refine our understanding of how they use qualitative representations. For example, how much of expert reasoning is analogical versus first principles? This is a fascinating open question.

V Summary and New Directions

Qualitative representations provide symbolic, structured representations of continuous phenomena. This book has argued that such representations play several essential roles in human cognition. Chapter 20 summarizes these roles and the arguments for them. Research on the cognitive implications of qualitative representations is only now coming into its own, and so several new directions are discussed in chapter 21.

20 Summary

This book has made the case for qualitative representations being central to human cognition. We have seen multiple arguments for this case, employing several kinds of evidence and covering a wide range of phenomena. Here we summarize these arguments for conciseness.

20.1 Bridge between Perception and Cognition

Perception encodes the world, providing information that we need to survive and thrive. To make sense of the massive stream of visual information, we appear to summarize this flood into sets of symbols, rooted in the quantitative descriptions computed by earlier stages of processing, which are more appropriate for conceptual reasoning. As part III described, these qualitative descriptions of shape and space are used to make predictions, do visual comparisons and other visual problem solving, and support learning spatial language. The evidence supporting the hypothesis that such qualitative representations exist and are used for reasoning and learning includes behavioral studies, linguistic analyses, neuroscience studies, and computational models.

20.2 Basis for Commonsense Reasoning

Commonsense reasoning is extraordinary in its breadth and flexibility. I have argued that qualitative representations play a key role in achieving these properties. Qualitative representations for quantity enable some kinds of intuitive reasoning to be performed without detailed numerical information (chapters 5 and 6). Qualitative process theory (chapters 7–10) provides a model for continuous causality that supports causal reasoning about quantities and processes, including feedback systems, which most models of causality proposed over the past two decades in cognitive psychology simply cannot handle (chapter 9). Qualitative representations enable knowledge about modeling itself to be expressed and reasoned with (chapter 11). Qualitative simulation provides a model for mental simulation about the continuous world. Qualitative representations provide a level of description that supports analogical reasoning and analogical generalization. A combination of analogical and first-principles reasoning is probably the best account for human qualitative simulation, given what we currently know (chapter 12).

Qualitative representations provide a natural level of description for natural-language semantics, with the composability of qualitative relationships providing the support needed for the incremental accumulation of information that occurs in natural-language understanding (chapter 13). The semantics of spatial language can be captured in part by qualitative spatial representations (Mani & Pustejovsky, 2012).

Common sense is learned from experience. The notion of mechanism provided by qualitative process (QP) theory aids learning, in that it constrains the search for theories by constraining the form of the answers. QP theory has been used to model conceptual change in several domains (chapter 17) and has been used to model numerical estimation, conceptual metaphors, and several types of social reasoning (chapter 18). Qualitative spatial representations have been used to help model early, perceptually grounded aspects of conceptual change (chapter 17), as well as to model aspects of human learning of spatial language (chapter 16).

20.3 Foundation for Expert Reasoning

Professional reasoning rests on the foundation of our commonsense knowledge. The ontologies of quantity, process, components, and fields all gracefully extend to quantitative information. Qualitative analysis provides the framing of problems, providing direct answers for simple questions and revealing the questions that need more information to resolve, thereby guiding the use of quantitative knowledge. Compositional modeling provides a formalization of the art of building models for specific purposes (chapter 11). The compositional capabilities of qualitative representations can also be extended to quantitative representations. Thus, compositional modeling has been used to create systems that can create models for engineering tasks, learn models from numerical data to support scientific discovery, and provide new tools for design, diagnosis, and monitoring (chapter 19).

21 New Directions

There are many exciting new directions suggested by progress in qualitative modeling. A few that I think are the most exciting are described next.

21.1 Formalizing Discrete Processes and Their Interactions with Continuous Processes

As noted in chapter 11, the same phenomenon can sometimes be viewed as both discrete and continuous, at different levels of analysis. Examples include Weld's (1986) notion of aggregation, which generated continuous process representations from discrete actions in reasoning about molecular genetics, and Hinrichs et al.'s (2011) use of continuous processes to ground discrete action representations in military simulation. Some attempts have been made to formalize discrete processes, typically by modifications of action representations (e.g., Simmons, 1983; Yin et al., 2010a), but much work remains to be done. Developing formalisms of discrete processes that provide the same explanatory range that qualitative process (QP) theory does and formalizing the conditions under which discrete/continuous-level shifts should be done would deepen our understanding of human reasoning and provide new tools for model-checking software, which is widely used in verifying designs and code.

21.2 Qualitative Vision

That qualitative distinctions could be used to segment video data was first realized by Badler (1976). Subsequent work in *cognitive vision* has led to systems that use qualitative representations for interesting tasks. For example, tracking interacting moving objects (Bennett, Magee, Cohn, & Hogg, 2004) and learning games via visual observation (Needham et al., 2005) are facilitated by using qualitative spatial calculi as an intermediate representation for interpreting visual sequences, because that level of abstraction leads to very robust interpretations. By using qualitative spatial calculi to decompose video data into segments, CogSketch's qualitative relationship computations can be combined with analogical generalization to recognize human behavior from Kinect data (Chen & Forbus, 2018). Qualitative descriptions of shape and color, as well as space, are also useful for scene recognition (Falomir & Olteteanu, 2015).

21.3 Qualitative Representations in Other Modalities

The use of qualitative representations in visual and spatial thinking suggests investigating whether other sensory modalities, too, use qualitative representations. Auditory scene analysis is a potentially fruitful example. People can infer a surprising amount about what happened from what they hear from events. Consider the difference in sound between dropping a pencil, a fork, and a glass on a hardwood floor, for example. Do qualitative representations play roles in smell, taste, and touch? These are completely open questions currently, to the best of my knowledge.

21.4 Qualitative Representations in Semantics

Chapters 13 and 15 explored roles of qualitative representation in natural-language semantics, and chapter 16 briefly examined reasoning about depiction. In addition, Mani and Pustejovsky (2012) make a convincing case for the utility of qualitative spatial representations for expressing the semantics of motion. Although much remains to be done, the evidence so far suggests that using qualitative representations to provide more precise accounts of semantics is very promising. Already this has led us to develop type-level qualitative representations, which better express the semantics of generic statements, but also provide a far more concise level of representation for complex constructive dynamical domains (Hinrichs & Forbus, 2012).

Analyses of gesture, another powerful form of communication, may also benefit from modeling via qualitative representations. For example, in describing continuous causal systems, gesture is used to construct spatialized representations of relational descriptions, which can be used by analogy for explanation and learning (Cooperrider, Gentner, & Goldin-Meadow, 2017).

21.5 Qualitative Representations in Robotics

Given that robots need a bridge between perception and cognition, there have already been a number of applications and extensions of qualitative spatial reasoning in robotics. Conveying tactics to robot teams in Robocup software by sketching them (Gspandl, Reip, Steinbauer, & Wotawa, 2010) is one example. Qualitative descriptions of pose constraints and configurations have been used to simplify manipulator planning (Berenson, Srinivasa, & Kuffner, 2011). Troha and Bratko (2011) show that qualitative representations can be used to enable robots to rapidly learn how to push novel objects to reach a desired location. By first learning a qualitative model, reinforcement learning within qualitative states can be applied to learn the quantitative parameters needed for effective action (Wiley, Sammut, Hengst, & Bratko, 2015). Qualitative representations are used by Hawes, Klenk, Lockwood, Horn, and Kelleher (2012) to define context-specific regions (e.g., "front of the room") in ways that enable a robot to use analogy to recognize the equivalent region in a new room. Walega, Zawidzki, and Mozaryn (2017) show that qualitative reasoning can be used to evaluate the stability of plans for unloading boxes (a real-world extension of the classic AI domain of blocks world to warehouses). The RACE robot architecture for learning from experiences uses a version of the metric diagram/place vocabulary model to integrate metric information needed for navigation and manipulation with qualitative representations as a bridge to conceptual knowledge (Rockel et al., 2013). Beetz's group at Bremen has used natural-language understanding of recipes, translated into qualitative process theory, to provide inputs for their cooking robot (Tenorth & Beetz, 2012). These examples suggest a promising future for qualitative representations in the minds of robots, just as they seem to be a key part of our own minds.

21.6 Cataloging the Range of Human Mental Models and Ontologies

Understanding how much people tend to know and what their theories and models look like is an important aspect of studying human cognition. Attempts to model human reasoning and learning by cognitive psychologists, learning scientists, and AI researchers has led to models of some aspects of human knowledge, such as motion, force, circulatory systems, and thermodynamics. With the possible exception of motion, where significant effort has been made to explore student misconceptions and ways to ameliorate them, none of these areas has been completely understood, and even the range of models of motion has not been fully formalized. Now multiply this across the range of domains in science and engineering, and the scope of the problem becomes clearer. Include in it qualitative reasoning as used in everyday life, and the problem becomes even more vast. A study of fourth-grade science exams suggests that qualitative dynamics models suffice to cover at least 29 percent of the material, with another 37 percent involving visual reasoning (including some qualitative spatial reasoning and depiction reasoning) and the remaining 35 percent involving more general world knowledge (Crouse & Forbus, 2016). Thus, creating cognitive systems that can gather and use large bodies of knowledge effectively is important even for understanding elementary school science.

Until recently, progress in compiling a catalog of human knowledge has been slow for three reasons. The first is that it has been far from clear which representations should be used. For the case of knowledge about continuous aspects of the world (physical, social, mental), the state of the art in qualitative reasoning is, I think, already sufficient to serve as a basis for such efforts. Attempting such a compilation at scale is probably one of the best ways to find out where there are gaps in our understanding of such representations. The second is that most domain theories have been built by hand, making it a labor-intensive enterprise. Rapid progress in natural-language processing and vision research is changing this. The third is that there hasn't been sufficient motivation. Scientific interest, although enough of a reason for some, doesn't always capture the imagination of those who write the checks. By contrast, the Human Genome project (and now the Proteomics project) was sold on the basis of eventual tangible benefits to society. Now we have a strong case for such benefits: building more humanlike cognitive systems represents a massive benefit to society, potentially as important as discovering fire. For example, a compendium of typical student models across the range of subjects taught in school could revolutionize the range of intelligent tutoring systems and learning environments that could be created and thereby radically increase science literacy, which is crucial for an informed citizenry.

21.7 Qualitative Representations for Social Science

I believe that qualitative reasoning provides an especially appropriate level of representation for reasoning about social causality, as chapter 18 illustrates. Such theories typically are expressed in terms of continuous parameters, such as "amount of intention" and "degree of foreknowledge." Unfortunately, there tend to not be principled ways of moving from qualitative ideas to quantitative models and numerical values for such parameters. For such domains, qualitative modeling is actually a more rigorous way to proceed because it makes fewer ad hoc assumptions. Similarly, ordinal fitting with human data is a more robust measure than numerical curve fitting, which requires many numerical parameters for which there are little or no empirical or theoretical ways to ascertain their values in advance.

21.8 Qualitative Representations in Cognitive Architecture

Cognitive architectures are cognitive simulations that attempt to capture broader swaths of mental capabilities compared to most simulations, which focus on only one or two psychological processes (Anderson, 2009; Forbus et al., 2009; Laird, 2012; Newell, 1994). Given the hypothesized centrality of qualitative representations, one would expect to find them widely used in cognitive architectures. That is not the case so far, mostly because of the choices of topics that have been explored with them. Our Companion cognitive architecture (Forbus et al., 2009; Forbus & Hinrichs, 2017) is an exception, as many of the simulations discussed in the rest of the book indicate. The visual processing module in SOAR (Laird, 2012) basically provides a metric diagram that is used for spatial reasoning. As cognitive architectures start to scale up to tackling more commonsense reasoning, I believe that they will start using qualitative representations more heavily.

However, I also think that cognitive architecture research raises new and interesting questions for qualitative representation. For example, strategic thinking might be better modeled as a set of continuous processes that an agent intends to make occur (Hinrichs & Forbus, 2015). One novel task for qualitative representations raised by cognitive architectures is self-modeling. We continually estimate our own capabilities to decide what to do and how to do it. Do I have enough time to cook breakfast before leaving the house, or do I need to grab something on the way to work? Can we lift that sofa, or should we find a cart? How long will it take me to respond to this email message? How much time will it take me to catch up on the literature on this phenomenon? These questions and many more like them all seem to need a combination of qualitative representations, to organize experience into meaningful units, and analogical generalization, to support estimation.

The Companion cognitive architecture is based on the hypothesis that qualitative representations and analogical processing are central to human cognition. As chapter 17 shows, Companions have been used to model conceptual change in several domains, lending credence to this hypothesis.

21.9 Multimodal Science Learning and Teaching

One of the early visionary papers in intelligent tutoring was Collins and Stevens's (1982) examination of what computational support would be required for creating software Socratic tutors. Two of the missing ingredients for such systems are now available: qualitative representations that can express human mental models in ways that are natural for people and manipulable by AI systems, as well as analogical reasoning that enables comparing and contrasting causal models and situations. The rise of cognitive architectures and increasing capabilities in natural modalities for interacting with people suggests that such open-ended Socratic tutors are becoming possible. One path to achieve this would be to first build *multimodal science learners*, cognitive systems that can accumulate general knowledge and build specific models of people they interact with over time, using natural modalities. This would build reusable libraries of both domain knowledge and communication strategies and skills. As both improve, such systems can be put to work as *multimodal science tutors*, helping students of all ages understand science better. (There are never enough people in education, so the usual concern about job loss via automation should not apply here.) Being able to discuss any topic, at any time, with a being that builds a relationship with you over time could revolutionize education.

21.10 In Conclusion

While the qualitative reasoning community has made considerable progress and laid some firm foundations, we are a long way from a complete understanding of qualitative representations and the roles that they play in human cognition. I hope some of you reading this will join us in this exciting endeavor.

Notes

1 Introduction

1. See "A Private Universe," a video produced by the Harvard-Smithsonian Center for Astrophysics. Clips are available on YouTube (e.g., https://youtu.be/p0wk4qG2mIg).

2. The coefficient of restitution indicates how elastic something is. A convenient form is to take it as the fraction of energy retained in each collision, so that a value of 0 means something is inelastic, and a value of 1 means something is perfectly elastic.

3. Publishing real examples from the test would violate its security, which is necessary for its use as a psychometric instrument.

4. Sixty-three percent of the participants chose to not redirect the truck. Twelve percent responded that they would not redirect even if only one person in the first camp would die.

5. Marr's choice of the term *computational* (i.e., what is being computed) is in some ways unfortunate. Since then, the term *computational model* has typically been used for process-level and implementation-level models as well as computational-level models.

2 Representation

1. For an especially good summary of the issues, see Markman and Dietrich (2000).

2. There are some in AI who identify "theory" with "formal theory." I believe that is a mistake. By that light, physics, chemistry, and biology do not have theories. Even mathematics, for most of its history, was not fully formalized—formalizing the differential calculus was an achievement of the nineteenth century, but differential equations had been productively used for several centuries before that. So while I respect and root for efforts to build fully formal theories of different aspects of reasoning, I think there is ample precedent to proceed without them.

3. Lisp is a programming language whose simplicity of syntax has made it a workhorse for AI research, because it simplifies building systems that can reason about

their own knowledge. Statements of knowledge are not programming language statements, of course, but similar advantages apply.

4. There are some heuristics that some systems have used, essentially doing morphological analysis on predicate names, to guess more of their properties. Although useful in some circumstances, this presumes that the predicate names were generated by hand and named in a useful way. Neither of these presumptions is safe these days.

5. And, in a technical sense, impossible because there is always one additional model: the model where the entities are the terms and sentences of the axioms themselves. These are called *Herbrand models*, and they represent a solipsistic perspective on the world that, interestingly, can never be formally ruled out.

6. http://www.cyc.com/documentation/ontologists-handbook/.

7. Tailorability describes the degree to which the results of a cognitive model depend on decisions that are not theoretically or empirically motivated. Tailorability needs to be minimized so that the results of a simulation actually depend on the theory it is intended to test, rather than other factors.

8. Collections are not sets, which avoids paradoxes. Consider the set of all sets that do not contain themselves. Does this set contain itself? Bertrand Russell managed to throw a monkey wrench into several lines of research with this paradox (Irvine & Deutsch, 2013).

9. This means that structural relations are higher order, in the logical sense, or, equivalently, first order with reification, meaning that for each predicate, there is a special entity introduced to represent it.

10. The ResearchCyc contents we were using in 2016 contained just under five million facts.

5 Quantity

1. In the domain of military simulation, a head-to-head comparison found that qualitative simulation was both more accurate and required orders of magnitude less computation than Monte Carlo simulation (Hinrichs et al., 2011).

2. Fans of thermodynamics will know that this is an approximation: phase change boundaries depend on pressure as well as temperature. The qualitative mathematics described in the next chapter enables such factors to be added incrementally, thereby allowing reasoning with partial knowledge and supporting learning.

3. A quantity space where the relationships always suffice to yield a total order is called a *value space* (Kuipers, 1994). Value spaces are useful for plotting qualitative

values (e.g., Bredeweg et al., 2009) and determining some dynamical properties, as discussed in chapter 9.

4. FOG stands for "Formalisation du raisonnement sur l'Ordre de Grandeur."

6 Relationships between Quantities

1. Traditionally, d/dt, but this is typographically challenging.

7 Qualitative Process Theory

1. Gentner and Stevens's notion of a mental model is much closer to the approach taken here than Johnson-Laird's, which focuses more on counting properties of discrete objects. Much of the early work on qualitative representations and reasoning comes out of the Gentner and Stevens's tradition of mental models.

2. Another reason, historically, for the focus on generality is a level confusion arising in some early causal reasoning research. To be composable requires that a model fragment not rely on tacit assumptions about other parts of the system. This was not always appreciated, hence the articulation of de Kleer and Brown's (1984) *no function in structure* principle. One solution is to construct domain theories such that their tacit assumptions are always aligned with the assumptions used in the kinds of reasoning to be performed. Another is to make such assumptions explicit and reason about them, as described in chapter 11.

3. Readers with some knowledge of thermodynamics may bristle at my seemingly cavalier use of the word *heat* instead of the more technically correct term, *internal energy*. Despite the continued efforts of generations of professors to eradicate the term *heat*, it continues to be used by working engineers in practice, seemingly without harm.

4. In the original descriptions of QP theory, these were called *individual views*.

5. Here it is constant because there is no influence on it—the connection between heat and temperature has been included only for gases. That disconnection is one way to model heat sources. Another method of modeling sources and sinks in QP theory is to add an additional process to an entity that replenishes or absorbs the extensive parameter, for example, heat or mass (Collins & Forbus, 1989).

6. Fans of this paradox will be pleased to know that forms of it come up again, in two places, in this book.

7. Readers familiar with the minimal model change literature (e.g., Winslett, 1988) will recognize that approach as a version of this algorithm, but for discrete actions instead of continuous processes.

8 Examples Using QP Theory

1. Interestingly, infants have some knowledge about differences between solids and nonsolid substances as early as five months (Hespos, Ferry, Anderson, Hollenbeck, & Rips, 2016).

2. There have been several more detailed formalizations of container geometry. For example, Kim's (1993) *bounded stuff* ontology describes locations inside containers, which is needed for representing systems such as lift pumps, flush toilets, and internal combustion engines.

3. This assumption can be abused, as the neglect of the impact of human activities on climate change illustrates.

4. This impart process handles applied forces applied over a period of time. Handling sharp blows, like a kick, requires extending QP theory with impulses (Kim, 1993).

9 Causality

1. QP theory deliberately ignores agency (e.g., intentional actions taken by some being) in order to focus on continuous phenomena. The impact of such actions on the continuous world can be modeled by changes that affect the conditions of model fragments. These can include changes in connectivity (e.g., opening a valve or setting a kettle on a stove) or discontinuous changes in an ordinal relationship (e.g., kicking a ball).

2. A STRIPS operator is a discrete model of action, describing what happens via lists of facts to be added and deleted in the current model of a situation to represent the effects of that action. Using STRIPS operators requires identifying a set of facts that should be regarded as primitive, with all others being derived from them, so that indirect consequences of actions will be correctly handled (Fikes & Nilsson, 1981).

3. Norton and Thevenin equivalents, specifically.

4. Interestingly, even numerical circuit simulators like SPICE use the disturbance model in organizing their computations.

5. Or have never taken apart high-voltage equipment containing capacitors with unpleasant consequences.

11 Modeling

1. Salt water cannot be used because it is corrosive: the chemistry of water in the boiler must be carefully controlled, because even a small amount of mineral deposits on its surfaces (i.e., the "scale" one sees in a tea kettle that needs cleaning) radically reduces thermal conductivity and hence the efficiency of the system.

2. Some real boilers are just large containers. Others, for efficiency, divide the working fluid into many small tubes to increase the surface area for thermal transfer.

3. For example, although turbofan engines are critically important in modern aircraft, accurate models of the consequences of failures such as fan blades breaking have not been developed, because it would require risking pilots' lives to gather the data needed to construct them.

4. The particular representation conventions were worked out jointly with the Educational Testing Service and Cycorp (Klenk & Forbus, 2009a).

12 Analogy in Dynamics

1. This is assuming a program running at 1 GHz, which is on the slow end for today's gaming rigs.

2. This is a simplification; multiple retrievals are one way to generate multiple possible behaviors.

13 Dynamics in Language

1. Here we depart from FrameNet's tradition and follow Minsky, in allowing frames to take other frames as arguments. We believe that when connecting linguistic to conceptual information, this approach is simpler.

2. As of July 20, 2012.

14 Qualitative Spatial Reasoning

1. In physics terms, their coefficient of restitution is equal to zero (i.e., perfectly inelastic). A perfectly elastic object has a coefficient of restitution of 1.

2. This representation leaves out corners deliberately for simplicity. A more complex version involving corners was constructed later, but the same principles hold.

3. A total order means that there can be no ambiguity: one number is either smaller than, greater than, or equal to another.

15 Qualitative Spatial Calculi

1. The formalization uses a combination of ideas from *mereology*, which studies theories of parthood, and topology. See Cohn and Renz (2007) for a summary.

2. Computing what is called *arc consistency* in a planar network of discrete variables is linear in the number of variables (Mackworth & Freuder, 1985).

3. See Galton (2001) for more details.

4. As discussed in chapter 3, this assumes serial processing. For parallel processing, the time complexity can often be reduced, but at the cost of an exponentially increasing number of processors based on size of problem. Given that organisms have finite amounts of hardware, parallelism is typically not a feasible solution to exponential complexity, unless the size of inputs is kept very small and bounded.

16 Understanding Sketches and Diagrams

1. For example, in one experiment involving a military sketch recognition system, participants were required to specify in advance what sorts of names would be used for different sorts of entities (i.e., color names for phase lines, as in "Phase Line Blue").

2. The knowledge base contents are derived from Cycorp's OpenCyc KB, extracted from it, and combined with our own group's extensions.

3. Although some devices provide additional information, such as pressure of a stroke or *Z* distance from the screen when hovering is supported, for portability, CogSketch throws such extra information away. This limits its ability to, for example, model the use of a pen hovering over a glyph to focus attention on it. We substitute selecting the glyph, as one would select a word or sentence in a word processor, as a means of indicating focus of attention.

4. It is more subtle in practice: CogSketch filters the nearby candidates based on the semantics of the relationship, and for convenience, users may override its judgments about bindings.

5. These embellishments often result in unrealistic geometric distortions of the street, leading to the somewhat counterproductive warning on many of them: "Not to be used for navigation."

6. For some problems, Evans's program was able to compute symbolic descriptions from coordinate data. Again, quite a feat, given that punch cards were the medium used to enter programs and data!

7. The other two are a simpler test intended for children and the Advanced test, which is the same kinds of problems as the Standard test but including some that are substantially harder.

17 Learning and Conceptual Change

1. This may be how diSessa's (1983) *p-prims* arise.

2. Because SAGE is order-dependent, in a subsequent experiment (Friedman, Taylor, & Forbus, 2009), we used cross-fold validation to examine what models were formed given different orders of experiences, which produced similar results.

3. If you are computationally oriented, you can think of this as a generalization of the justification structure used in truth maintenance systems (Forbus & de Kleer, 1994).

18 Commonsense Reasoning

1. Feigenson (2007) argues that there is converging evidence for a uniform underlying mental representation for quantities, based on evidence that discrimination thresholds are surprisingly constant across domains, which might seem to contradict the evidence for variability. But assuming that what is varying is what quantities are measured and how, this is actually compatible with variability in learned judgments across domains.

2. k-means clustering is a statistical method for creating groups, where one of the inputs is k, the number of clusters to be created.

3. See Marinier, Laird, and Lewis (2009) and Wilson, Forbus, and McLure (2013) for some initial forays.

19 Expert Reasoning

1. It is one of the most successful ideas in human history, responsible for our current standards of living. It is also, sadly, implicated in the negative impacts that humanity is having on our biosphere. But as such concerns become integrated into engineering practice, it is likely to be a key component in how we solve the problems we have brought upon ourselves.

2. At least in domains and problems where such simulation makes sense. A domain where it doesn't make sense is industrial design, where concerns include ergonomics and aesthetics as well as functionality. An extreme example is creating an environment that an autistic person will find soothing, which is all about understanding the psychology of its intended users.

References

Abbott, K. (1988). Robust operative diagnosis as problem solving in a hypothesis space. In T. M. Mitchell & R. G. Smith (Eds.), *Proceedings of the Seventh National Conference on Artificial Intelligence* (pp. 369–374). Menlo Park, CA: AAAI Press.

Allen, J. F. (1994). *Natural language understanding* (2nd ed.). Redwood City, CA: Benjamin/Cummings.

Amorapanth, P., Kranjec, A., Bromberger, B., Lehet, M., Widick, P., Woods, A., ... Chatterjee, A. (2012). Language, perception, and the schematic representation of spatial relations. *Brain and Language, 120,* 226–236.

Anderson, J. (2009). *How can the human mind occur in the physical universe?* Oxford, UK: Oxford University Press.

Antonelli, G. (2012). Aldo, "non-monotonic logic." In E. N. Zalta (Ed.), *The Stanford encyclopedia of philosophy*. Retrieved from http://plato.stanford.edu/archives/win2012/entries/logic-nonmonotonic/.

Baader, F., Calvanese, D., McGuinness, D., Nardi, D., & Patel-Schneider, P. (2010). *The description logic handbook* (2nd ed.). Cambridge, UK: Cambridge University Press.

Badler, N. (1976). Conceptual descriptions of physical activities. *American Journal of Computational Linguistics, 35,* 70–83.

Bailey-Kellogg, C., & Ramakrishnan, N. (2004). Spatial aggregation for qualitative assessment of scientific computations. In G. Ferguson & D. McGuinness (Eds.), *Proceedings of the Nineteenth National Conference on Artificial Intelligence* (pp. 585–591). Menlo Park, CA: AAAI Press.

Bailey-Kellogg, C., & Zhao, F. (2003). Qualitative spatial reasoning: Extracting and reasoning with spatial aggregates. *AI Magazine, 24*(4), 47–60.

Baillargeon, R. (1994). How do infants learn about the physical world? *Current Directions in Psychological Science, 3*(5),133–140.

Baillargeon, R. (1998). A model of physical reasoning in infancy. *Advances in Infancy Research, 3*, 305–371.

Baillargeon, R. (2002). The acquisition of physical knowledge in infancy: A summary in eight lessons. In U. Goswami (Ed.), *The Wiley-Blackwell handbook of childhood cognitive development* (pp. 47–83). Malden, MA: Blackwell.

Banach, S., & Tarski, A. (1924). Sur la décomposition des ensembles de points en parties respectivement congruentes [On decomposition of point sets into respectively congruent parts]. *Fundamenta Mathematicae, 6*, 244–277.

Baron, J., & Spranca, M. (1997). Protected values. *Organizational Behavior and Human Decision Processes, 70*, 1–16.

Bartlett, F. (1932). *Remembering: A study in experimental and social psychology*. Cambridge, UK: Cambridge University Press.

Barton, G., Berwick, R., & Ristad, E. (1987). *Computational complexity and natural language*. Cambridge, MA: MIT Press.

Batt, G., Besson, B., Ciron, P., de Jong, H., Dumas, E., Geiselmann, J., ... Ropers, D. (2012). Genetic Network Analyzer: A tool for the qualitative modeling and simulation of bacterial regulatory networks. In J. van Helden, A. Toussaint, & D. Thieffry (Eds.), *Bacterial molecular networks* (pp. 439–462). New York: Humana.

Battaglia, P. W., Hamrick, J. B., & Tenenbaum, J. B. (2013). Simulation as an engine of physical scene understanding. *Proceedings of the National Academy of Sciences, 110*(45), 18327–18332.

Bechtel, W., & Abrahamsen, A. (2005). Explanation: A mechanistic alternative. *Studies in History and Philosophy of the Biological and Biomedical Sciences, 36*, 421–441.

Bell, D., Bobrow, D., Falkenhainer, B., Fromherz, M., Saraswat, V., & Shirley, M. (1994). RAPPER: The Copier Modeling Project. In T. Nishida (Ed.), *Proceedings of the Eighth International Workshop on Qualitative Reasoning about Physical Systems* (pp. 1–12). Nara, Japan.

Bennett, B., Magee, D., Cohn, A., & Hogg, D. (2004). Using spatio-temporal continuity constraints to enhance visual tracking of moving objects. In R. Lopez de Mantaras & L. Saitta (Eds.), *Proceedings of the Sixteenth European Conference on Artificial Intelligence* (pp. 922–926). Valencia, Spain: IOS Press.

Berenson, D., Srinivasa, S., & Kuffner, J. (2011). Task space regions: A framework for pose-constrained manipulation planning. *International Journal of Robotics Research, 30*(12), 1435–1460.

Biswas, G., Schwartz, D., Bransford, J., & The Teachable Agents Group at Vanderbilt. (2001). Technology support for complex problem solving: From SAD environments to AI. In K. Forbus & P. Feltovich (Eds.), *Smart machines in education: The coming revolution in educational technology* (pp. 71–97). Cambridge, MA: MIT Press.

Bizer, C., Lehmann, J., Kobilarov, G., Auer, S., Becker, C., Cyganiak, R., & Hellmann, S. (2009). DBPedia—A crystallization point for the web of data. *Journal of Web Semantics, 7*, 154–165.

Blass, J., & Forbus, K. (2017). Analogical chaining with natural language instruction for commonsense reasoning. In S. Singh & S. Markovitch (Eds.), *Proceedings of the Thirty-First AAAI Conference on Artificial Intelligence* (pp. 4357–4363). Palo Alto, CA: AAAI Press.

Bobrow, D. (Ed.). (1985). *Qualitative reasoning about physical systems*. Cambridge, MA: MIT Press.

Bobrow, D., Falkenhainer, B., Farquhar, A., Fikes, R., Forbus, K, Gruber, T., … Kuipers, B. (1996). A compositional modeling language. In Y. Iwasaki & A. Farquhar (Eds.), *Proceedings of the Tenth International Workshop for Qualitative Reasoning* (pp. 12–21). Menlo Park, CA: AAAI Press.

Borrett, S., Bridewell, W., Langley, P., & Arrigo, K. (2007). A method for representing and developing process models. *Ecological Complexity, 4*, 1–12.

Bowdle, B., & Gentner, D. (1997). Informativity and asymmetry in comparisons. *Cognitive Psychology, 34*, 244–286.

Bowdle, B., & Gentner, D. (2005). The career of metaphor. *Psychological Review, 112*, 193–216.

Brachman, R., & Levesque, H. (2004). *Knowledge representation and reasoning*. San Francisco, CA: Morgan-Kaufmann.

Bradley, E., Easley, M., & Stolle, R. (2001). Reasoning about nonlinear system identification, *Artificial Intelligence, 133*, 139–188.

Bratko, I., & Suc, D. (2003). Learning qualitative models. *AI Magazine, 24*(4), 107.

Bredeweg, B., Linnebank, F., Bouwer, A., & Liem, J. (2009). Garp3—workbench for qualitative modelling and simulation. *Ecological Informatics, 4*(5–6), 263–281.

Bredeweg, B., Salles, P., Bouwer, A., Liem, J., Nuttle, T., … Zitek, A. (2008). Towards a structured approach to building qualitative reasoning models and simulations. *Ecological Informatics*, 3, 1–12.

Brown, D. (1994). Facilitating conceptual change using analogies and explanatory models. *International Journal of Science Education, 16*(2), 201–214.

Brown, J. S., Collins, A., & Duguid, P. (1989). Situated cognition and the culture of learning. *Educational Researcher, 18*(1), 32–42.

Brown, N. R., & Siegler, R. S. (2001). Seeds aren't anchors. *Memory and Cognition, 29*(3), 405–412.

Buckley, S. (1979). *Sun up to sun down*. New York: McGraw-Hill.

Bundy, A. (1983). *The computer modeling of mathematical reasoning*. New York: Academic Press.

Bundy, A., Sasnauskas, G., & Chan, M. (2015). Solving guesstimation problems using the semantic web: Four lessons from an application. *Semantic Web, 6*(2), 1–14.

Burstein, M. H. (1983). A model of learning by incremental analogical reasoning and debugging. In M. R. Genesereth (Ed.), *Proceedings of the Third National Conference on Artificial Intelligence* (pp. 45–48). Menlo Park, CA: AAAI Press.

Bylander, T. (1991). A theory of consolidation for reasoning about devices. *International Journal Man-Machine Studies, 35*, 467–489.

Camblin, C., Gordon, P., & Swaab, T. (2007). The interplay of discourse congruence and lexical association during sentence processing: Evidence from EPRs and eye tracking. *Journal of Memory and Language, 56*(1), 103–128.

Carey, S. (2011). *The origin of concepts*. Oxford, UK: Oxford University Press.

Carraher, T. N., Carraher, D. W., & Schliemann, A. D. (1985). Mathematics in the streets and in schools. *British Journal of Developmental Psychology, 3*, 21–29.

Cassimatis, N., Bello, P., & Langley, P. (2008). Ability, breadth, and parsimony in computational models of higher-order cognition. *Cognitive Science, 32*, 1304–1322.

Catino, C., Grantham, S., & Ungar, L. (1991). Automatic generation of qualitative models of chemical process units. *Computers & Chemical Engineering, 15*(8), 583–599.

Catrambone, R., Craig, D., & Nersessian, N. (2006). The role of perceptually represented structure in analogical problem solving. *Memory and Cognition 34*(5), 1126–1132.

Chang, M., & Forbus, K. (2015). *Towards interpretation strategies for multimodal instructional analogies*. Paper presented at the 28th International Workshop on Qualitative Reasoning (QR2015), Minneapolis, MN.

Chen, K., & Forbus, K.D. (2018). Action recognition from skeleton data via analogical generalization over qualitative representations. *Proceedings of AAAI 2018*. New Orleans, LA: AAAI Press

Cheng, P., & Holyoak, K. (1985). Pragmatic reasoning schemas. *Cognitive Psychology, 17*, 391–416.

Chi, M., Bassok, M., Lewis, M., Reimann, P., & Glaser, R. (1989). Self-Explanations: How students study and use examples in learning to solve problems. *Cognitive Science, 13*, 145–182.

Chi, M., de Leeuw, N., Chiu, M., & LaVancher, C. (1994). Eliciting self-explanations improves understanding. *Cognitive Science, 18*, 439–477.

Chi, M. T. H., Feltovich, P. J., & Glaser, R. (1981). Categorization and representation of physics problems by experts and novices. *Cognitive Science, 5*, 121–152.

Chi, M. T. H., Slotta, J. D., & de Leeuw, N. (1994). From things to processes: A theory of conceptual change for learning science concepts. *Learning and Instruction, 4,* 27–43.

Choi, S. (2006). Influence of language-specific input on spatial cognition: Categories of containment. *First Language, 26,* 187–205.

Chou, T., & Winslett, M. (1994). A model-based belief revision system. *Journal of Automated Theorem Proving, 12*(2), 157–208.

Christie, S., & Gentner, D. (2010). Where hypotheses come from: Learning new relations by structural alignment. *Journal of Cognition and Development, 11*(3), 356–373.

Christie, S., & Gentner, D. (2014). Language helps children succeed on a classic analogy task. *Cognitive Science, 38,* 383–397.

Cioaca, E., Linnebank, F., Bredeweg, B., & Salles, P. (2009). A qualitative reasoning model of algal bloom in the Danube Delta Biosphere Reserve (DDBR). *Ecological Informatics, 4*(5–6), 282–298.

Clark, A. (1987). From folk psychology to naïve psychology. *Cognitive Science, 11,* 139–154.

Clement, C. A., & Gentner, D. (1991). Systematicity as a selection constraint in analogical mapping. *Cognitive Science, 15,* 89–132.

Clement, J. (1983). A conceptual model discussed by Galileo and used intuitively by physics students. In D. Gentner & A. Stevens (Eds.), *Mental models* (pp. 325–339). Hillsdale, NJ: Lawrence Erlbaum.

Clementini, E., Di Felice, P., & Hernandez, D. (1997). Qualitative representation of positional information. *Artificial Intelligence, 95,* 317–356.

Cohen, P., Johnston, M., McGee, D., Oviatt, S., Pittman, J., Smith, I., … Clow, J. (1997). QuickSet: Multimodal interaction for distributed applications. *Proceedings of the Fifth ACM International Conference on Multimedia,* (pp. 31–40). New York: ACM.

Cohn, A., Magee, D., Galata, A., Hogg, D., & Hazirika, S. (2008). Towards an architecture for cognitive vision using qualitative spatio-temporal representations and abduction. In C. Freksa, W. Brauer, C. Habel, & K. F. Wender (Eds.), *Spatial cognition III, LNAI 2685* (pp. 232–248). Berlin, Germany: Springer-Verlag.

Cohn, A., & Renz, J. (2007). Qualitative spatial representation and reasoning. In F. van Harmelen, V. Lifshitz, & B. Porter (Eds.), *Handbook of knowledge representation* (pp. 551–596). San Diego: Elsevier Science.

Cohn, A., & Renz, J. (2008). Qualitative spatial representation and reasoning. In F. van Hermelen, V. Lifschitz, & B. Porter (Eds.), *Handbook of knowledge representation* (pp. 551–596). San Diego: Elsevier Science.

Collins, A., & Gentner, D. (1987). How people construct mental models. In D. Holland & N. Quinn (Eds.), *Cultural models in language and thought* (pp. 243–265). Cambridge, UK: Cambridge University Press.

Collins, A., & Quillian, R. (1969). Retrieval time from semantic memory. *Journal of Verbal Learning and Verbal Behavior, 9*, 432–438.

Collins, A., & Stevens, A. (1982). Goals and strategies for inquiry teachers. In R. Glaser (Ed.), *Advances in instructional psychology* (Vol. 2, pp. 65–119). Hillsdale, NJ: Erlbaum.

Collins, A., Warnock, E., Aiello, N., & Miller, M. (1975). Reasoning from incomplete knowledge. In D. Bobrow & A. Collins (Eds.), *Representation and understanding*. New York: Academic Press.

Collins, J. (1993). *Process-based diagnosis: An approach to understanding novel failures.* Evanston, IL: Institute for the Learning Sciences.

Collins, J., & Forbus, K. (1987). Reasoning about fluids via molecular collections. In K. Forbus & H. Shrobe (Eds.), *Proceedings of the Sixth National Conference on Artificial Intelligence* (pp. 590–594). Menlo Park, CA: AAAI Press.

Collins, J., & Forbus, K. (1989). *Building qualitative models of thermodynamic processes.* Retrieved from http://www.qrg.northwestern.edu/papers/files/fsthermo(searchable).pdf.

Cooperrider, K., Gentner, D., & Goldin-Meadow, S. (2017). Analogical gestures foster understanding of causal systems. In G. Gunzelmann, A. Howes, T. Tenbrink, & E. Davelaar (Eds.), *Proceedings of the Thirty-Ninth Annual Meeting of the Cognitive Science Society* (pp. 240–245). London.

Coventry, K. R., & Garrod, S. C. (2004). *Saying, seeing and acting: The psychological semantics of spatial prepositions.* New York: Psychology Press, Taylor & Francis.

Cox, P., Plimmer, B., & Rodgers, P. (2012). *Diagrammatic representation and inference: Seventh International Conference, Diagrams 2012.* Berlin, Germany: Springer.

Crawford, L. S., Do, M. B., Ruml, W., Hindi, H., Eldershaw, C., Zhou, R., … Larner, D. L. (2013). Online reconfigurable machines. *AI Magazine, 34*(3), 73–88.

Crouse, M., & Forbus, K. (2016). *Elementary school science as a cognitive system domain: How much qualitative reasoning is required?* Paper presented at the Fourth Annual Conference on Advances in Cognitive Systems, Evanston, IL.

Crowley, K., & Siegler, R. S. (1999). Explanation and generalization in young children's strategy learning. *Child Development, 70*, 304–316.

Crupi, V. (2013). Confirmation. In E. N. Zalta (Ed.), *The Stanford encyclopedia of philosophy*. Retrieved from http://plato.stanford.edu/archives/win2013/entries/confirmation/.

Cui, Z., Cohn, A. G., & Randell, D. A. (1992). Qualitative simulation based on a logical formalism of space and time. *Proceedings AAAI-92* (pp. 679–684). Menlo Park, CA: AAAI Press.

Culicover, P. W., & Jackendoff, R. (1999). The view from the periphery: The English comparative correlative. *Linguistic Inquiry, 30*, 543–571.

Curtis, J., Baxter, D., Wagner, P., Cabral, J., Schneider, D., & Witbrock, M. (2009). Methods of rule acquisition in the TextLearner system. In S. Nirenburg & T. Oates (Eds.), *2009 AAAI Spring Symposium on Learning by Reading and Learning to Read* (pp. 22–28). Menlo Park, CA: AAAI Press.

Dauge, P. (1993). Symbolic reasoning with relative orders of magnitude. *Proceedings of the Thirteenth IJCAI* (pp. 1509–1515). Chambery, France.

Davis, E. (1987). *Order of magnitude reasoning in qualitative differential equations* (Technical Report 312). New York: NYU Computer Science Department.

Davis, E. (1990). *Representations of commonsense knowledge.* San Mateo, CA: Morgan-Kaufmann.

Davis, E., & Marcus, G. (2016). The scope and limits of simulation in automated reasoning. *Artificial Intelligence, 233*, 60–72.

Day, S., & Gentner, D. (2007). Nonintentional analogical inference in text comprehension. *Memory and Cognition, 35*, 39–49.

deCoste, D. (1991). Dynamic across-time measurement interpretation. *Artificial Intelligence, 51*(1), 273–341.

deCoste, D., & Collins, J. W. (1991). IQE: An incremental qualitative envisioner. *Proceedings of the Fifth International Workshop on Qualitative Reasoning about Physical Systems* (pp. 58–70). Austin, TX: QR-91.

Dehaene, S., Izard, V., Pica, P., & Spelke, E. (2006). Core knowledge of geometry in an Amazonian indigene group. *Science, 311*, 381–384.

Dehghani, M., Sachdeva, S., Ekhtiari, H., Gentner, D., & Forbus, K. (2009). The role of cultural narratives in moral decision-making. In N. Taatgen & H. van Rijn (Eds.), *Proceedings of the Annual Meeting of the Cognitive Science Society* (pp. 1912–1917). Amsterdam.

Dehghani, M., Tomai, E., Forbus, K., & Klenk, M. (2008). *An integrated reasoning approach to moral decision-making.* Paper presented at the Twenty-Third AAAI Conference on Artificial Intelligence (AAAI), Chicago, IL.

Dehghani, M., Unsworth, S., Lovett, A., & Forbus, K. (2007). Capturing and categorizing mental models of food webs using QCM. *Proceedings of the Twenty-First International Workshop on Qualitative Reasoning.* Aberystwyth, UK.

de Jong, H. (2008). Qualitative modeling and simulation of bacterial regulatory networks. In A. Uhrmacher & M. Heiner (Eds.), *Computational methods in systems biology (CMSB-08)*. Berlin, Germany: Springer-Verlag.

de Kleer, J. (1984). How circuits work. *Artificial Intelligence, 24,* 205–280.

de Kleer, J., & Brown, J. (1984). A qualitative physics based on confluences. *Artificial Intelligence, 24,* 7–83.

de Koning, K., Bredeweg, B., Breuker, J., & Wielinga, B. (2000). Model-based reasoning about learner behavior. *Artificial Intelligence, 117,* 173–229.

Derbinsky, N., Laird, J. E., & Smith, B. (2010, August). Towards efficiently supporting large symbolic declarative memories. In D. Salvucci & G. Gunzelmann (Eds.), *Proceedings of the Tenth International Conference on Cognitive Modeling* (pp. 49–54). Philadelphia, PA.

diSessa, A. (1983). Phenomenology and the evolution of intuition. In D. Gentner & A. Stevens (Eds.), *Mental models* (pp. 15–33). Hillsdale, NJ: Lawrence Erlbaum.

diSessa, A. (1988). Knowledge in pieces. In G. Forman & P. Pufall (Eds.), *Constructivism in the computer age* (pp. 49–70). Hillsdale, NJ: Lawrence Erlbaum.

diSessa, A. (1993). Toward an epistemology of physics. *Cognition and Instruction, 10*(2–3), 105–225.

diSessa, A. (2008). A bird's eye view of "pieces" versus "coherence" controversy. In S. Vosniadou (Ed.), *Handbook of conceptual change research* (pp. 35–60). Mahwah, NJ: Lawrence Erlbaum.diSessa, A., Gillespie, N., & Esterly, J. (2004). Coherence versus fragmentation in the development of the concept of force. *Cognitive Science, 28,* 843–900.

Donlon, J., & Forbus, K. (1999). Using a Geographic Information System for qualitative spatial reasoning about trafficability. *Proceedings of QR99* (pp. 62–72). Loch Awe, Scotland.

Doyle, R. J., Chien, S. A., Fayyad, U. M., & Wyatt, E. J. (1993). Focused real-time systems monitoring based on multiple anomaly models. In Daniel Weld (Ed.), *Proceedings of QR93* (pp. 75–82). Orcas Island, WA.

Drabble, B. (1993). EXCALIBUR: A program for planning and reasoning with processes. *Artificial Intelligence, 6291,* 1–40.

Dylla, F., Lee, J., Mossakowski, T., Schneider, T., Van Delden, A., Van De Ven, J., & Wolter, D. (2017). A survey of qualitative spatial and temporal calculi: Algebraic and computational properties. *ACM Computing Surveys, 50*(1), Article 7.

Egenhofer, M. J., & Franzosa, R. D. (1991). Point-set topological spatial relations. *International Journal of Geographical Information Systems, 5*(2), 161–174.

Egenhofer, M. J., & Mark, D. M. (1995). Naive geography. In A. U. Frank & W. Kuhn (Eds.), *Spatial information theory: A theoretical basis for GIS. International Conference, COSIT 95* (pp. 1–15). Berlin, Germany: Springer.

Elio, R., & Anderson, J. R. (1981). The effect of category generalizations and instance similarity on schema abstraction. *Journal of Experimental Psychology: Human Learning and Memory, 7*, 397–417.

Elkan, C. (1994, August). The paradoxical success of fuzzy logic. *IEEE Expert*, pp. 3–8.

Epley, N., & Gilovich, T. (2005). When effortful thinking influences judgmental anchoring: Differential effects of forewarning and incentives on self-generated and externally-provided anchors. *Journal of Behavioral Decision Making, 18*, 199–212.

Evans, T. (1968). A program for the solution of geometric-analogy intelligence test questions. In M. Minsky (Ed.), *Semantic information processing*. Cambridge, MA: MIT Press.

Falkenhainer, B. (1987). An examination of the third stage in the analogy process: Verification-based analogical learning. In John P. McDermott (Ed.), *Proceeding of IJCAI* (pp. 260–264). Milan, Italy.

Falkenhainer, B. (1990). A unified approach to explanation and theory formation. In J. Shrager & P. Langley (Eds.), *Computational models of scientific discovery and theory formation* (pp. 157–196). Los Altos, CA: Morgan Kaufmann.

Falkenhainer, B. (1992). Modeling without amnesia: Making experience-sanctioned approximations. In R. Leitch (Ed.), *Proceedings of the Sixth International Workshop on Qualitative Reasoning about Physical Systems* (pp. 44–55). Edinburgh, UK: Heriot-Watt University.

Falkenhainer, B., & Forbus, K. (1991). Compositional modeling: Finding the right model for the job. *Artificial Intelligence, 51*(1–3), 95–143.

Falkenhainer, B., Forbus, K. D., & Gentner, D. (1989). The structure-mapping engine: Algorithm and examples. *Artificial Intelligence, 41*(1), 1–63.

Falomir, Z., & Olteteanu, A. (2015). Logics based on qualitative descriptors for scene understanding. *Neurocomputing, 161*, 3–16.

Faltings, B. (1992). A symbolic approach to qualitative kinematics. *Artificial Intelligence, 56*(2–3), 139–170.

Fan, J., Ferrucci, D., Gondek, D., & Kalyanpur, A. (2010). *PRISMATIC: Inducing knowledge from a large scale lexicalized relation resource*. Paper presented at the NAACL Workshop on Formalisms and Methodology for Learning by Reading, Los Angeles, CA.

Fan, J., Kalyanpur, A., Gondek, D., & Ferruci, D. (2012). Automatic knowledge extraction from documents. *IBM Journal of Research & Development, 56*(3/4), 5:1–5:10.

Faries, J. M., & Reiser, B. J. (1988). Access and use of previous solutions in a problem solving situation. In V. Patel & G. Groen (Eds.) *Proceedings of the Tenth Annual Meeting of the Cognitive Science Society* (pp. 433–439). Hillsdale, NJ: Erlbaum.

Feigenson, L. (2007). The quality of quantity. *Trends in Cognitive Science, 11*(5), 185–187.

Feltovich, P., Coulson, R., & Spiro, R. (2001). Learners' (mis)understanding of important and difficult concepts. In K. Forbus & P. Feltovich (Eds.), *Smart machines in education* (pp. 349–375). Cambridge, MA: /MIT Press.

Fikes, R., & Nilsson, N. (1981). STRIPS: A new approach to the application of theorem proving to problem solving. *Artificial Intelligence, 2*, 189–208.

Fillmore, C. (1976). Frame semantics and the nature of language. *Annals of the New York Academy of Sciences, 280*, 20–32.

Fillmore, C., & Atkins, S. (1994). Starting where dictionaries stop: The challenge for computational lexicography. In S. Atkins & A Zampolli (Eds.), *Computational approaches to the lexicon* (pp. 349–393). Oxford, UK: Oxford University Press.

Forbus, K. (1980). Spatial and qualitative aspects of reasoning about motion. In R. M. Balzer (Ed.), *Proceedings of the First National Conference on Artificial Intelligence* (pp. 170–173). Menlo Park, CA: AAAI Press.

Forbus, K. (1983). Qualitative reasoning about space and motion. In D. Gentner & A. Stevens (Eds.), *Mental models*. Hillsdale, NJ: Lawrence Erlbaum.

Forbus, K. (1984). Qualitative process theory. *Artificial Intelligence, 24*, 85–168.

Forbus, K. (1989). Introducing actions into qualitative simulation. *Proceedings of the Eleventh International Joint Conference on Artificial Intelligence* (pp. 1273–1278). Detroit, MI.

Forbus, K. (2001). Exploring analogy in the large. In D. Gentner, K. Holyoak, & B. Kokinov (Eds.), *The analogical mind: Perspectives from cognitive science*. Cambridge, MA: MIT Press.

Forbus, K., Carney, K., Sherin, B., & Ureel, L. (2004, July). *VModel: A visual qualitative modeling environment for middle-school students*. Paper presented at the 16th Innovative Applications of Artificial Intelligence Conference, San Jose, CA.

Forbus, K., Chang, M., McLure, M., & Usher, M. (2017). The cognitive science of sketch worksheets. *Topics in Cognitive Science, 9*(4), 921–942.

Forbus, K., & de Kleer, J. (1994). *Building problem solvers*. Cambridge, MA: MIT Press.

Forbus, K., Ferguson, R., & Gentner, D. (1994, August). Incremental structure-mapping. In A. Ram & K. Eiselt (Eds.), *Proceedings of the Cognitive Science Society*. Hillsdale, NJ: Lawrence Erlbaum Associates.

Forbus, K., Ferguson, R., Lovett, A., & Gentner, D. (2017). Extending SME to handle large-scale cognitive modeling. *Cognitive Science, 41*, 1152–1201.

Forbus, K., Garnier, B., Tikoff, B., Marko, W., Usher, M. & Mclure, M. (2018). Sketch worksheets in STEM classrooms: Two deployments. Deployed Application Prize paper. *Proceedings of IAAI 2018*. New Orleans, LA: AAAI Press.

Forbus, K., & Gentner, D. (1986a, August). *Causal reasoning about quantities*. Paper presented at the *Eighth Annual Conference of the Cognitive Science Society*, Amherst, MA.

Forbus, K., & Gentner, D. (1986b). Learning physical domains: Towards a theoretical framework. In R. Michalski, J. Carbonell, & T. Mitchell (Eds.), *Machine learning: An artificial intelligence approach* (Vol. 2). Palo Alto, CA: Tioga Press.

Forbus, K., & Gentner, D. (1989). Structural evaluation of analogies: What counts? *Proceedings of the Eleventh Annual Conference of the Cognitive Science Society*. Ann Arbor, MI.

Forbus, K., & Gentner, D. (1997, June). *Qualitative mental models: Simulations or memories?* Paper presented at the Eleventh International Workshop on Qualitative Reasoning, Cortona, Italy.

Forbus, K., Gentner, D., & Law, K. (1995). MAC/FAC: A model of similarity-based retrieval. *Cognitive Science, 19*(2), 141–205.

Forbus, K., & Hinrichs, T. (2017). Analogy and relational representations in the companion cognitive architecture. *AI Magazine, 38*(4), 34–42.

Forbus, K., Klenk, M., & Hinrichs, T. (2009). Companion cognitive systems: Design goals and lessons learned so far. *IEEE Intelligent Systems, 24*(4), 36–46.

Forbus, K., & Kuehne, S. (2005, May). *Towards a qualitative model of everyday political reasoning*. Paper presented at the 19th International Qualitative Reasoning Workshop, Graz, Austria.

Forbus, K., Nielsen, P., & Faltings, B. (1991). Qualitative spatial reasoning: The CLOCK Project. *Artificial Intelligence, 51*(1–3), 417–471.

Forbus, K., Riesbeck, C., Birnbaum, L., Livingston, K., Sharma, A., & Ureel, L. (2007). Integrating natural language, knowledge representation and reasoning, and analogical processing to learn by reading. In R. C. Holte & A. Howe (Eds.), *Proceedings of the Twenty-Second AAAI Conference on Artificial Intelligence* (pp. 1542–1547). Menlo Park, CA: AAAI Press.

Forbus, K., Usher, J., & Chapman, V. (2003). *Qualitative spatial reasoning about sketch maps*. Paper presented at the Fifteenth Annual Conference on Innovative Applications of Artificial Intelligence, Acapulco, Mexico.

Forbus, K., Usher, J., Lovett, A., Lockwood, K., & Wetzel, J. (2011). CogSketch: Sketch understanding for cognitive science research and for education. *Topics in Cognitive Science, 3*, 648–666.

Forbus, K., Usher, J., & Tomai, E. (2005). Analogical learning of visual/conceptual relationships in sketches. In M. Veloso & S. Kambhampati (Eds.), *Proceedings of the Twentieth National Conference on Artificial Intelligence* (pp. 202–208). Menlo Park, CA: AAAI Press.

Forbus, K., Whalley, P., Everett, J., Ureel, L., Brokowski, M., Baher, J., & Kuehne, S. (1999). CyclePad: An articulate virtual laboratory for engineering thermodynamics. *Artificial Intelligence, 114*, 297–347.

Forrester, J. W. (1961). *Industrial dynamics*. Waltham, MA: Pegasus Communications.

Freksa, C. (1991). Conceptual neighborhood and its role in temporal and spatial reasoning. In M. Singh & L. Trave-Massuyes (Eds.), *Proceedings of the IMACS Workshop on Decision Support Systems and Qualitative Reasoning* (pp. 181–187). Amsterdam, Holland.

Freksa, C. (1992). Using orientation information for qualitative spatial reasoning. In A. U. Frank, I. Campari, & U. Formentini (Eds.), *Theories and methods of spatio-temporal reasoning in geographic space* (pp. 162–178). Berlin: Springer.

Friedman, S. E. (2012). *Computational conceptual change: An explanation-based approach*. PhD dissertation, Northwestern University, Department of Electrical Engineering and Computer Science, Evanston, IL.

Friedman, S. E., & Forbus, K. (2008). Learning causal models via progressive alignment and qualitative modeling: A simulation. In B. C. Love, K. McRae, & V. M. Sloutsky (Eds.), *Proceedings of the Thirtieth Annual Conference of the Cognitive Science Society* (pp. 1123–1128). Austin, TX: Cognitive Science Society.

Friedman, S. E., & Forbus, K. (2010). *An integrated systems approach to explanation-based conceptual change*. Paper presented at the Twenty-Fourth AAAI Conference on Artificial Intelligence, Atlanta, GA.

Friedman, S. E., & Forbus, K. (2011). *Repairing incorrect knowledge with model formulation and metareasoning*. Paper presented at the 22nd International Joint Conference on Artificial Intelligence, Barcelona, Spain.

Friedman, S. E., Forbus, K. D., & Sherin, B. (2011a). Constructing and revising commonsense science explanations: A metareasoning approach. *Proceedings of the AAAI Fall Symposium on Advances in Cognitive Systems*. Arlington, VA: AAAI Press.

Friedman, S. E., Forbus, K. D., & Sherin, B. (2011b). *How do the seasons change? Creating & revising explanations via model formulation & metareasoning*. Paper presented at the 25th International Workshop on Qualitative Reasoning, Barcelona, Spain.

Friedman, S. E., Forbus, K., & Sherin, B. (2018). Representing, running, and revising mental models: A computational model. *Cognitive Science, 42*(4), 1110–1145.

Friedman, S. E., Taylor, J., & Forbus, K. (2009). Learning naïve physics models by analogical generalization. *Proceedings of the Second International Analogy Conference* (pp. 168–177). Sofia, Bulgaria.

Fromherz, M. P. J., Bobrow, D. G., & de Kleer, J. (2003). Model-based computing for design and control of reconfigurable systems. *AI Magazine, 24*(4), 120–130.

Funt, B. (1980). Problem-solving with diagrammatic representations. *Artificial Intelligence, 13*(3), 201–230.

Gagnier, K., Atit, K., Ormand, C., & Shipley, T. (2016). Comprehending 3D diagrams: Sketching to support spatial reasoning. *Topics in Cognitive Science, 9*(4), 883–901.

Galton, A. (2001). *Qualitative spatial change.* Oxford, UK: Oxford University Press.

Gardin, F., & Meltzer, B. (1989). Analogical representations of naïve physics. *Artificial Intelligence, 38*(2), 139–159.

Genesereth, M., & Nilsson, N. (1987). *Logical foundations of artificial intelligence.* San Mateo, CA: Morgan-Kaufmann.

Gentner, D. (1983). Structure-mapping: A theoretical framework for analogy. *Cognitive Science, 7,* 155–170.

Gentner, D. (1988). Metaphor as structure mapping: The relational shift. *Child Development, 59,* 47–59.

Gentner, D. (1989). The mechanisms of analogical learning. In S. Vosniadou & A. Ortony (Eds.), *Similarity and analogical reasoning* (pp. 199–241). London, UK: Cambridge University Press. (Reprinted in *Knowledge Acquisition and Learning,* 1993, 673–694.)

Gentner, D. (2003). Why we're so smart. In D. Gentner & S. Goldin-Meadow (Eds.), *Language in mind: Advances in the study of language and thought* (pp. 195–235). Cambridge, MA: MIT Press.

Gentner, D. (2010). Bootstrapping the mind: Analogical processes and symbol systems. *Cognitive Science, 34*(5), 752–775.

Gentner, D., Bowdle, B., Wolff, P., & Boronat, C. (2001). Metaphor is like analogy. In D. Gentner, K. J. Holyoak, & B. N. Kokinov (Eds.), *The analogical mind: Perspectives from cognitive science* (pp. 199–253). Cambridge, MA: MIT Press.

Gentner, D., & Bowerman, M. (2009). Why some spatial semantic categories are harder to learn than others: The typological prevalence hypothesis. In J. Guo, E. Lieven, N. Budwig, S. Ervin-Tripp, K. Nakamura, & S. Ozcaliskan (Eds.), *Crosslinguistic approaches to the psychology of language: Research in the tradition of Dan Isaac Slobin* (pp. 465–480). New York: Psychology Press.

Gentner, D., Brem, S., Ferguson, R. W., Markman, A. B., Levidow, B. B., Wolff, P., & Forbus, K. D. (1997). Analogical reasoning and conceptual change: A case study of Johannes Kepler. *The Journal of the Learning Sciences, 6*(1), 3–40.

Gentner, D., & Kurtz, K. (2005). Relational categories. In W. K. Ahn, R. L. Goldstone, B. C. Love, A. B. Markman, & P. W. Wolff (Eds.), *Categorization inside and outside the lab* (pp. 151–175). Washington, DC: APA.

Gentner, D., Loewenstein, J., & Hung, B. (2007). Comparison facilitates children's learning of names for parts. *Journal of Cognition and Development, 8,* 285–307.

Gentner, D., & Markman, A. B. (1997). Structure mapping in analogy and similarity. *American Psychologist, 52,* 45–56.

Gentner, D., & Namy, L. L. (2006). Analogical processes in language learning. *Current Directions in Psychological Science, 15*(6), 297–301.

Gentner, D., & Rattermann, M. J. (1991). Language and the career of similarity. In S. A. Gelman & J. P. Byrnes (Eds.), *Perspectives on language and thought: Interrelations in development* (pp. 225–275). London, UK: Cambridge University Press.

Gentner, D., Rattermann, M. J., & Forbus, K. D. (1993). The roles of similarity in transfer: Separating retrievability from inferential soundness. *Cognitive Psychology, 25*, 524–575.

Gentner, D., Rattermann, M. J., Markman, A. B., & Kotovsky, L. (1995). Two forces in the development of relational similarity. In T. J. Simon & G. S. Halford (Eds.), *Developing cognitive competence: New approaches to process modeling* (pp. 263–313). Hillsdale, NJ: Lawrence Erlbaum.

Gentner, D., & Stevens, A. (1983). *Mental models*. Hillsdale, NJ: Lawrence Erlbaum.

Gentner, D., & Toupin, C. (1986). Systematicity and surface similarity in the development of analogy. *Cognitive Science, 10*, 277–300.

Gigerenzer, G., Todd, P., & the ABC Research Group. (2000). *Simple heuristics that make us smart*. Oxford, UK: Oxford University Press.

Giunchiglia, E., Lee, J., Lifschitz, V., McCain, N., & Turner, H. (2004). Nonmonotonic causal theories. *Artificial Intelligence, 153*(1–2), 49–104.

Goel, A. (2013). A 30-year case study and 15 principles: Implications of an artificial intelligence methodology for functional modeling. *Artificial Intelligence for Engineering Design, Analysis and Manufacturing, 27*, 203–215.

Gopnik, A., & Schulz, L. (2007). *Causal learning: Psychology, philosophy and computation*. Oxford, UK: Oxford University Press.

Goswami, U., & Brown, A. L. (1989). Melting chocolate and melting snowmen: Analogical reasoning and causal relations. *Cognition, 35*, 69–95.

Gratch, J., & Marsella, S. (2004). A domain-independent framework for modeling emotion. *Journal of Cognitive Systems Research, 5*(4), 269–306.

Griffiths, T. L., Chater, N., Kemp, C., Perfors, A., & Tenenbaum, J. B. (2010). Probabilistic models of cognition: Exploring the laws of thought. *Trends in Cognitive Sciences, 14*, 357–364.

Gspandl, S., Reip, M., Steinbauer, G., & Wotawa, F. (2010). From sketch to plan. In J. de Kleer & K. Forbus (Eds.), *Proceedings of QR-2010*. Portland, OR.

Guerrin, F. (1995). *Dualistic algebra for qualitative analysis*. Paper presented at the 9th International Workshop on Qualitative Reasoning, Amsterdam, Holland.

Halloun, I., & Hestenes, D. (1985). Common-sense concepts about motion. *American Journal of Phsyics, 53*, 1056.

Halpern, J. (2016). *Actual causality*. Cambridge: MIT Press.

Hammond, T., & Davis, R. (2005). LADDER: A sketching language for user interface developers. *Computers and Graphics, 29*, 518–532.

Hanks, S., & McDermott, D. (1987). Nonmonotonic and temporal projection. *Artificial Intelligence, 33*(3), 379–412.

Harte, J. (1988). *Consider a spherical cow: A course in environmental problem solving*. Sausalito, CA: University Science Books.

Hawes, N., Klenk, K., Lockwood, K., Horn, G., & Kelleher, J. (2012). Towards a cognitive system that can recognize spatial regions based on context. In J. Hoffman & B. Selman (Eds.), *Proceedings of the Twenty-Sixth AAAI Conference on Artificial Intelligence* (pp. 200–206). Palo Alto, CA: AAAI Press.

Hayes, P. (1979). The naive physics manifesto. In D. Michie (Ed.), *Expert systems in the micro-electronic age*. Edinburgh, UK: Edinburgh University Press.

Hayes, P. (1985a). The logic of frames. In R. Brachman & H. Levesque (Eds.), *Readings in knowledge representation*. San Francisco, CA: Morgan-Kaufmann.

Hayes, P. (1985b). Naive Physics 1: Ontology for liquids. In R. Hobbs & R. Moore (Eds.), *Formal theories of the commonsense world*. Norwood, NJ: Ablex.

Hayes-Roth, R., Waterman, D., & Lenat, D. (1983). *Building expert systems*. Boston, MA: Addison-Wesley.

Hespos, S., & Baillargeon, R. (2001). Infants' knowledge about occlusion and containment events: A surprising discrepancy. *Psychological Science, 12*(2), 141–147.

Hespos, S., Ferry, A., Anderson, E., Hollenbeck, E., & Rips, J. (2016). Five-month-old infants have general knowledge of how nonsolid substances behave and interact. *Psychological Science, 27*(2), 244–256.

Hestenes, D., Wells, M., & Swackhamer, G. (1992). Force Concept Inventory. *The Physics Teacher, 30*, 141–158.

Hinrichs, T., & Forbus, K. (2012, July). Toward higher-order qualitative representations. *Proceedings of the Twenty-Sixth International Workshop on Qualitative Reasoning*. Los Angeles, CA.

Hinrichs, T., & Forbus, K. (2015). *Qualitative models for strategic planning*. Paper presented at the 3rd Annual Conference on Advances in Cognitive Systems, Atlanta, GA.

Hinrichs, T., Forbus, K., de Kleer, J., Yoon, S., Jones, E., Hyland, R., & Wilson, J. (2011). *Hybrid qualitative simulation of military operations*. Paper presented at the 23rd Innovative Applications for Artificial Intelligence Conference, San Francisco, CA.

Hobbs, J. R. (2004). Abduction in natural language understanding. In L. Horn & G. Ward (Eds.), *Handbook of pragmatics* (pp. 724–741). Malden, MA: Blackwell.

Hogge, J. (1987). Compiling plan operators from domains expressed in qualitative process theory. In K. Forbus & H. Shrobe (Eds.), *Proceedings of the Sixth National Conference on Artificial Intelligence* (pp. 229–233). Menlo Park, CA: AAAI Press.

Holden, M. P., Curby, K. M., Newcombe, N. S., & Shipley, T. F. (2010). A category adjustment approach to memory to memory for spatial location in natural scenes. *Journal of Experimental Psychology: Learning, Memory, & Cognition, 36*, 590–604.

Horvitz, E. (2001). Principles and applications of continual computation. *Artificial Intelligence, 126*(1–2), 159–196.

Huttenlocher, J., Hedges, L., & Duncan, S. (1991). Categories and particulars: Prototype effects in estimating spatial location. *Psychological Review, 98*, 352–376.

Ioannides, C., & Vosniadou, S. (2002). The changing meanings of force. *Cognitive Science Quarterly, 2*, 5–61.

Ironi, L., & Tentoni, S. (2007). Automated detection of qualitative spatio-temporal features in electrocardiac activation maps. *Artificial Intelligence in Medicine, 39*, 99–111.

Ironi, L., & Tentoni, S. (2011). Interplay of spatial aggregation and computational geometry in extracting diagnostic features from cardiac activation data. *Computer Methods and Programs in Biomedicine, 107*(3), 456–467.

Irvine, A., & Deutsch, H. (2013). Russell's paradox. In E. Zalta (Ed.), *The Stanford encyclopedia of philosophy*. Retrieved from http://plato.stanford.edu/archives/win2013/entries/russell-paradox/.

Iwasaki, Y., & Simon, H. (1994). Causality and model abstraction. *Artificial Intelligence, 67*, 143–194.

Johnson-Laird, P. (1983). *Mental models*. Cambridge, MA: Harvard University Press.

Jones, M., & Love, B. (2011). Bayesian fundamentalism or enlightenment? On the explanatory status and theoretical contributions of Bayesian models of cognition. *Behavioral and Brain Sciences, 34*(4), 169–188.

Kahneman, D., Slovic, P., & Tversky, A. (Eds.). (1982). *Judgment under Uncertainty: Heuristics and Biases*. Cambridge, UK: Cambridge University Press.

Kamp, H., & Reyle, U. (1993). *From discourse to logic*. Dordrecht, the Netherlands: Kluwer.

Kandaswamy, S., Forbus, K., & Gentner, D. (2014). Modeling learning via progressive alignment using interim generalizations. In P. Bellow, M. Guarini, M. McShane, & B. Scassellati (Eds.) *Proceedings of the Cognitive Science Society* (pp. 2471–2476). Québec City, Canada.

Kareev, Y., Lieberman, I., & Lev, M. (1997). Through a narrow window: Sample size and perception of correlation. *Journal of Experimental Psychology: General, 126*(3), 278–287.

Karp, P. (1993). A qualitative biochemistry and its application to the regulation of the tryptophan operon. In L. Hunter (Ed.), *Artificial intelligence and molecular biology* (pp. 289–324). Menlo Park, CA: AAAI Press.

Keane, M. T. (1995). On order effects in analogical mapping: Predicting human error using IAM. In J. D. Moore & J. F. Lehmann (Eds.), *Proceedings of the Seventeenth Annual Conference of the Cognitive Science Society*. Hillsdale, NJ: Lawrence Erlbaum.

Keisler, H. J. (1976). *Elementary calculus: An approach using infinitesimals*. Boston, MA: Prindle Weber & Schmidt.

Kempton, W. (1986). Two theories of home heat control. *Cognitive Science, 10*, 75–90.

Keppens, J., & Shen, Q. (2002). Compositional modeling repositories via dynamic constraint satisfaction with order-of-magnitude preferences. *Journal of AI Research, 21*, 499–550.

Kim, H. (1993). *Qualitative reasoning about fluids and mechanics*. PhD dissertation and ILS Technical Report, Northwestern University, Evanston, IL.

King, R., Garrett, S., & Coghill, G. (2005). On the use of qualitative reasoning to simulate and identify metabolic pathways. *Bioinformatics, 21*(9), 2017–2026.

Klenk, M., de Kleer, J., Bobrow, D., Yoon, S., Handley, J., & Janssen, B. (2012). DRAFT: Guiding and verifying early design using qualitative simulation. *ASME 2012 International Design Engineering Technical Conferences and Computers and Information in Engineering Conference* (pp. 1097–1103). Houston: American Society of Mechanical Engineers.

Klenk, M., & Forbus, K. (2007). Cognitive modeling of analogy events in physics problem solving from examples. In D. S. McNamara & J. G. Trafton (Eds.), *Proceedings of the Twenty-Ninth Annual Cognitive Science Society* (pp. 1163–1168). Austin, TX: Cognitive Science Society.

Klenk, M., & Forbus, K. (2009a). Analogical model formulation for AP physics problems. *Artificial Intelligence, 173*(18), 1615–1638.

Klenk, M., & Forbus, K. (2009b). Domain transfer via cross-domain analogy. *Cognitive Systems Research, 10*(3), 240–250.

Klenk, M., & Forbus, K. (2013). Exploiting persistent mappings in cross-domain analogical learning of physical domains *Artificial Intelligence, 195*, 398–417.

Klenk, M., Friedman, S., & Forbus, K. (2008). *Learning modeling abstractions via generalization*. Paper presented at the 22nd International Workshop on Qualitative Reasoning, Boulder, CO.

Klippel, A., Li, R., Yang, J., Hardisty, F., & Xu, S. (2013). The Egenhofer-Cohn hypothesis, or topological relativity? In M. Raubal & A. Frank (Eds.), *Cognitive and linguistic aspects of geographic space*. Berlin, Germany: Springer-Verlag.

Knauff, M., Rauh, R., & Renz, J. (1997). A cognitive assessment of topological spatial relations: Results from an empirical investigation. *Proceedings of the Third International Conference on Spatial Information Theory (COSIT'97)* (pp. 193–206). Berlin, Heidelberg: Springer.

Kolodner, J. (1993). *Case-based reasoning*. San Mateo, CA: Morgan-Kaufmann.

Kosslyn, S. (1994). *Images and brain: The resolution of the imagery debate*. Cambridge, MA: MIT Press.

Kosslyn, S., Koenig, O., Barrett, A., Cave, C., Tang, J., & Gabrieli, J. (1989). Evidence for two types of spatial representations: Hemispheric specialization for categorical and coordinate relations. *Journal of Experimental Psychology: Human Perception and Performance, 15*(4), 723–735.

Kosslyn, S., & Schwartz, S. (1977). A simulation of visual imagery. *Cognitive Science, 1*, 265–295.

Kotovsky, L., & Gentner, D. (1996). Comparison and categorization in the development of relational similarity. *Child Development, 67*, 2797–2822.

Kuehne, S. (2004). *Understanding natural language descriptions of physical phenomena*. PhD dissertation, Northwestern University, Evanston, IL.

Kuehne, S., & Forbus, K. (2002). *Qualitative physics as a component in natural language semantics: A progress report*. Paper presented at the Twenty-Fourth Annual Meeting of the Cognitive Science Society, George Mason University, Fairfax, VA.

Kuehne, S., & Forbus, K. (2004, August). *Capturing QP-relevant information from natural language text*. Paper presented at the 18th International Qualitative Reasoning Workshop, Evanston, IL.

Kuehne, S., Gentner, D., & Forbus, K. (2000, August). Modeling infant learning via symbolic structural alignment. *Proceedings of the Twenty-Second Annual Conference of the Cognitive Science Society*. Philadelphia, PA.

Kuipers, B. (1994). *Qualitative simulation: Modeling and simulation with incomplete knowledge*. Cambridge, MA: MIT Press.

Kuipers, B., & Kassirer, J. (1984). Causal reasoning in medicine: Analysis of a protocol. *Cognitive Science, 8*, 363–385.

Kunda, M., McGreggor, K., & Goel, A. K. (2013). A computational model for solving problems from the Raven's Progressive Matrices intelligence test using iconic visual representations. *Cognitive Systems Research, 22–23*, 47–66.

Laird, J. (2012). *The SOAR cognitive architecture*. Cambridge, MA: MIT Press.

Lakoff, G., & Johnson, M. (1980). *Metaphors we live by*. Chicago, IL: University of Chicago Press.

Lakoff, G., & Nunez, R. (2000). *Where mathematics comes from: How the embodied mind brings mathematics into being*. New York: Basic Books.

Lange-Kuttner, C., & Vinter, A. (Eds.). (2008). *Drawing and the non-verbal mind: A life-span perspective*. Cambridge, UK: Cambridge University Press.

Langley, P. (1981). Data-driven discovery of physical laws. *Cognitive Science, 5*, 31–54.

Langley, P., Shiran, O., Shrager, J., Todorovski, L., & Pohorille, A. (2006). Constructing explanatory process models from biological data and knowledge. *Artificial Intelligence in Medicine, 37*, 191–201.

Larkin, J., & Simon, H. (1987). Why a diagram is (sometimes) worth ten thousand words. *Cognitive Science, 11*(1), 65–100.

Lassaline, M. (1996). Structural alignment in induction and similarity. *Journal of Experimental Psychology: Learning, Memory, and Cognition, 22*, 754–770.

Leake, D. (Ed.). (2000). *Case-based reasoning: Experiences, lessons, and future directions*. Cambridge, MA: MIT Press.

Le Clair, S., Abrams, F., & Matejka, R. (1989). Qualitative process automation: Self directed manufacture of composite materials. *Artificial Intelligence in Engineering Design & Manufacturing, 3*(2), 125–136.

Lim, C. S., & Baron, J. (1997). *Protected values in Malaysia, Singapore, and the United States*. Unpublished manuscript, Department of Psychology, University of Pennsylvania.

Linder, B. (1991). *Understanding estimation and its relation to engineering education*. PhD dissertation, MIT, Department of Mechanical Engineering, Cambridge, MA.

Lockwood, K., & Forbus, K. (2009). Multimodal knowledge capture from text and diagrams. *Proceedings of KCAP-2009*, Redondo Beach, CA.

Lockwood, K., Lovett, A., & Forbus, K. (2008). Automatic classification of containment and support spatial relations in English and Dutch. *Proceedings of Spatial Cognition 2008*. Berlin: Springer.

Lockwood, K., Lovett, A., Forbus, K., Dehghani, M., & Usher, J. (2008). A theory of depiction for sketches of physical systems. *Proceedings of the Twenty-Second International Workshop on Qualitative Reasoning*. Boulder, CO.

Lovett, A. (2012). *Spatial routines for sketches: A framework for modeling spatial problem-solving*. PhD dissertation, Northwestern University, Department of Electrical Engineering and Computer Science, Evanston, IL.

Lovett, A., Dehghani, M., & Forbus, K. (2008). Building and comparing qualitative descriptions of three-dimensional design sketches. *Proceedings of the Twenty-Second International Qualitative Reasoning Workshop*. Boulder, CO.

Lovett, A., & Forbus, K. (2011). Cultural commonalities and differences in spatial problem-solving: A computational analysis. *Cognition, 121*(2), 281–287.

Lovett, A., & Forbus, K. (2012). *Modeling multiple strategies for solving geometric analogy problems*. Paper presented at the 34th Annual Conference of the Cognitive Science Society, Sapporo, Japan.

Lovett, A., & Forbus, K. (2013). Modeling spatial ability in mental rotation and paper-folding. *Proceedings of the Thirty-Fifth Annual Conference of the Cognitive Science Society*. Berlin, Germany.

Lovett, A., & Forbus, K. (2017). Modeling visual problem solving as analogical reasoning. *Psychological Review, 124*(1), 60–90.

Lovett, A., Forbus, K., & Usher, J. (2010). A structure-mapping model of Raven's Progressive Matrices. *Proceedings of the Thirty-Second Annual Meeting of the Cognitive Science Society*. Portland, OR.

Lovett, A., Gentner, D., Forbus, K., & Sagi, E. (2009). Using analogical mapping to simulate time-course phenomena in perceptual similarity. *Cognitive Systems Research, 10*, 216–228.

Lovett, A., Tomai, E., Forbus, K., & Usher, J. (2009). Solving geometric analogy problems through two-stage analogical mapping. *Cognitive Science, 33*(7), 1192–1231.

Lucke, D., Mossakowski, T., & Moratz, R. (2011). *Streets to the OPRA—finding your destination with imprecise knowledge*. Paper presented at the Workshop on Benchmarks and Applications of Spatial Reasoning, IJCAI-2011, Barcelona, Spain.

Mackie, J. (1980). *The cement of the universe: A study of causation*. Oxford, UK: Oxford University Press.

Mackworth, A. (1977). Consistency in networks of relations. *Artificial Intelligence, 8*(1), 99–118.

Mackworth, A., & Freuder, E. (1985). The complexity of some polynomial network consistency algorithms for constraint satisfaction problems. *Artificial Intelligence, 25*(1), 65–74.

Macleod, C., Grishman, R., & Meyers, A. (1998). *COMLEX Syntax Reference Manual, Version 3.0*. Philadelphia: Linguistic Data Consortium, University of Pennsylvania.

Mani, I., & Pustejovsky, J. (2012). *Interpreting motion: Grounded representations for spatial language*. Oxford, UK: Oxford University Press.

Mao, W. (2006). *Modeling social causality and social judgment in multi-agent interactions*. PhD dissertation, University of Southern California, Los Angeles.

Marcus, G., & Davis, E. (2013). How robust are probabilistic models of higher-level cognition? *Psychological Science, 24*(12), 2351–2360.

Marcus, G. F., Vijayan, S., Rao, B., & Vishton, P. (1999) Rule-learning in seven-month-old infants. *Science, 283*, 77–80.

Marinier, R., Laird, J., & Lewis, R. (2009). A computational unification of cognitive behavior and emotion. *Cognitive Systems Research, 10*, 48–69.

Mark, D. M., & Egenhofer, M. J. (1994a). Calibrating the meanings of spatial predicates from natural language: Line-region relations. In T. C. Waugh & R. G. Healey (Eds.), *Advances in GIS Research, 6th International Symposium on Spatial Data Handling* (pp. 538–553). Edinburgh, UK: International Geographical Union Commission on GIS, Association for Geographic Information.

Mark, D. M., & Egenhofer, M. J. (1994b). Modeling spatial relations between lines and regions: Combining formal mathematical models and human subject testing. *Cartography and Geographic Information Systems, 21*(3), 195–212.

Markman, A. (1997). Constraints on analogical inference. *Cognitive Science, 21*(4), 373–418.

Markman, A. (1998). *Knowledge representation.* Hove, UK: Psychology Press.

Markman, A., & Dietrich, E. (2000). In defense of representation. *Cognitive Psychology, 40*(2), 138–171.

Markman, A., & Gentner, D. (1993). Splitting the differences: A structural alignment view of similarity. *Journal of Memory and Language, 32*, 517–535.

Markman, A., & Medin, D. L. (2002). Decision making. In *Stevens handbook of experimental psychology: Vol. 2. Memory and cognitive processes* (3rd ed.). New York: Wiley.

Marr, D. (1982). *Vision.* New York: W. H. Freeman & Co.

McCloskey, M. (1983). Naïve theories of motion. In D. Gentner & A. Stevens (Eds.), *Mental models* (pp. 299–324). Hillsdale, NJ: Lawrence Erlbaum.

McDermott, D. (1976). Artificial intelligence meets natural stupidity. *ACM SIGART Bulletin, 57*, 4–9.

McFate, C. J., & Forbus, K. (2015). *Frame semantics of continuous processes.* Paper presented at the 28th International Workshop on Qualitative Reasoning, Minneapolis, MN.

McFate, C., & Forbus, K. (2016, August). *An analysis of frame semantics of continuous processes.* Paper presented at the 38th Annual Meeting of the Cognitive Science Society, Philadelphia, PA.

McFate, C. J., Forbus, K., & Hinrichs, T. (2014). *Using narrative function to extract qualitative information from natural language texts.* Paper presented at the Twenty-Eighth AAAI Conference on Artificial Intelligence, Québec City, Québec, Canada.

McLure, M., Friedman, S., & Forbus, K. (2010). *Combining progressive alignment and near-misses to learn concepts from sketches*. Paper presented at the 24th International Workshop on Qualitative Reasoning, Portland, OR.

McLure, M. D., Friedman S. E., & Forbus, K. D. (2015). Extending analogical generalization with near-misses. In B. Bonet & S. Koenig (Eds.), *Proceedings of the Twenty-Ninth AAAI Conference on Artificial Intelligence* (pp. 565–571). Palo Alto,CA: AAAI Press.

McLure, M. D., Friedman, S. E., Lovett, A., & Forbus, K. D. (2011). *Edge-cycles: A qualitative sketch representation to support recognition*. Paper presented at the 25th International Workshop on Qualitative Reasoning, Barcelona, Spain.

Medin, D. L., Goldstone, R. L., & Gentner, D. (1993). Respects for similarity. *Psychological Review, 100*(2), 254–278.

Meltzoff, A. N. (2007). 'Like me': A foundation for social cognition. *Developmental Science, 10*(1), 126–134.

Milch, B., Marthi, B. & Russell, S. G. (2004). BLOG: Relational modeling with unknown objects. *Proceedings of the ICML-2004 Workshop on Statistical Relational Learning and Its Connections to Other Fields*. Banff, Alberta, Canada.

Minsky, M. (1974, June). A framework for representing knowledge. MIT AI Lab Memo #306. (Reprinted in Winston, P. (Ed.). (1975). *The psychology of computer vision*. New York: McGraw-Hill).

Minsky, M. (2007). *The emotion machine: Commonsense thinking, artificial intelligence, and the future of the human mind*. New York: Simon & Schuster.

Moratz, R., Dylla, F., & Frommberger, L. (2005). *A relative orientation algebra with adjustable granularity*. Paper presented at the Workshop on Agents in Real-Time and Dynamic Environments (IJCAI), Edinburgh, Scotland.

Mueller, E. (2014). *Commonsense reasoning: An event calculus based approach* (2nd ed.). San Mateo, CA: Morgan-Kaufmann.

Muggleton, S. (1992). *Inductive logic programming* (Vol. 38). San Mateo, CA: Morgan Kaufmann.

Mulholland, T. M., Pellegrino, J. W., & Glaser, R. (1980). Components of geometric analogy solution. *Cognitive Psychology, 12*, 252–284.

Murdock, J. W. (2011). Structure mapping for *Jeopardy!* clues. *Proceedings of ICCBR 2011*, London, UK: Springer-Verlag.

Muscettola, N., Nayak, P., Pell, B., & Williams, B. (1998). Remote agent: To boldly go where no AI system has gone before. *Artificial Intelligence, 103*, 5–47.

Mustapha, S., Jen-Sen, P., & Zain, S. (2002). Application of qualitative process theory to qualitative simulation and analysis of inorganic chemical reaction. *Proceedings of the Sixteenth International Workshop on Qualitative Reasoning*. Barcelona, Catalonia, Spain.

Nagel, E., Newman, J., & Hofstadter, D. (2001). *Gödel's proof.* New York: NYU Press.

Nayak, P. (1994). Causal approximations. *Artificial Intelligence, 70*, 277–334.

Needham, C., Santos, P., Magee, D., Devin, V., Hogg, D., & Cohn, A. (2005). Protocols from perceptual observations. *Artificial Intelligence, 167*, 103–136.

Newcombe, N. C., & Uttal, D. H. (2006). Whorf versus Socrates, Round 10. *Trends in Cognitive Science, 10*(9), 394–396.

Newell, A. (1994). *Unified theories of cognition*. Cambridge, MA: Harvard University Press.

Norman, D., Rumelhart, D., & the LNR Research Group. (Eds.). (1975). *Explorations in cognition*. San Francisco, CA: Freeman.

Novak, J. (1990). Concept mapping: A useful tool for science education. *Journal of Research in Science Teaching, 27*(10), 937–949.

Novick, L. R. (1988). Analogical transfer, problem similarity, and expertise. *Journal of Experimental Psychology: Learning, Memory, and Cognition, 14*, 510–520.

Noy, N., & Hafner, C. (1998). Representing scientific experiments: Implications for ontology design and knowledge sharing. In J. Mostow & C. Rich (Eds.), *Proceedings of the Fifteenth National Conference on Artificial Intelligence* (pp. 615–622). Menlo Park, CA: AAAI Press.

Ord-Hume, A. W. J. G. (2006). *Perpetual motion: The history of an obsession*. Kempton, IL: Adventures Unlimited Press, Kempton, IL.

Ortony, A., Clore, G. L., & Collins, A. (1988). *The cognitive structure of emotions*. New York: Cambridge University Press.

Osmani, A. (2004). Introduction to reasoning about cyclic intervals. *Proceedings of the Twelfth International Conference on Industrial and Engineering Applications of Artificial Intelligent and Expert Systems: Multiple Approaches to Intelligent Systems* (pp. 698–706). Berlin: Springer.

Ouyang, T., & Forbus, K. (2006). Strategy variations in analogical problem solving. In Y. Gil & R. J. Mooney (Eds.), *Proceedings of the Twenty-First AAAI Conference on Artificial Intelligence* (pp. 446–451). Menlo Park, CA: AAAI Press.

Palmer, S. E. (1978). Fundamental aspects of cognitive representation. In E. Rosch & B. L. Lloyd (Eds.), *Cognition and categorization* (pp. 259–302). Hillsdale, N.J.: Erlbaum.

Palmer, S. E. (1999). *Vision science: Photons to phenomenology*. Cambridge, MA: MIT Press.

Paritosh, P. K. (2004). *Symbolizing quantity*. Paper presented at the 26th Cognitive Science Conference, Chicago, IL.

Paritosh, P. K. (2007). *Back of the envelope reasoning for robust quantitative problem solving* (Tech. Rep. No. NWU-EECS-07-11). PhD dissertation, Northwestern University, Department of Electrical Engineering and Computer Science, Evanston, IL.

Paritosh, P. K., & Forbus, K. (2005). *Analysis of strategic knowledge in back of the envelope reasoning*. Paper presented at the 20th National Conference on Artificial Intelligence (AAAI-05), Pittsburgh, PA.

Paritosh, P. K., & Klenk, M. E. (2006). *Cognitive processes in quantitative estimation: Analogical anchors and causal adjustment*. Paper presented at the 28th Annual Conference of the Cognitive Science Society, Vancouver, Canada.

Park, C., Bridewell, W., & Langley, P. (2010). Integrated systems for inducing spatio-temporal process models. In M. Fox & D. Poole (Eds.), *Proceedings of the Twenty-Fourth AAAI Conference on Artificial Intelligence* (pp. 1555–1560). Menlo Park, CA: AAAI Press.

Parsons, T. (1990). *Events in the semantics of English*. Cambridge, MA: MIT Press.

Pearl, J. (2009). *Causality* (2nd ed.). Cambridge, UK: Cambridge University Press.

Pearl, J. & Mackenzie, D. (2018). *The book of why: The new science of cause and effect*. New York: Basic Books.

Petrinovich, L., O'Neill, P., & Jorgensen, M. (1993). An empirical study of moral intuitions: toward and evolutionary ethics. *Journal of Personality and Social Psychology, 64*(3), 467–478.

Piaget, J. (1952). *The origin of intelligence in children*. Madison, CT: International Universities Press.

Piaget, J., & Inhelder, B. (1956). *The child's conception of space*. London, UK: Routledge & Kegan Paul.

Pisan, Y. (1996). Using qualitative representations in controlling engineering problem solving. *Proceedings of the Tenth International Workshop on Qualitative Reasoning* (pp. 190–197). Stanford, CA.

Pisan, Y. (1998). *An integrated architecture for engineering problem solving*. PhD dissertation, Northwestern University, Evanston, IL (UMI No. 733042431).

Price, C. J. (2000). AutoSteve: Automated electrical design analysis. *Proceedings ECAI-2000* (pp. 721–725). Berlin, Germany.

Raiman, O. (1991). Order of magnitude reasoning. *Artificial Intelligence, 51*, 11–38.

Rasch, R., Kott, A., & Forbus, K. (2002, July). *AI on the battlefield: An experimental exploration*. Paper presented at the 14th Innovative Applications of Artificial Intelligence Conference, Edmonton, Canada.

Rashid, A., Shariff, B., Egenhofer, M., & Mark, D. (1998). Natural-language spatial relations between linear and areal objects: The topology and metric of English language terms. *International Journal of Geographical Information Science, 12*(3), 215–246.

Rassbach, L., Bradley, E., & Anderson, K. (2011). Providing decision support for cosmogenic isotope dating. *AI Magazine, 32*, 69–78.

Rattermann, M. J., & Gentner, D. (1998). The effect of language on similarity: The use of relational labels improves young children's performance in a mapping task. In K. Holyoak, D. Gentner, & B. Kokinov (Eds.), *Advances in analogy research: Integration of theory & data from the cognitive, computational, and neural sciences* (pp. 274–282). Sophia: New Bulgarian University.

Raven, J., Raven, J. C., & Court, J. H. (2000). *Manual for Raven's Progressive Matrices and Vocabulary Scales*. Oxford, UK: Oxford Psychologists Press.

Reiter, R., & Mackworth, A. (1989). A logical framework for depiction and image interpretation. *Artificial Intelligence, 41*, 125–155.

Richardson, M., & Domingos, P. (2006). Markov logic networks. *Machine Learning, 62*(1–2), 107–136.

Richland, L. E., Morrison, R. G., & Holyoak, K. J. (2006). Children's development of analogical reasoning: Insights from scene analogy problems. *Journal of Experimental Child Psychology, 94*, 249–271.

Rickel, J., & Porter, B. (1994). Automated modeling for answering prediction questions: Selecting the time scale and system boundary. In B. Hayes-Roth & R. E. Korf (Eds.), *Proceedings of the Twelfth National Conference on Artificial Intelligence* (pp. 1191–1198). Menlo Park, CA: AAAI Press.

Rieger, C., & Grinberg, M. (1977). The declarative representation and procedural simulation of causality in physical mechanisms. *IJCAI-1977, 1*, 250–256.

Riesbeck, C., & Schank, R. (1989). *Inside case-based reasoning*. Hove, UK: Psychology Press.

Ritov, I., & Baron, J. (1999). Protected values and omission bias. *Organizational Behavior and Human Decision Processes, 79*(2), 79–94.

Rockel, S., Newumann, B., Zhang, J., Dubba, K., Cohn, A., Konecny, S., … Hotz, L. (2013). *An ontology-based multi-level robot architecture for learning from experiences*. Paper presented at the 2013 AAAI Spring Symposium on Designing Intelligent Robots: Reintegrating AI, Palo Alto, CA.

Röhrig, R. (1994). A theory for qualitative spatial reasoning based on order relations. In B. Hayes-Roth & R. E. Korf (Eds.), *Proceedings of the Twelfth National Conference on Artificial Intelligence* (pp. 1418–1423). Menlo Park, CA: AAAI Press.

Rousu, J., & Aarts, R. (2001). An integrated approach to biorecipe design. *Integrated Computer-Aided Engineering, 8*, 363–373.

Rozenblit, L., & Keil, F. (2002). The misunderstood limits of folk science: An illusion of explanatory depth. *Cognitive Science, 26*, 521–562.

Russell, S., & Norvig, P. (2009). *Artificial intelligence: A modern approach* (3rd ed.). Upper Saddle River, NJ: Prentice Hall.

Russell, S., & Wefald, E. (1991). Principles of metareasoning. *Artificial Intelligence, 49*(1–3), 361–395.

Sachenbacher, M., & Struss, P. (2005). Task-dependent qualitative domain abstraction. *Artificial Intelligence, 162*, 121–143.

Sachenbacher, M., Struss, P., & Weber, R. (2000). *Advances in design and implementation of OBD functions for diesel injection systems based on a qualitative approach to diagnosis*. Paper presented at the SAE World Congress, Detroit, MI.

Sacks, E., & Joscowicz, L. (2010). *The configuration space method for kinematic design of mechanisms*. Cambridge, MA: MIT Press.

Sagi, E., Gentner, D., & Lovett, A. (2012). What difference reveals about similarity. *Cognitive Science, 36*(6), 1019–1050.

Salles, P., & Bredeweg, B. (2003). Qualitative reasoning about population and community ecology. *AI Magazine, 24*(4), 77–90.

Saund, E., Fleet, D., Larner, D., & Mahoney, J. (2003). Perceptually-supported image editing of text and graphics. *Proceedings of the Sixteenth Annual ACM Symposium on User Interface Software and Technology (UIST 03)*. Vancouver, British Columbia, Canada.

Schank, R. (1972). Conceptual dependency: A theory of natural language understanding. *Cognitive Psychology, 3*(4), 532–631.

Scheiter, K., Schleinschock, K., & Ainsworth, S. (2017). Why sketching may aid learning from science texts: Contrasting sketching with written explanations. *Topics in Cognitive Science, 9*(2), 866–882.

Schwering, A., Kuhnberger, K-U., Krumnack, U., & Gust, H. (2009). Spatial cognition of geometric figures in the context of proportional analogies. *Proceedings of COSIT-09*. Aber Wrac'h, France

Schwering, A., Wang, J., Chipofya, M., Jan, S., Li, R., & Broelemann, K. (2014). SketchMapia: Qualitative representations for the alignment of sketch and metric maps. *Spatial Cognition and Computation, 14*(3), 220–254.

Seidenberg, M., & Ellman, J. (1999). Networks are not hidden rules. *Trends in Cognitive Science, 3*, 288–289.

Sgouros, N. (1998). Interaction between physical and design knowledge in design from physical principles. *Engineering Applications of Artificial Intelligence, 11*, 449–459.

Shah, P., Schneider, D., Matuszek, C., Kahlert, R. C., Aldag, B., Baxter, D., … Curtis, J. (2006). *Automated population of Cyc: Extracting information about named-entities from the web*. Paper presented at the Nineteenth International FLAIRS Conference, Melbourne Beach, FL.

Shapiro, S. (2013). Classical logic. In E. N. Zalta (Ed.), *The Stanford encyclopedia of philosophy*. Retrieved from http://plato.stanford.edu/archives/win2013/entries/logic-classical/.

Shaver, K. (1985). *The attribution theory of blame: Causality, responsibility and blameworthiness*. New York: Springer-Verlag.

Shen, Q., & Leitch, R. (1993). Fuzzy qualitative simulation. *IEEE Transactions on Systems, Man, and Cybernetics, 23*(4), 1038–1064.

Shepard, R. N. (1987). Toward a universal law of generalization for psychological science. *Science, 237*, 1317–1323.

Shepard, R. N., & Cooper, L. A. (1982). *Mental images and their transformations*. Cambridge, MA: MIT Press.

Sheredos, B., & Bechtel, B. (2017). Sketching biological phenomena and mechanisms. *Topics in Cognitive Science, 9*(4), 970–985.

Sherin, B., Krakowski, M., & Lee, V. (2012). Some assembly required: How scientific explanations are constructed during clinical interviews. *Journal of Research in Science Teaching, 49*(2), 166–198.

Shimomura, Y., Tanigawa, S., Umeda, Y., & Tomiyama, T. (1995). Development of self-maintenance photocopiers. *Proc. IAAI-95* (pp. 171–180). Montreal, Québec, Canada: AAAI Press.

Simmons, R. (1983). *Representing and reasoning about change in geologic interpretation*. PhD dissertation, MIT, Cambridge, MA.

Simon, H. (1953). Causal ordering and Identifiability. In W. C. Hood & T. C. Koopmans (Eds.), *Studies in econometric methods* (pp. 49–74). New York: Wiley.

Simon, H., & Iwasaki, Y. (1988). Causal ordering, comparative statics, and near decomposability. *Journal of Econometrics, 39*, 149–173.

Sinclair, F., & Walker, D. (1998). Acquiring qualitative knowledge about complex agroecosystems. Part 1: Representation as natural language. *Agricultural Systems, 56*, 341–393.

Skorstad, G. (1992). Towards a qualitative Lagrangian theory of fluid flow. In P. Rosenbloom & P. Szolovits (Eds.), *Proceedings of the Tenth National Conference on Artificial Intelligence* (pp. 691–696). Menlo Park, CA: AAAI Press.

Sloman, S. (2009). *Causal models: How people think about the world and its alternatives*. Oxford, UK: Oxford University Press.

Snow, R. E., Kyllonen, P. C., & Marshalek, B. (1984). The topography of learning and ability correlations. In R. J. Sternberg (Ed.), *Advances in the psychology of human intelligence* (Vol. 2, pp. 47–103). Hillsdale, NJ: Lawrence Erlbaum.

Spelke, E. (2003). What makes us smart? In D. Gentner & S. Goldin-Meadow (Eds.), *Language in mind* (pp. 277–312). Cambridge, MA: MIT Press.

Spelke, E., & Kinzler, K. (2007). Core knowledge. *Developmental Science, 10*(1), 89–96.

Spellman, B. A., & Holyoak, K. J. (1992). If Saddam is Hitler then who is George Bush? Analogical mapping between systems of social roles. *Journal of Personality and Social Psychology, 62*, 913–933.

Sprio, R., Feltovich, P., Coulson, R., & Anderson, D. (1989). Multiple analogies for complex concepts: Antidotes for analogy-induced misconception in advanced knowledge acquisition. In S. Vosniadou & A. Ortony (Eds.), *Similarity and analogical reasoning* (pp. 498–531). Cambridge, UK: Cambridge University Press.

Sternberg, R. J. (1977). *Intelligence, information processing, and analogical reasoning*. Hillsdale, NJ: Erlbaum.

Strasser, C., & Aldo, A. (2015). Non-monotonic logic. In E. N. Zalta (Ed.), *The Stanford encyclopedia of philosophy*. Retrieved from http://plato.stanford.edu/archives/fall2015/entries/logic-nonmonotonic/.

Swartz, C. (2003). *Back-of-the-envelope physics*. Baltimore, MD: Johns Hopkins University Press.

Suc, D., & Bratko, I. (2002). Qualitative reverse engineering. *Proceedings of the International Conference on Machine Learning*. Sydney, Australia.

Talmy, L. (2000). *Toward a cognitive semantics*. Cambridge, MA: MIT Press.

Tenenbaum, J. B., & Griffiths, T. L. (2001). Generalization, similarity, and Bayesian inference. *Behavioral and Brain Sciences, 24*(4), 629–640.

Tenorth, M., & Beetz, M. (2012). *A unified representation for reasoning about robot actions, processes, and their effects on objects*. Paper presented at the IEEE/RSJ International Conference on Intelligent Robots and Systems (IROS). Vilamoura, Algarve, Portugal.

Thomson, J. (1985). The trolley problem. *Yale Law Journal, 94*, 1395–1415.

Tomai, E., & Forbus, K. (2008). *Using qualitative reasoning for the attribution of moral responsibility*. Paper presented at the 30th Annual Conference of the Cognitive Science Society, Washington, DC.

Tomai, E., & Forbus, K. (2009). *EA NLU: Practical language understanding for cognitive modeling*. Paper presented at the 22nd International Florida Artificial Intelligence Research Society Conference, Sanibel Island, FL.

Trelease, R., & Park, J. (1996). Qualitative process modeling of cell-cell-pathogen interactions in the immune system. *Computer Methods and Programs in Biomedicine, 51*, 171–181.

Troha, M., & Bratko, I. (2011). Qualitative learning of object pushing by a robot. *Proceedings of the Twenty-Fifth International Workshop on Qualitative Reasoning*. Barcelona, Catalonia, Spain.

Tversky, A., & Kahneman, D. (1974). Judgment under uncertainty: Heuristics and biases, *Science, 185*, 1124–1131.

Ullman, S. (1984). Visual routines. *Cognition, 18*, 97–159.

Uttal, D., Meadow, N., Tipton, E., Hand, L., Alden, A., Warren, C., & Newcombe, N. (2013). The malleability of spatial skills: A meta-analysis of training studies. *Psychological Bulletin, 139*(2), 352–402.

Valentine, S., Vides, F., Lucchese, G., Turner, D., Kim, H., Li, W., ... Hammond, T. (2012). Mechanix: A sketch-based tutoring system for statics courses. *Proceedings of IAAI 2012*. Toronto, Ontario, Canada: AAAI Press.

Van Sommers, P. (1984). *Drawing and cognition: Descriptive and experimental studies of graphic production processes*. Cambridge, UK: Cambridge University Press.

Waldmann, M. R., & Dieterich, J. (2007). Throwing a bomb on a person versus throwing a person on a bomb: Intervention myopia in moral intuitions. *Psychological Science, 18*(3), 247–253.

Walega, P., Zawidzki, M., & Mozaryn, J. (2017). Qualitative evaluation of stability in disassembling block structures with robot manipulator. *Proceedings of the Thirtieth International Workshop on Qualitative Reasoning*. Melbourne, Australia.

Wallgrun, J., Frommberger, L., Wolter, D., Dylla, F., & Freksa, C. (2007). Qualitative spatial representation and reasoning in the SparQ Toolbox. In T. Barkowsky, M. Knauff, G. Ligozat, & D. R. Montello (Eds.), *Spatial cognition V, LNAI 4387* (pp. 39–58). Berlin, Germany: Springer-Verlag.

Wason, P. C. (1968). Reasoning about a rule. *Quarterly Journal of Experimental Psychology, 20*(3), 273–281.

Weisberg, D., Keil, F., Goodstein, J., Rawson, E., & Gray, J. (2008). The seductive allure of neuroscience explanations. *Journal of Cognitive Neuroscience, 20*, 470–477.

Weld, D. (1986). The use of aggregation in causal simulation. *Artificial Intelligence, 30*(1), 1–34.

Weld, D. (1990). *Theories of comparative analysis*. Cambridge, MA: MIT Press.

Weld, D., & de Kleer, J. (1990). *Readings in qualitative reasoning about physical systems*. San Mateo, CA: Morgan-Kaufmann.

Wiley, T., Sammut, C., Hengst, B., & Bratko, I. (2015). A multi-strategy architecture for on-line learning of robotic behaviours using qualitative reasoning. *Proceedings of the Third Annual Conference on Advances in Cognitive Systems*, pp. 1–16. Atlanta, GA.

Williams, B. (1984). Qualitative analysis of MOS circuits. *Artificial Intelligence, 24*(1–3), 281–346.

Williams, B. (1991). A theory of interactions: Unifying qualitative and quantitative algebraic reasoning. *Artificial Intelligence, 51*, 39–94.

Wilson, J., Forbus, K., & McLure, M. (2013). Am I really scared? A multiphase computational model of emotions. *Proceedings of Advances in Cognitive Systems*. Baltimore, MD.

Winslett, M. (1988). Reasoning about action using a possible models approach. In T. M. Mitchell & R. G. Smith (Eds.), *Proceedings of the Seventh National Conference on Artificial Intelligence* (pp. 89–93). Menlo Park, CA: AAAI Press.

Winston, P. H. (1980). Learning and reasoning by analogy. *Communications of the ACM, 23*(12), 689–703.

Witbrock, M., Pittman, K., Moszkowicz, J., Beck, A., Schneider, D., & Lenat, D. (2015). Cyc and the big C: Reading that produces and uses hypotheses about complex molecular biology mechanisms. *AAAI Workshop on Scholarly Big Data: AI Perspectives, Challenges, and Ideas*. Menlo Park, CA: AAAI Press.

Wolff, P. (2007). Representing causation. *Journal of Experimental Psychology: General, 136*, 82–111.

Wolff, P., & Gentner, D. (2011). Structure-mapping in metaphor comprehension. *Cognitive Science, 35*, 1456–1448.

Wu, M., & Gentner, D. (1998). Structure in category-based induction. *Proceedings of the Twentieth Annual Conference of the Cognitive Science Society* (pp. 1154–1158). Madison, WI.

Yan, J., & Forbus, K. (2005). Similarity-based qualitative simulation. *Proceedings of the Twenty-Seventh Annual Meeting of the Cognitive Science Society*. Stressa, Italy.

Yan, J., Forbus, K., & Gentner, D. (2003). A theory of re-representation in analogical matching. *Proceedings of the Twenty-Fifth Annual Meeting of the Cognitive Science Society*. Boston, MA.

Yaner, P., & Goel, A. (2008). Analogical recognition of shape and function in design drawings. *Artificial Intelligence for Engineering Design, Analysis, and Manufacturing, 22*(2), 117–128.

Yin, P., Chang, M., & Forbus, K. (2010). Sketch-based spatial reasoning in geologic interpretation. *Proceedings of the Twenty-Fourth International Workshop on Qualitative Reasoning*. Portland, OR.

Yin, P., Forbus, K., Usher, J., Sageman, B., & Jee, B. (2010). Sketch worksheets: A sketch-based educational software system. *Proceedings of the Twenty-Second Annual Conference on Innovative Applications of Artificial Intelligence*. Portland, OR: AAAI Press.

Yip, K. (1991). *KAM: A system for intelligently guiding numerical experimentation by computer*. Cambridge, MA: MIT Press.

Yip, K., & Zhao, F. (1996). Spatial aggregation: Theory and applications. *Journal of Artificial Intelligence Research, 5*, 1–26.

Zabkar, J., Janez, T., Mozina, M., & Bratko, I. (2010). Active learning of qualitative models with Pade. *Proceedings of the Twenty-Fourth International Workshop on Qualitative Reasoning*. Portland, OR.

Zadeh, L. (1996). Fuzzy logic = computing with words. *IEEE Transactions on Fuzzy Systems, 4*(2), 103–111.

Zhao, F., Bailey-Kellogg, C., Huang, X., & Ordomez, I. (2007). Structure discovery from massive spatial data sets using intelligent simulation tools. In S. Dzeroski & L. Todorovski (Eds.), *Computational discovery LNAI 4660* (pp. 158–174). Berlin, Germany: Springer-Verlag.

Index